We hope you enjoy this boo'
Please return or renew i*
You can renew it at ' ....Jraries
or by using our free ...ıse you can
phone **0344 800 802u.** ....ɔ your library
card and pin ready.
You can sign up for emaiı ...minders too.

# GLADIUS

## Living, Fighting and Dying
in the Roman Army

GUY DE LA BÉDOYÈRE

ABACUS

First published in Great Britain in 2020 by Little, Brown
This paperback edition published in 2021 by Abacus

1 3 5 7 9 10 8 6 4 2

Copyright © Guy de la Bédoyère 2020

The moral right of the author has been asserted.

Maps by John Gilkes

A CIP catalogue record for this book
is available from the British Library.

Paperback ISBN 978-0-3491-4391-0

Typeset in Spectrum by M Rules
Printed and bound in Great Britain by
Clays Ltd, Elcograf S.p.A.

Papers used by Little, Brown are from well-managed forests
and other responsible sources.

Abacus
An imprint of
Little, Brown Book Group
Carmelite House
50 Victoria Embankment
London EC4Y 0DZ

An Hachette UK Company
www.hachette.co.uk

www.littlebrown.co.uk

This book is dedicated to my four sons,
Hugh, Thomas, Robert and William

*Titus Flaminius . . . of Legio XIIII Gemina,*
*served as a soldier for 22 years, and now here*
*I am. Read this and be more or less lucky in*
*your lifetime.*

Tombstone of a legionary, found at
Wroxeter, Britain. Mid-first century AD

# CONTENTS

Roman Empire, West

□ Legionary fortress
■ Fort
⚔ Major battle
● Major settlement

AP Alpes Penninae
AC Alpes Cottiae
AM Alpes Maritimae

0    100   200   300 miles
0       200      400      600 km

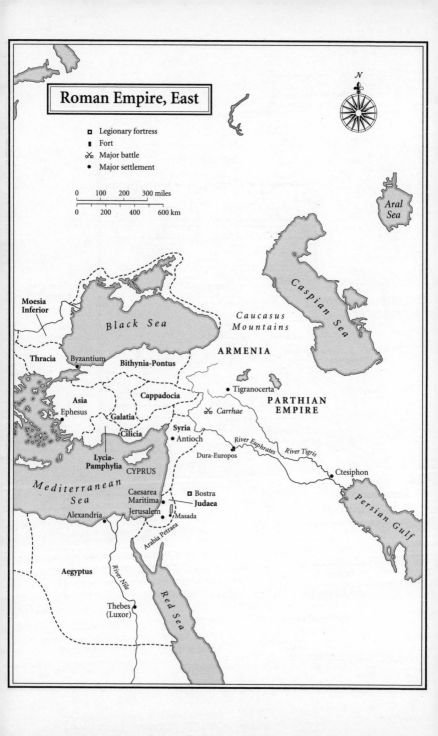

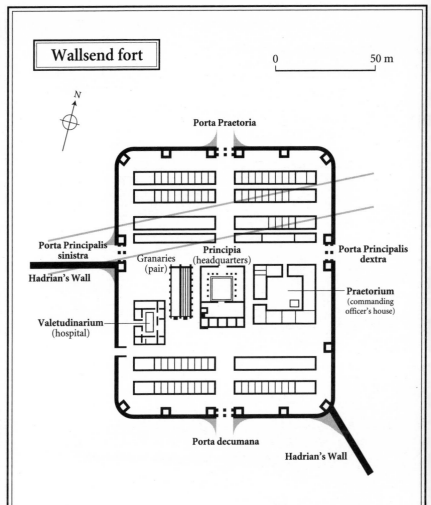

**Wallsend fort**

0          50 m

*N*

Porta Praetoria

Porta Principalis
sinistra

Hadrian's Wall

Granaries
(pair)

Principia
(headquarters)

Porta Principalis
dextra

Praetorium
(commanding
officer's house)

Valetudinarium
(hospital)

Porta decumana

Hadrian's Wall

Plan of the auxiliary fort (1.66 h, 4.1 acres) at Wallsend (Segedunum) on
Hadrian's Wall in Britain in the late second century (after Daniels). The
oblique lines represent a modern road. Most of the fort's area was filled
with barracks, some of which were used as stables.

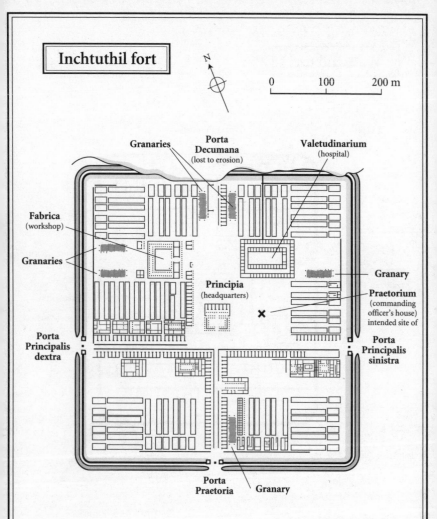

**Inchtuthil fort**

0    100    200 m

Granaries

**Porta Decumana**
(lost to erosion)

**Valetudinarium**
(hospital)

**Fabrica**
(workshop)

Granaries

**Principia**
(headquarters)

**Granary**

**Praetorium**
(commanding officer's house)
intended site of

**Porta Principalis dextra**

**Porta Principalis sinistra**

**Porta Praetoria**    Granary

Plan of the legionary fortress (22 ha, 53 acres) at Inchtuthil in Britain (Scotland) built and used *c.* 83–7 by Legio XX and perhaps VIIII Hispana during Agricola's campaign (after Richmond). The short-lived fortress was not finished before its demolition and clearance but remains one of the most complete legionary fortress plans known. To the west of the fortress were the temporary labour camps and to the east temporary compounds for the stores and officers, the latter including an elaborate officers' baths built after the officers moved into the fortress. These are not shown here.

# FOREWORD

G ladius is not a history of the Roman army, though it has a great deal of Roman army history in it. Nor is it a handbook of Roman military organization, equipment and fighting methods.[1] Instead, *Gladius* uses evidence from the Roman world to recreate a sense of what it was like to be a soldier in the army that brought the Romans their vast empire. During its many centuries of existence the Roman army fought its wars in places as far apart as northern Britain and Syria, North Africa and across the Danube. Roman troops faced enemies as diverse as Parthian cavalry and archers in the east, and Caledonian tribes who fought a guerrilla war from the forests and swamps of northern Britain. They were sometimes defeated but more often they prevailed. They came from every part of the Roman Empire and were stationed in isolated forts anywhere from the Arabian Desert to the Rhine, in great legionary fortresses like Lambaesis and Xanten (Vetera), in the Castra Praetoria in Rome, or took their turn in lonely watchtowers on remote frontiers. They variously sharpened their swords, were bullied by centurions, erected forts, built aqueducts and bridges, made weapons and equipment, policed civilians, collected taxes, sought promotion, wrote letters, had families, petitioned the emperor, marched on campaign, committed acts of great valour, participated in atrocities and worshipped their gods. Some died in service from disease, enemy action or accidents. Others lived to sign on again as veterans or retired to find their way in civilian life, with a few reaching remarkably advanced ages. The Roman army was the greatest force, perhaps in most respects the only force, the Roman state had to exert its power and influence over the Empire and beyond. Soldiers and veterans were present in every community in the Roman world.

The word *gladius* – or *gladius Hispaniensis*, to give its full name – was the name of the standard Roman infantry sword. The so-called 'Spanish sword' appeared in the days of the Second Punic War and remained in use for centuries. Few other examples of military weaponry endured so long. Although the exact shape of the tapering iron blade and its length varied over the period, all examples of the gladius shared a similar blade and a carefully shaped handgrip made of wood, bone or sometimes ivory, with a wooden pommel. Some of the most elaborate examples had decorative scabbards made of wood with metal skins embossed with designs. Although not all Roman soldiers used the gladius – cavalry troops had a longer version known as a *spatha* – it seems from a letter found at Carlisle that in practice Roman soldiers used the word *gladius* as a generic term for 'regulation swords'.[2] There can be no other piece of Roman military weaponry that so effectively symbolizes the Roman army. It therefore seemed only logical to name this book after the gladius.

*

Compiling this book has been an absorbing and very interesting experience. Indeed, the research could have gone on indefinitely. The flow of stories about soldiers, their units, their lives, families, preoccupations, and their failures and achievements, seemed to be unstoppable. Inevitably, compressing such a vast subject into so small a space has involved a great deal of selection, frustrating though it has been to have to leave so much out. The result is a primarily and unashamedly anecdotal text, made up of evidence from inscriptions, original letters and other documents, and the writings of Roman historians or commentators who composed personal accounts of what they had seen or used other sources which no longer exist.

Some instances and episodes were automatic inclusions, such as the Varian disaster of AD 9 or the military letters and records found at Vindolanda, a fort on Britain's northern frontier. In many other cases it was a question of making a choice, often a difficult one.

The book is also unavoidably unbalanced. This is a consequence of the material we have to hand. Rome's great age of warfare and conquest was the last three centuries BC during the late Republic, a period beloved of some of Rome's greatest historians like Livy, Polybius and Appian who provide us with detailed accounts of the period and in particular the generals, the campaigns and the battles. However, at that time the Roman army was not a standing force and we also know very little about individual soldiers. Under the emperors the picture turns almost volte face. Fighting was less sustained and wars of conquest were infrequent. Gaps in the sources also mean that apart from Tacitus, who only covers the first century AD (and even then not all of it), we have little to match what we know about the earlier period. Paradoxically this is when the ordinary soldiers, the *milites gregarii*, emerge from the shadows in the haul of tombstones and documents of imperial date recovered from sites across the Empire to provide us with a wealth of information about their lives.

The ancient texts have invariably been consulted rather than relying on modern secondary sources. These days with so much published online the task has been made much easier, whether that means reading a papyrus found in Egypt or an inscription from a legionary fortress on the Rhine. All those used here are fully referenced so that the reader can follow any of them up. The process is very instructive. The original wording of these texts is often somewhat more ambiguous than a secondary source might imply, but can also reveal additional information or nuances omitted from a secondary source. The many different ways in which a sword was referred to is an excellent case in point.

The army of the Romans belongs to the ages. But to read a letter from a soldier in Egypt or northern Britain, study the life of a centurion and his family on the Rhine frontier from his tombstone, or learn about the successes and disasters of legionaries and auxiliaries across the centuries in countless places from Syria to Spain, is to make the Roman army live again. No other military institution of the ancient world offers us this remarkable privilege.

It is important to make clear that it is the laborious work of untold numbers of scholars over the last two centuries or more that has made a book like this possible. The Further Reading section, inadequate though it is, should be considered a testimony to their efforts as much as anything else.

## A NOTE ABOUT STYLE

There is today, surprisingly, no agreed format for the names of Roman military units, even amongst Roman military specialists. A reader might find a particular legion referred to in any number of books variously as Legio III Augusta (its Latin name) or Third Legion Augusta, Third Augustan Legion, III Legion Augusta or Legion III Augusta. Auxiliary units are even more challenging: Cohors II Thracum might appear for example as Second Cohort of Thracians or II Cohort of Thracians; an auxiliary cavalry *ala* might be called an Ala, or a cavalry regiment or wing. With this in mind the decision was made early on in this book's preparation to use the Latin form as far as possible to avoid ambiguity. It is, incidentally, worth noting that the Roman army did not normally use IV for 'Fourth', IX for 'Ninth' or XIV for 'Fourteenth'. Instead IIII, VIIII and XIIII were more commonly used.* This book follows Roman military practice. Dates throughout are AD (CE) unless specified as BC (BCE). Place names are normally given in modern form unless the location has no modern equivalent, such as Vindolanda and Dura-Europos.

Sometimes Roman sources supply details of distances in Roman

---

* Needless to say, even that was not consistent. The Eighteenth Legion is attested, for instance, as Legio XIIX, i.e. 'ten plus ten less two'.

miles. I have not normally bothered to include laborious comparisons with modern miles; such precision is not really relevant, because the Roman mile was not measured with any high degree of accuracy or consistency (reflecting Roman society in general). A Roman mile was theoretically about 4,850 ft, or approximately 92 per cent of a modern mile (1.47 km). So long as the reader bears this in mind when coming across such references he or she will not be misled.

It is hoped those who read the book from start to finish will forgive occasional repetitions of information and dates. These are to help others who have chosen to read the book by dipping into individual chapters.

Guy de la Bédoyère
Welby by Ermine Street, a road built by Legio
VIIII Hispana in the mid-first century AD
*Lincolnshire 2020*

# ONE

# INTRODUCTION

## The Army of the Emperors

*The Romans have conquered the whole world by no other means than thorough training in the use of weapons, strict discipline in military forts, and practice at war.*

VEGETIUS[3]

The Roman army has been a source of fascination ever since antiquity. No other military force has lasted so long or achieved so much. The focus in *Gladius* is on the individual soldier, on real people whose personal experience of the Roman army has reached us through a variety of routes. The information comes from Roman historians, from inscriptions set up by the units, from official documents, and from the soldiers themselves or their families on their tombstones, religious dedications or even personal letters. These include everything from the records of individual soldiers to the stories of Rome's great battles, whether victories or defeats, and the physical remains of forts and military equipment. This material has been found all over the Roman world, and includes writing tablets preserved in bogs in northern Britain, inscriptions in North

Africa, papyri in Egypt, the physical remains of forts on the frontiers, and of the civil engineering projects undertaken by the army. Taken together they paint a remarkable picture of the largest permanent organization in the western ancient world, one without which the Roman Empire could not have existed, let alone been created; of the men who made it work and the women and children who shared their lives.

Imagine the whole Roman army, which numbered from around a quarter to not much less than half a million at its height (see below for how this has been arrived at), all gathered together on one vast parade ground. Now multiply that many times over to take into account all the soldiers who ever served in it over the centuries. If just 25,000 new men were needed annually to compensate for the discharge of veterans and those dying in service from illness or in war, then over the three centuries from the reign of Augustus to the accession of Diocletian it is clear that millions of men served in the Roman army at some time during its history. Obviously it is impossible to work out the actual total. Now imagine that vast parade ground emptied out abruptly. All that remains are a few scattered broken swords, fragments of shields, the odd helmet, as well as a few hundred scattered, ripped and damaged documents blowing about in the wind, some battered tombstones and dedications to gods. Apart from the derelict and usually buried ruins of forts, that is what we are left with today in terms of physical relics of the mighty Roman army. Of course we also have the anecdotes and accounts recorded by Roman historians, virtually all of which survive only in copies of copies of copies, usually incomplete and frequently short on essential detail, which made it through to the Middle Ages in monastic libraries.

Despite all that, the Roman army has survived into our own times as one of the most vivid of all ancient institutions, not least because the evidence for it is stronger than for any other part of the Roman world. Thanks also to cinema and television (despite all their reckless liberties with accuracy), books, re-enactors and education, the Roman soldier is often regarded today as the defining image of the Roman world. This is not at all inappropriate. The Romans 'surround the Empire with great armies, and garrison the whole stretch of land and sea like a shingle stronghold', said Appian.[4] Even the word Rome itself was derived from the Greek word ρωμη ('rome̅'), which means 'strength' or 'might'.

Writing in the early second century AD, the historian Florus stressed that until the reign of Augustus Rome had been almost permanently at war. Until 29 BC, he said, the doors of the temple of Janus had been symbolically closed only twice to mark times of peace. Valour and Fortune (*Virtus et Fortuna*) had competed to create the Roman Empire, 'so widely have their armed forces spread through the whole world' in the face of toils and peril. He was disparaging about what he called the 'inertia of the Caesars' compared to the endless wars of the Republic, but was delighted that the soldier emperor Trajan had reinvigorated Rome with war in his own time.[5] In the late fourth century an unknown poet harked back to the Republic too and was certain that adversity and war had driven the Roman people to success, concluding that 'therefore a protracted and oppressive peace is the ruin of Romulus' people'.[6] This was the cultural backdrop to every Roman soldier's life.

Roman soldiers not only built and lived in forts and went on campaign, but they also served in limitless capacities in everyday Roman life on behalf of the state. This was especially so under the emperors when the army had become a standing force and the Empire enjoyed long periods of internal peace. Soldiers were among the most literate members of the ordinary Roman population. More written material survives from the Roman army than from any other sector of society. The vast majority of soldiers' inscriptions or surviving documents are in Latin in the Western Empire and Greek in the East (though occasionally Latin appears in the East and Greek in the West). This shows that, regardless of where they came from, soldiers were accustomed to using the official languages of the Roman world, even if privately they still spoke to their fellow countrymen in their native tongue. Soldiers used writing all the time, whether it was to pursue grievances, claim pay or expenses, record loans, request leave, write letters, prepare their wills or appeal to their gods.

The army was the Roman world's biggest bureaucracy. Although most documents are long lost and rotted away, some remarkable material has been found, especially in Egypt and Britain, that shows how inclined to recording and cataloguing everything Roman military administrators were. Roman army units commemorated their achievements and activities in inscriptions. Individual soldiers were also far more likely than civilians, especially in frontier provinces, to record themselves on religious dedications or tombstones.

The Roman army is the best-documented military force of the ancient world. But that does not mean it is consistently or even well recorded. For example, the historian Appian explained that an individual report was made of every soldier's character. When Mark Antony wanted his tribunes to bring all the troublemakers in his army before him so they could be punished, the tribunes were able to do so by consulting these documents.[7] Such records are virtually non-existent today in any form for any part of the Roman army.

Inevitably, the evidence varies enormously across both time and place, not least because the details of Roman military history varied enormously. The time from the Second Punic War to the end of the Republic saw Rome's most sustained and extensive wars of territorial conquest, but it is also a period from which little evidence of individual soldiers has survived. Under the emperors there was a great deal of warfare but much of it involved Roman civil wars, and the defence or consolidation of frontiers. The campaigns to conquer Britain, Dacia and Parthia were exceptions. Conversely, the first to early third centuries AD are amongst the best recorded, albeit erratically, and include by far and away the most extensive evidence for the lives of individual soldiers from inscriptions, letters and other documents.

We have therefore to make the best of what there is. Although the works of, to take one instance, the Roman historian Tacitus are invaluable, we do not have all his *Annals*, which covered the period AD 14–68. His description of the invasion of Britain in 43 by Claudius, for example, is lost. His *Histories*, which carried the story on to the end of the first century, are only extant for the Civil War period of 68–9 and the early months of the following year. Military themes often dominated his history, but Tacitus had many other topics to cover and was not concerned with providing the level of detail modern historians, especially military ones, crave. Similar problems afflict the accounts written by other Roman historians such as Livy, Appian, Plutarch, and Cassius Dio (sometimes known as Dio Cassius).

These historians were also often writing about events that occurred generations or even centuries before they lived, and had to use documents and histories written by their predecessors that no longer exist. On the other hand we have, for example, some of the correspondence of Pliny the Younger from the early second century AD. He included references to officers and soldiers he knew personally, especially in connection with his letters to the emperor Trajan while he was governor of Bithynia. As a young man, before being elevated to senatorial status, Pliny had moreover served as an equestrian tribune with Legio III Gallica in Syria.[8] However, there is no comparable collection of correspondence from any other governor to any other emperor. To Pliny and Trajan's letters we can add anecdotes and asides that pop up in a myriad other places, such as the collection of past deeds and sayings compiled, with accompanying commentary, by Valerius Maximus in the reign of Tiberius; or Pliny the Elder's *Natural History*, which happens to include a description of the famous strongman praetorian Vinnius Valens among other military references. Valerius Maximus and Pliny provide some fascinating insights, but what they include is inherently rather random.[9] Much the same applies to Valerius Maximus and Aulus Gellius. Their compilations of anecdotes cover a huge variety of topics. In among what they recorded are snippets of information about military history and the Roman army, sometimes preserving references to much longer works that no longer exist.

Not all of these written sources are reliable; indeed they are all unreliable in some way. It is one thing to make that observation but quite another to know what to do about it. Evaluating the relative reliability of sources, especially where they contradict each other, is challenging enough. If every source is treated with outright scepticism then the only alternative is for the modern historian to substitute his or her opinion instead, a tactic that is unlikely to be any more reliable. However, there are certain principles which need to be considered, especially with military matters. All Roman historians were given to a greater or lesser degree to using rhetorical devices such as stock depictions of villainy, atrocities, brutality and heroism, or inventing stories that were transposed retrospectively onto accounts of the past. They were also inclined to provide numbers for armies and casualties that were patently rounded up or down, depending on the agenda. These figures should also be seen as rhetorical devices, rather than as statements of fact. There are many instances of this habit which had a long tradition, stretching back into the Republic. For example, Sulla said his army had killed 20,000 soldiers under Marius the Younger's command at Signia in 83 BC with the loss of only 23 of his men; or so Plutarch reported after reading Sulla's autobiography.[10]

In his account of the Civil War of 68–9, Tacitus described complex and fast-moving events several decades after they occurred, using sources about which we know next to nothing and which were unlikely to have been based on notes taken at the times and places concerned. Most of the time, we only have one source for an event. When we have two, such as Herodian or Cassius Dio for the Severan period, they often diverge in crucial detail.

Some of the biographies of second- and third-century emperors in the *Historia Augusta* are unreliable in important respects. Modelled on the *Twelve Caesars* of Suetonius which covered Julius Caesar to Domitian, they were attributed to six different authors such as Aelius Spartianus and Flavius Vopiscus, but these names were probably made up. The texts were compiled during the late third and early fourth centuries from official records and other unspecified sources, and are anecdotal in style in their treatment of emperors such as Hadrian, Septimius Severus and Probus. Although they make for entertaining reading, the texts frequently include what are obviously invented quotes from fabricated letters and speeches, as well as interpolations made by later contributors. The result is a series of pastiches incorporating a jumble of truths, half-truths and outright falsehoods. Deciding which is which is far from straightforward.

Some Roman military manuals such as those by Frontinus, Onasander and Vegetius have survived. These can provide some very useful information but present other problems. Vegetius, for example, wrote his military manual in the later fourth or early fifth century, a very long time after the period he was writing about. He may not have had reliable information to hand, and he was also inclined to idealise the world of 'Augustus and his illustrious successors' as he called it.[11] Another work on building a military encampment was once thought to have been written by a surveyor called Hyginus in the early second century but now seems to have been written a century or more later by someone whose identity is unknown.

Military sculptures can provide an exceptional visual record of the Roman army at war. Trajan's Column and the Column of Marcus Aurelius in Rome both have running friezes that record campaigns. They show everything from foraging and camp building to the crossing of rivers, fighting and parading. Triumphal arches were occasionally erected both by generals in the Republic and by emperors from Augustus' time on to commemorate great victories. Appropriately enough they featured sculptures of the campaigns, the despoliation of the vanquished enemy and other symbols of military success. In Rome today, three still stand: the arches of Titus, Septimius Severus and Constantine the Great, but there were once dozens more not only in Rome but also across the Empire.

As for the personal record of individual soldiers, there are few letters and documents and no diaries we can read through to find out about their experiences as we might do for, say, the Napoleonic Wars, the American Civil War or the First and Second World Wars. Unless they were copied into carved inscriptions, written documents of Roman date tend only to survive in waterlogged or arid conditions. The two provinces that have produced the most such documents are Britain and Egypt. Two places which could hardly have been further apart or more different, they were also not necessarily representative of the rest of the Roman Empire. The documents are usually damaged, difficult to read, and entirely random in their nature, but are often fascinating records.

There are also the far more numerous inscriptions that record imperial edicts, the construction of buildings, religious dedications in the name of a unit or individual soldiers, and tombstones. The texts on funerary memorials and personal offerings to deities, the vast majority on stone, provide the most tantalizing glimpses of military lives. They represent a tiny fraction of the number of soldiers who once served in the Roman legions or auxiliary forces. Almost all belong only to the first to third centuries AD, though few carry any precise dating information. The edict of Domitian in 94 declaring that veterans were to be exempt from certain taxes (see Chapter 15) illustrates just how random the record we have is. The text only survives because a veteran copied it down on a writing tablet found in Egypt.

Funerary inscriptions are by far and away the most important evidence for the lives of individual soldiers, whether as the deceased or as those responsible for commemorating their comrades, friends, children, wives or parents. A soldier was entitled to write a will and leave instructions as to how he was to be remembered, either by his parents, his family or his comrades (see the Epilogue). Tombstones of soldiers have been found at military sites across the Empire, and also in civilian communities where some were stationed or where as veterans they made second careers, though inevitably their survival is down mainly to chance. They record invaluable information which can include where the soldier came from, the units in which he served, his age at death, and details of his wife and children, freedmen, and friends.

For all their shortcomings, these inscriptions and documents of the Roman army of the emperors are a universe apart from the records of other ancient or medieval wars, where nothing comparable has survived or perhaps ever existed. To take the English Wars of the Roses, fought across the period 1455–85 as an example, there is not a single instance of any equivalent personal record of the thousands of ordinary soldiers who fought in that conflict.

The standard of modern publication varies hugely too – Britain, Egypt and Germany have been well served, some other provinces of the Roman Empire rather less so.

*Gladius* inevitably includes material covering a long period of time (about five centuries). It features a mixture of individuals from lowly recruits to exalted generals, and a number of specific battles, campaigns and rebellions, placing them in the broader context of life in the Roman army. The reader will find lists of Roman emperors and of Rome's wars at the end of the book, together with other appendices, tables and a glossary, which will make it easy to place any part of the main text in historical context as well as acting as a reference for Roman army organization.

The position with the evidence may sound bleak, but in fact the Roman army has been the subject of such intense scrutiny and study that a remarkable amount is now known, even if that usually means relying on information gathered from different times and places in what must seem sometimes to be a haphazard way. This is unavoidable but there is always a danger that in the absence of anything comparable, a body of evidence – the Vindolanda letters are a good example – will be taken as representative of the Roman army at all times and places. Usually there is no way of knowing. However, there is reasonable consistency in the evidence, which suggests that most parts of the Roman army operated in a fairly similar way. The way almost all forts known today from excavation and extant remains resembled one another is a plausible basis for assuming that. Nonetheless, perhaps the most remarkable aspect of the Roman army – one that has become more and more conspicuous – is how much local variation there was in almost everything from basic equipment to the way army units were organized.

## THE EMPERORS AND MILITARY COMMAND

The emperors had supreme command of the army, but it was typical of the Roman world that their military power was cloaked in nebulous and tenuous links to the days of the Republic. When Augustus took power he embarked on a charade of having restored the Republic in order to avoid any public admission that he ruled in the capacity of a monarch. For the Romans any whiff of kingship was completely unacceptable but everyone was prepared to play along with him for the sake of peace and stability.[12] A successful Republican general could be acclaimed *imperator* ('commander') by his army. Such an honour could be conferred on more than one general at any one time and crucially did not denote precedence of any one over another.

Imperator was itself linked to *imperium*, the power of military command. Under the Republic, imperium was granted by the Senate on a strictly temporary basis to the consuls when a war had to be fought, and could not be held within the sacred boundary (*pomerium*) of Rome. In exceptional cases it could be awarded to a more junior man of senatorial rank. For example, Scipio was awarded imperium in 212 BC when he was only twenty-four because his military skills had become a matter of life or death for Rome in the Second Punic War.[13] This all changed under Augustus in 23 BC, when the Senate voted that he could not only hold imperium in Rome but also did not need to have it renewed. In addition, his imperium was deemed to exceed in authority that held by any consul or former consul and was known as *maius imperium proconsulare*. In addition, the Senate annually voted him and his successors the privileges and powers of a tribune of the plebs which allowed him to pose as the protector of the plebs.[14] Although there was no Latin word for emperor, the fact that Augustus' special imperium had become an integral legal component of his supreme power led seamlessly in the future to the title imperator becoming synonymous with that supreme power and is the basis of our word emperor.

Under the emperors it was briefly possible for a general to be acclaimed imperator by his legions in the manner of the old days as a special favour. Augustus allowed this several times prior to 27 BC, Tiberius just once in AD 22.[15] Thereafter only the emperors held the title, often enumerated on their coins, but from then on the legions were still quite capable of making a unilateral acclamation of their general as imperator. When this happened they were effectively declaring him supreme ruler in place of the incumbent emperor. This was what happened in 68 when Galba's legions declared for him and initiated the Civil War of 68–9, again when civil war broke out in 193 in the aftermath of the murder of Commodus, and on and off throughout the third century.

## LEGIONS, AUXILIARY INFANTRY AND CAVALRY, AND IRREGULARS

Writing about the state of the Roman world at the death of Augustus in AD 14 Tacitus said 'the Empire was enclosed by the Ocean or far-off rivers. Legions, provinces, fleets – all were interconnected'.[16] Even so, the Roman army under the emperors was a remarkably varied collection of units, ranging from the legions to irregular bands of infantry or cavalry hired from frontier tribes. The Roman army reached its most coherent and consistent form under the emperors, especially from the time of Augustus until the middle of the third century AD. But it never operated as a single institution, and its overall size and the disposition of its units changed continuously.

Accounts of the army's origins in Republican times provide us with a great deal of evidence for how its various elements came into existence. Great generals of the Republic, such as Scipio Africanus and Scipio Aemilianus, were greatly venerated in later centuries.[17] The Augustan poet Virgil even had Aeneas discover the two Scipios during his visit to the Underworld, when they were pointed out to him as 'the two thunderbolts of war'.[18] Stories about the Scipios were included in admiring anecdotes and references in all sorts of later works, for example Valerius Maximus' collection of 'memorable deeds and sayings'. Disastrous battles of the Republic, like Cannae in 216 BC, were also remembered, but as warnings.

There were important differences between the armies of the Republic and the army under the emperors. In the Republic armies were raised on an as-needed basis and were made up predominantly of Roman citizens from Rome and the surrounding area, supplemented later by Italian allies. This reinforced a sense that their purpose was to defend the homeland, especially during the First and Second Punic Wars, even though by the second century BC the army was more likely to be used in predatory wars to seize more territory. Under the emperors the army was not only permanent but was also distributed predominantly in the frontier provinces. In a process that started in the late Republic, its manpower was largely drawn from much further afield, such as legionaries from Gaul and Spain, or Thracian, Batavian and Sarmatian auxiliaries. These soldiers were liable to feel a sense of loyalty to their units and commanding officers, or to their own homes, rather than to Rome – a place which few of them had probably ever seen.

The best-known parts of the Roman army today are the legions. The infantry legions were dominated by men from Italy, Gaul and Spain, but by no means exclusively so. They were supported by provincial auxiliary units (*auxilia*) which ranged from crack cavalry wings (*alae*) to basic infantry cohorts. The Praetorian Guard (*Cohortes Praetorianae*), the imperial bodyguard, was the elite section of the army and the most privileged. It was based in Rome where it held sway along with its associated cavalry force, the *equites singulares Augusti*. The Guard was commanded by one of the most senior equestrian prefects in the Roman world, the *praefectus praetorio* (praetorian prefect), and was supplemented by the urban cohorts (*Cohortes Urbanae*), effectively a city police force created by Augustus in around 13 BC and commanded by the *praefectus urbi* (Prefect of Rome). There was also the night watch or fire brigade (*Cohortes Vigilum*, or *Vigiles Urbani*) under the *praefectus vigilum*, established by Augustus in AD 6. Although the Guard and the urban cohorts were nominally based at the Castra Praetoria in Rome under the emperors from the time of Tiberius on, in practice they were often sent out to various places across the Empire, either as cohorts or in the form of individual soldiers detached on imperial duties. There were, for example, additional urban cohorts stationed at Lyon and Carthage. Scattered around the Empire were various fleets, such as the *Classis Britannica* and the *Classis Germanica*, also placed under individual equestrian prefectures.

The numbers of legions, auxiliary units, Praetorian Guard, urban cohorts, Vigiles and fleets all varied constantly, and it is impossible now to do more than estimate how many there were of each at any specific time. Only rarely do we have any evidence for the army's size at any given date, especially under the Republic. When Livy researched the size of the army assembled in late 217 BC he found his various unnamed sources disagreed about the numbers and types of additional recruits. 'I should hardly to venture to declare (these) with any certainty', said Livy, before explaining how the figures differed, drastically affecting the possible number of legions and men involved.[19] He made no attempt to decide which was most likely to be accurate. (This incidentally illustrates why figures for battle casualties given by ancient historians are virtually meaningless, other than as metaphors for the scale of relative losses.)

The position was not a great deal better under the emperors. Tacitus provided a reasonably detailed description of the army's disposition in 23 during the reign of Tiberius. Even so, he did not write this down until around seventy years later, using records he must have found in the imperial archives. There were, he said, eight legions on the Rhine, three in Spain, two in North Africa and another two in Egypt, four in Syria and the Middle East, and four on the Danube with an additional two behind them in reserve, making a total of 25 legions. Rome had its own army, made up of nine praetorian cohorts (Tacitus does not say how big these cohorts were) and three urban cohorts, with the whole of Italy defended by a fleet on the west coast at Misenum and on the east at Ravenna. He added that the men in the auxiliary units of cavalry and infantry came in total to not much less than those of the legions, but was at pains to stress their numbers constantly fluctuated according to requirements.[20] From this we can gather that the legions amounted to about 125,000 men, and if Tacitus was right the auxiliaries came to roughly the same again. The Roman army in 23, then, numbered approximately a quarter of million soldiers of various types, scattered across a vast Empire but mainly on the frontiers, distributed in provinces where the need was greatest. Their distribution varied over time as legions were raised and lost, or moved, and the same applied to auxiliaries. Tantalizingly, an inscription found in Rome lists all the legions, and the provinces in which they were stationed, but can only be approximately dated to the second century AD, with additions made after 160.[21]

One survey of attested auxiliary units, both cavalry and infantry, known to have been in existence under Hadrian around 130, has provided an estimate of just under 218,000 auxiliaries in service by then, with a ratio of about 2:1 in favour of infantry. Another survey generated a figure of just under 181,000 auxiliaries. The numbers are nominal, prone to all sorts of caveats (see Chapter 5 for surviving strength reports of individual units), and the difference shows how susceptible the evidence is to variant interpretations. However, both estimates suggest that the auxiliary forces had grown considerably since the reign of Tiberius.[22]

Writing in the early third century, Cassius Dio looked back at the army's size in AD 5 under Augustus so that he could make a comparison with his own time. He reckoned there had been 23 or 25 legions in those days, although like Livy he said his sources did not agree with one another. Of those, 19 still existed when he wrote, but he explained that a further 13 had been raised since Nero's time, making a grand total of 32, or approximately 160,000 men (see Table 1 for a summary). Dio was uncertain about the auxiliaries: 'I cannot give the exact figure', he said. But if the estimates for the auxiliaries under Hadrian are at least approximately true it seems likely the Roman army numbered about 340,000 to 380,000 men by the reign of Septimius Severus (193–211). It may even have been considerably larger; some modern estimates take it up to around 450,000 plus another 30,000 or so in the fleets.[23]

Dio added that the Praetorian Guard was made up of 10,000 praetorians in ten cohorts, in addition to an unspecified number of veterans who had signed on again, in the army of Augustus in AD 5.[24] There is some confusion over whether he had transposed figures from his own time concerning the Praetorian Guard, as obviously the figure does not match the nine cohorts mentioned by Tacitus for 23. Other evidence shows the Guard varied in size, with the result that we cannot be certain of its exact numbers or organization at any one time. The same applies to the Cohortes Urbanae, for which Dio gives no figure. Neither Dio's account nor that of Tacitus (both of which can be found in the Appendices) refers to other parts of the army that we know to have existed, for example the equites singulares Augusti, the 300 *speculatores* who served as the emperor's personal mounted bodyguard, or the special units of German bodyguards called the *Germani corporis custodes* hired by some emperors.

The evidence shows that, whatever the position at one date, it was different at others, and usually in ways we cannot now resolve. It is extremely unlikely that the Romans themselves ever knew the army's size with any precision. It would have been impossible to produce a wholly reliable census of the army's strength, which in any case would have been out of date the moment it had been compiled. Legions also operated largely independently of one another. The men might bestow their loyalty on their commanding officer or the provincial governor if the emperor fell short of their expectations. This was what made it relatively easy for rival would-be emperors, who were frequently provincial governors, to mount rebellions leading to civil wars, for example in the turmoil following the disastrous reigns of Nero and Commodus.

The Romans moreover used inconsistent numbering systems to label both legions and cohorts, whether praetorian or auxiliary. Disbanded or lost legions, or auxiliary units, disappeared from the numerical sequence, leaving gaps (see below).[25] Legions themselves were not organized in a broadly consistent way until the first century BC, though they had been a key part of the army for much longer than that. By the time of the emperors, the structure of a legion still bore some resemblance to that of the army described by Polybius in the mid-second century BC, while having changed significantly into a form that lasted until well into the third century AD.

Each legion from Augustus' time until the late third century AD theoretically numbered around 4,800 men, divided into ten cohorts of 480 men each, plus centurions, optiones, officers and at least in some cases 120 cavalry (whether all legions had 120 cavalry is unknown. See Chapter 6). By the late first century, the first cohort was doubled in size, but not necessarily in every legion, elevating the total to 5,120 men plus the extras.[26] In practice there is not a single reference in the sources to the exact size of any one legion. The general assumption is that a legion's strength was probably between 5,000 and 6,000 soldiers most of the time. In time this changed. In the early fifth century Vegetius described what he called the 'ancient establishment of the legion', but said a legion consisted of 6,100 infantry and 726 cavalry in ten cohorts. Although he implied this was how many men a legion had had in earlier times it is a clearly a mixture of old detail and arrangements in legions closer to his own time, especially the much larger number of cavalry.[27]

Each legionary cohort was made up of six 'centuries' of 80 men each, except for the first cohort which was made up of five double centuries of 160 men each (800 in total).* 'Century' meant what it sounds like: 100 men under the command of a centurion (*centurio*). But it was typical of the Roman world that an old word had been kept on when its original meaning had become obsolete. A century of 80 men was divided up into ten 'tent-parties' or 'squads' (*contubernia*) of eight men. Each *contubernium* (the word means 'with a tent') shared a tent on campaign, or rooms in a barrack block in the unit's fort. Pairs of facing barrack blocks in forts may represent the continuity for two centuries of the old term *manipulus*. The word means 'a handful' but came to be a colloquial military term for a group of soldiers. In the later Empire 'among the maniples' meant 'among the general soldiery'.[28] There were also the legion's cavalry contingent, plus various non-commissioned officers, the centurions and the optiones.

The Roman legion mirrored Roman social structure. Each was commanded by a senator of senior rank, the *legatus legionis*. Normally, the only exceptions were legions based in Egypt. Here senators were prohibited from entering or holding office: there was always a fear they might use the province's massive resources and mount a challenge to become emperor. Instead Egypt was governed by the *praefectus Aegyptii* rather than a senatorial legate and each Egyptian legion was commanded by an equestrian *praefectus*. Until the third century, one

---

* A normal-sized cohort's six centurions, each commanding a century of 80 men, were arranged in an ascending order of seniority:

*hastatus posterior* – rear man armed with a spear
*hastatus prior* – forward man armed with a spear
*princeps posterior* – rear captain
*princeps prior* – forward captain
*pilus posterior* – rear spear
*pilus prior* – forward spear (except for the *primus pilus* who was first centurion of the first cohort in the legion)

In the double-sized first cohort the centurions were *hastatus posterior, hastatus, princeps posterior, princeps* and *primus pilus*. Although little is known about first cohorts, the likelihood is that they were the pre-eminent force within each legion.

of the few exceptions to this general rule was when Commodus placed equestrians in certain legionary commands in 184 as part of his vendetta against senators.[29] During the third century equestrian legionary commands became the norm.

The legatus legionis was the emperor's personal delegate. He was a senator in his thirties who had reached the status of a *praetor*, a senior magistracy that entitled him to be awarded a military command. He was by now approaching the climax of his senatorial career, which might involve proceeding next to a provincial governorship, or to a consulship in Rome, the most senior magistracy, and then to the most senior provincial governorships as a proconsul. He might command the legion for only a few years. The legate's second in command was the *tribunus laticlavius*, which meant 'tribune with the broad purple stripe'. This was a reference to the fact that he was a senator too, albeit at an earlier stage of his career, and thus had the right to wear the senatorial toga, whose wide stripe of purple indicated his rank. He would move on in due course to a variety of positions which might include commanding a legion himself. These two men were the only men of senatorial rank in the legion.

After the two senior senatorial posts in a legion came the *praefectus castrorum*, 'prefect of the camp', a man promoted to equestrian status. The position was filled by a soldier who had been in the army all his working life and had made his way up through the centurionate to be *primus pilus*, 'first spear', the most senior centurion in a legion. Marcus Pompeius Asper was one such man; it is recorded on his elaborately carved tombstone that he had served as a centurion in Legio XV Apollinaris and Cohors III Praetoria, and as primus pilus of Legio III Cyrenaica, before his final promotion to praefectus castrorum of Legio XX in the late first or early second century.[30] As so often for a man of his seniority, his earlier career was considered trivial by comparison and not worth mentioning.

After the praefectus castrorum came five *tribuni angusticlavii*, 'tribunes of the narrow purple stripe', who were men of equestrian rank. These were officers who could be given any task involving leadership and might subsequently go on to command an auxiliary unit. Below the tribunes came the centurions. Each centurion had a second called an *optio*, a word that meant 'assistant'.

Legions were also often temporarily split into detachments known as *vexillationes* ('wings'), based in and fighting in different places. The word came from *vexillum*, 'flag', because each vexillation had its own flag. Veterans were sometimes organized into wings attached to legions, augmenting their numbers. A few surviving strength reports show that soldiers were liable to be detached either in groups or individually for an almost infinite variety of duties, such as serving on the governor's staff, keeping the peace in a city, or supervising at the mines. On the edge of London, the capital of Britain, a fort housed the governor's bodyguard. Soldiers from all over the province were detached from their legions or auxiliary units to serve there on his staff, recorded on inscriptions found in London. Similar secondments happened across the Roman world, such as to take part in building work. Construction of Legio XX's fortress at Chester in Britain seems to have been suspended for some time in the second century because so many of the men were away working on building Hadrian's Wall. Other absences at any fort were caused by sickness, or by leave for personal reasons. In short, were we able to travel through time and visit a legionary fortress there would be virtually no chance of finding anything like the whole legion there at any given time, and exactly the same would happen on a visit to an auxiliary fort.

Although legions numbered anywhere from I to XXII (First to Twenty-Second) – and, incongruously, XXX (Thirtieth) – have been identified from the evidence from Augustus' time onwards, there was no regular sequence in operation. After his victory at Actium in 31 BC Augustus had both his own army and Antony's, a total of 60 legions. He cut these down to 28, but the numbering does not reflect that. Six different legions styled I (the 'First') are known to have existed at some point; they are distinguished only by their epithets, which might identify where they were raised or mark some special quality they had, such as Legio I Germanica or Legio I

Flavia Minervia. A legion's life might end prematurely because the soldiers had been defeated, or even wiped out. Three legions were destroyed by German tribes in AD 9 in the humiliating Varian disaster. Consequently their numbers, XVII (Seventeenth), XVIII/XIIX (Eighteenth) and XVIIII (Nineteenth), were never used again. One of the later legions, Legio XXX Ulpia Victrix, was raised by Trajan in the early second century. The numbers XXIII (Twenty-Third) to XXVIIII (Twenty-Ninth) were omitted. Perhaps raising a Legio XXX helped compensate for the duplicated lower numbers and provided a more accurate overall impression of the army's size. However, another explanation is that Legio XXX was supposed to be the first of a new series of legions, starting with XXX.[31]

This peculiar numbering almost certainly had its origins in a much more coherent system that operated under the Republic. Lawrence Keppie has suggested that originally I–IIII were the numbers allocated to the four legions raised by two consuls if circumstances meant that they had to form an army during their year of office. Higher numbers were then allocated to additional legions raised for specific deployment on campaign.[32] The legions were also only numbered, rather than having more elaborate titles.[33] In 58 BC in Gallia Cisalpina and Transalpina, the garrison was made up of the VII–X legions when Caesar arrived to take up command. A centurion called Numerius Granonius from Luceria in Italy was serving in Legio 'XIIX' (the Eighteenth) in the East when he died in Athens in the mid-first century BC, suggesting it was one of another block of higher-numbered legions allocated to serve in the area. He had also served in Pompey's Legio II.[34] The inscription is, incidentally, exceptionally rare for the date and largely serves to illustrate that nothing comparable from anywhere else at that time has been found to clarify which legions were stationed where. All that can be said is that some of these legions might have survived the late Republic's civil wars, retaining their numbers and being joined by others that were formed along the way.

It is equally possible that the numbers are purely coincidental. Any geographical significance was long lost by the first century AD, which was why Britain had four legions numbered II, VIIII, XIIII and XX by then. When XIIII was withdrawn by 70 it was temporarily replaced with another one numbered II (Adiutrix, drawn from fleet troops). Likewise, the Praetorian Guard was known collectively as the *Cohortes Praetorianae*, 'the praetorian cohorts'. The number and size of the praetorian cohorts varied across the period, with mysterious gaps in the attested sequence. The same applied to the urban cohorts.[35]

Under the late Republic and the emperors, the legions, or at least those that remained in existence, developed strong identities, especially if they had enjoyed great success and were favoured, which played an important part in morale. Caesar, for example, was particularly fond of his Legio X.[36] Most acquired names that reflected their history or the nature of their formation, and these names can help distinguish them. Legio I Flavia Minervia was raised by the emperor Domitian of the Flavian dynasty, who had a special interest in the goddess Minerva. As early as the 30s BC, one of the legions numbered VI became known as Legio VI Ferrata, 'Ironsides', clearly a nickname acquired from its performance in war and linked to the word *ferrum*, used sometimes for swords. Similarly, Legio VI Victrix was called 'Victorious'. Legio XIIII Gemina Martia Victrix had one of the longest names. The Gemina ('twin') element commemorated how, shortly after the Battle of Actium, it was reformed out of one of Octavian's legions with the addition of soldiers drawn from Antony's disbanded army. Martia Victrix means 'the Warlike, the Victorious', titles awarded for its victory over the tribal leader Boudica in Britain in 60–1. It enjoyed those titles for the rest of its existence. One of the most unusual was Legio V Alaudae, using a Gaulish word meaning 'the crested larks'.[37] The legion was raised by Caesar in Gaul, and the only explanation for the name is perhaps that wings were worn on the legionaries' helmets (but see below for its incongruous emblem).

In the Greek-speaking eastern part of the Empire private correspondence used the Greek convention of letters for numbers, and substituted a Greek translation of the legion's titles. Legio II Adiutrix, 'the Helper', was formed by Vespasian in 70 from fleet soldiers, possibly as a reward for their support in the Civil War. It was sometimes referred to by its own men in Greek as λεγεών β βοηθός (Legeōn II Boēthos), a direct translation of the Latin.[38]

The auxiliary units defy attempts to place them in any kind of neat categories. Even Roman historians made no attempt to do so. Auxiliaries were generally units originally made up of soldiers raised from a provincial ethnic group. They served in theory for 25 years; in practice discharge was carried out in batches on specific dates. For example, troopers and soldiers were all discharged together from three cavalry alae and 21 infantry cohorts respectively in Moesia Superior on 8 May 100.[39] The auxiliary forces were frequently distinguished by ethnic labels, such as the Cohors VIIII Batavorum, but in reality the men serving at any one time could have come either from Batavia (a region on the lower Rhine and now part of the Netherlands) or any one of a myriad other places in the Empire (see Chapter 2). On completion of service, auxiliaries were made Roman citizens on honourable discharge. Some whole units were awarded Roman citizenship as a reward for valour while still in service, but that only applied to the individual men involved in the action which had earned them that honour and not subsequent recruits to the unit.

Auxiliary organization varied wildly too. Auxiliaries were only ever arranged into units that approximated in size to a legionary cohort or double cohort. Anything bigger might have risked creating a potential rebel force of dangerous size. In small units the auxiliaries were useful for manning frontier garrisons and outpost forts, or for being attached to legions as supplementary forces, particularly where auxiliary cavalry could compensate for the legion's minimal mounted component. Auxiliary units were usually led by men of equestrian status, who could be prefects or tribunes. Sometimes centurions detached from legions were used instead. Infantry were formed into cohorts of a nominal 500 men, a *cohors quingenaria*. These were made up of six centuries of 80 men each (480), with centurions and their assistants, the *optiones*. They were commanded by prefects. There was a 'double-sized' version known as the *cohors milliaria*, which literally meant '1,000 strong', though these were in fact made up of ten centuries (800 men). A milliaria cohort was usually commanded by a tribune. Some infantry cohorts were part-mounted, meaning there was a cavalry component, indicated by the addition of *equitata* to the unit's name. It is likely, but uncertain, that a *cohors equitata quingenaria* had about 120 cavalry and 480 infantry, while a *cohors equitata milliaria* had around 240 cavalry and 800 infantry.

There were also the cavalry units, known as *alae*, 'wings', which were organized differently to infantry. An ala was divided into *turmae* ('squadrons') of around 30 or 32 cavalrymen each, led by a *decurion* and his second, a *duplicarius*. An *ala quingenaria*, made up of 16 turmae (about 512 men), was commanded by a prefect; an *ala milliaria*, made up of 24 turmae (about 768 men), by a tribune. Again, the totals were nominal. Actual strength varied from unit to unit and from day to day. Virtually no detailed history of auxiliary units is known. They were in a constant state of flux. Tacitus explained that it was impossible to identify or enumerate them all for that reason. He also tells us that in exceptional instances, like civil war, gladiators could be recruited as extra but 'disreputable' auxiliaries.[40] The same numbering problem that affects legions applies also to the auxiliary units. In Britain a Cohors VIIII Batavorum is known. Cohortes I–IIII Batavorum are also attested, but V–VIII have never been identified and either never existed or had had only short lives before being disbanded.[41] One possibility is that the ethnic epithet is misleading and the numbering was based on something else. Cohors XX Palmyrenorum at Dura-Europos in Syria is the only such Palmyrene unit attested. The number may just refer to it being the twentieth cohort in Syria when it was formed.[42] Specialist units, such as archers or even camel riders (*dromedarii*), were also hired. And then there were the irregular units of cavalry (*equites*) and infantry (*numeri*), taken on when needed but whose men enjoyed few privileges beyond being paid.[43]

A papyrus found in Egypt and written in 63 offers a good example of the various legal provisions affecting different parts of the Roman army. It was concerned with the rights and entitlements of veterans from legions, auxiliary cavalry wings and infantry cohorts, and of men who had served in the fleet. The veterans were so disgruntled at the way they were being treated differently from other serving soldiers that they had decided to make a representation to the prefect of Egypt, Caius Caecina Tuscus. The legionary veterans seem to have accosted him 'on the camp road at the temple of Isis'. Caecina Tuscus admonished them and told them not to make 'an ungodly uproar'. He told them that no one was harassing them, and instructed them to write down their names and addresses so that he could make sure the local officials (*strategoi*) left them alone.

When the legionaries handed over the documents on 4 August, Caecina Tuscus asked them if they had done so individually, which they assured him they had. They greeted each other the next day, and on 7 August the legionaries went to hear Caecina Tuscus address them from the tribunal (the original Greek text describes this as being in an 'atrium' ('hall'), presumably part of a basilican building where the governor heard cases). Realizing that the men were so angry that a rebellion was possible, Caecina Tuscus pointed out that the legal rights differed for each branch of the service (legions, auxiliaries or fleets) and that this would affect what was due to each man. He reminded them that 'I told you (this) before'. He said he would see to it that those individual rights were established and guaranteed.[44]

The identity and loyalties of each legion and auxiliary unit were displayed on a variety of different types of standard which the soldiers were supposed to venerate and protect to the last man. Every legion had an *aquila* (eagle) standard in the charge of the aquilifer standard-bearers. They were made of, or at least plated with, gold and were stored in a shrine at the legion's winter quarters.[45] On campaign they were placed on a long pole with a sharp point at one end. It was the general Gaius Marius who in 104 BC came up with the idea of making eagles the sole symbol of the legions. Before that the legions had used not only displayed eagle standards but also ones depicting minotaurs, wolves, horses and bears. Marius ordered those to be given up permanently – though in fact they were already falling out of use – and thereafter 'a pair of eagles' was routinely displayed in the legion's winter camp (for Marius' wider changes to the army, see Chapter 2), according to Pliny the Elder.[46] A legionary eagle standard was not supposed to leave the unit's base unless the whole legion was on the march. It was carried on a long pole by one of the legion's aquilifers, who planted it in the ground when the legion made camp. When Titus Flavius Surillio, aquilifer of Legio II Adiutrix, died around the year 214 aged forty his tombstone was erected by his fellow aquilifer Aurelius Zanax.[47]

Every unit, legionary or auxiliary, had *imaginiferes* who carried a standard bearing an image of the emperor. Josephus described military standards with 'busts of the emperor' attached to them.[48] In the late first century Virssuccius was a trooper and imaginifer with the Cohors I Brittonum equitata when he died aged thirty-five in Pannonia Inferior. His tombstone was set up by his heirs Albanus, probably another soldier, and a fellow imaginifer called Bodiccius.[49]

Standard-bearers of all varieties, for whom the generic word was *signiferi* (from *signum*, another word for a standard), stuck together when it came to funerals and other activities. Attianus Coresi was a *vexillarius* (flag-bearer) of an unnamed unit in Germania Superior in 239. Together with an imaginifer called Fortionius Constitutus he invested in a shrine and stone votive tablet dedicated to the Imperial Divine House and the Genius *vexillariorum et imaginiferorum* ('of the flag and imperial image bearers').[50]

The vexilla flags were said by Dio to resemble sails, by which he meant they were hung from a cross-bar on a pole and provided a setting for symbols of the unit, including the name of the unit and the name of commander spelled out in purple letters. He was referring to the late Republican army of Crassus in 53 BC.[51] From Augustus on, the name of the emperor took precedence. Legions had motifs which appeared on their standards and sometimes also as sculptures, used to lead the cohorts and centuries on the march and in parades, and to rally the men on the battlefield. Legio XIIII's emblems included a capricorn symbol, associated either with the date Octavian took the title Augustus (16 January 27 BC) or when he was conceived. Legio II Augusta shared the same emblem and also used a figure of Pegasus. Caesar's Legio V Alaudae, despite its name, used an elephant, which represented the animals in Metellus Scipio's army, which the legion had defeated at the Battle of Thapsus in 45 BC. Appropriately, Legio XII Fulminata ('Lightning-Bearer') used a thunderbolt.

Standards played an essential visual role in coordinating fighting. In AD 15, during Germanicus' campaign in Germany, part of the army's baggage became bogged down. In the chaos the Roman formations were disrupted, and the soldiers could neither see their standards properly nor hear orders. The German tribal leader Arminius spotted his chance and attacked. The battle raged around the standards, which could not be carried forward in the hail of weapons. Nor could they be erected where they were. The battle was nearly lost and in the end only the German tribes' preference for chasing after plunder and the bravery of Legio I helped save the day.[52]

Planting standards where a victory had been won was a powerfully symbolic act. When Titus seized the citadel in Jerusalem in 70, 'the Romans erected their standards on the towers, clapping and singing to celebrate their victory with joy'.[53] Affiliation to a legion and loyalty to the legion's reputation and identity became an important part of a soldier's career. Even detachments of soldiers made up of a century or two sent out to build roads or bridges, or perform any other task, had banners or flags (*vexilla*) that they stuck into the ground wherever they were working to identify themselves and serve as rallying points.[54]

Standards also appeared in imperial parades, serving as imperial propaganda in a civilian context. Gallienus laid on one in Rome in around 261 as a publicity stunt to divert attention for the military disasters and troubles that had afflicted his reign. Along with the senators, equestrians, a troupe of 1,200 gladiators and other groups, the standards of the legions as well as statues from their sanctuaries, and the standards of auxiliary units, were displayed along with those of the guilds of the professions.[55]

Disgracing the legion (or indeed any unit), and especially losing its standards, by being defeated was a devastating blow both collectively and individually. Recovering lost standards was the only way to restore face. Those lost by Crassus at Carrhae in 53 BC were eventually brought back by Tiberius to great acclaim in 20 BC, and those lost by Varus in AD 9 were recovered one by one in AD 15, 16 and 41.*

Members of the Roman armed forces found themselves in an organization that was hugely complicated by a variety of legal arrangements, terms and conditions, custom and practice, precedent, and ad hoc circumstances. Evidence shows, for example, that while the children of auxiliaries, born while their fathers were serving, were enfranchised when their fathers were discharged and themselves enfranchised, the children of praetorians born in service were not, even though their fathers were already Roman citizens.

In short, anyone looking for a predictable and reliable system of organization in the Roman army will look in vain. Units, even legions, came and went. No one seems to have been troubled by the chaotic numbering system, the nominal nature of auxiliary ethnicity or the use of terminology that implied units were of fixed sizes. A nice neat picture of a Roman legion or cohort's deployment belies the everyday reality that saw various members of the units scattered across the province on different duties, or even engaged in illicit second jobs; instead the picture emerges of an organization that often operated on an ad hoc basis with semi-autonomous units. Yet, somehow, it usually worked brilliantly and made the Roman army the most powerful military force in antiquity.

Whatever the differences in detail, the Roman army pervaded every corner of the Roman world. The army was made up of men drawn from one end of the Empire to the other and beyond. With no central 'high command', this was an army based on numerous

---

* See Chapter 7.

individual units widely dispersed around the Roman Empire, each with its own commanding officers and overseen by a provincial governor. Many soldiers had wives, legal or otherwise, and children. Most soldiers spent at least some of their time fulfilling duties on behalf of the state, but there is a great deal of evidence that soldiers also engaged in private business during their term of service.

This book focuses mainly on the Roman army up to the time of Constantine I (307–37), even if some of the sources post-date his reign. The late Roman army was a different institution from the one described here. Probus (276–82), one of the most successful soldier emperors, was said in his 'biography' to have promised a 'golden age' to his people in which there would be no wars, no forts, soldiers could leave the army and become farmers or students, and weapons would not need to be made.[56] Whether Probus ever said anything like that it is impossible now to say, but it reflects an era weary of endless civil wars and militarization.

What happened to the Roman army by the fourth century? The answer is that it changed dramatically, largely in response to the way the Empire had been forced onto the defensive.[57] One of the Empire's darkest days came in 260 when Valerian I, desperate to avoid fighting Shapur I's Persians with his own plague-ridden army, tried to talk terms and buy them off. Valerian set off to negotiate with Shapur but was captured by him and ended his life in a state of degradation and humiliation unprecedented for an emperor. The historian Zosimus dismissed Valerian as 'effeminate and indolent'.[58]

One of the few significant records we possess for the late Roman army is the *Notitia Dignitatum*, a document compiled in the late fourth and early fifth centuries that records government offices and military units across the Western and Eastern Empires. It was made up from various sources but its exact date and author are unknown, and survives only in medieval copies. It tells us little more than the names of units, where they were stationed and the title of the official in overall charge of groups of units such as 'the Count of the Saxon

Shore in Britain'. Some of the old names, such as those of the legions, survived, but it is clear from the limited evidence that they were differently organized and of variable sizes though we know almost nothing about either. Legio II Augusta, for example, was said by the Notitia to be based in Richborough, a coastal fort in east Kent in Britain that was much smaller than its former long-time legionary fortress at Caerleon.[59]

Diocletian had begun the process of change at the end of the third century, splitting the army into the *limitanei* and the *comitatenses*. The limitanei were the fixed frontier defence forces, manned by Roman conscripts and hired barbarians. The lives of these soldiers blurred into the civilian communities that grew up around them. The comitatenses were mixed units of infantry and cavalry, but the mounted soldiers were far more important and provided Diocletian and each of his imperial colleagues with a highly mobile force that could race to trouble spots. In 312, Constantine I abolished both the Praetorian Guard and their mounted contingent, the equites singulares, for backing his enemy Maxentius. They were replaced with new types of bodyguard, including the *protectores*, which had first appeared half a century earlier or more, and the *scholae palatinae*, a mounted body-guard created by Diocletian. We have immeasurably less evidence for the lives of individual soldiers from this time because the habit of creating inscriptions waned dramatically. However, legal decisions involving soldiers at this date often do survive in the emperor Justinian's sixth-century Codex and Digest; a number of these are cited in this book.

Regardless of the reorganization, the army was increasingly on the back foot, the victim of circumstances more often than defining them. Losses in frontier wars were only kept at bay by relying on the induction of barbarian tribesmen as confederates into the Roman army. In 378 Valens was trying to organize a major resettlement of Visigoths into the Empire. It went spectacularly wrong. Roman officials took bribes instead of relieving the Visigoths of their weapons. Next they extorted the tribesmen with overpriced food. The

Visigoths rose up and were joined by other tribes. Valens set out against them but failed to bring enough soldiers. He was defeated and killed on 9 August 378 at Adrianople, making him the first Roman emperor to be killed in battle by a barbarian force. Two-thirds of his army was annihilated. The historian Ammianus Marcellinus considered the disaster the greatest massacre of a Roman army since Cannae 594 years earlier.[60]

The death of Valens was not the end for Rome, at least not yet, but by the end of the fourth century the great days of the Roman army had long been just a memory. In 396 the priest and theologian Jerome said that 'the Roman army, conqueror and ruler of the world, is now conquered by [the mounted Huns], and are terrified at the sight of these men who cannot walk on foot'.[61]

# TWO

# STRENGTH AND HONOUR

## Signing On in Caesar's Army

*Sextus Valerius Genialis, a trooper of the Thracian cavalry*
*regiment, a member of the Frisiavones tribe, in the squadron*
*of Genialis, lived forty years and served twenty. He is buried*
*here. His heir erected this.*

Tombstone, Cirencester, Britain[1]

In AD 68 the emperor Galba, subscribing to the view that he wanted the best, famously boasted 'I choose my soldiers, not buy them.'[2] Tacitus nostalgically called it an 'impeccable statement of public policy'. In reality Galba's brag was monumental hubris. Every Roman soldier had his price. A few months later, in January 69, Galba found himself being murdered by some of his own soldiers who had been bought by his rival Otho. In reality the common soldiery, the *milites gregarii*, of the Roman army were drawn from a vast range of men of all sorts and from across the Empire. Soldiers were enticed into the service with the prospect of regular pay, theoretically enviable conditions of service, bonuses paid by emperors on their succession, retirement grants of land and money, and even bribes.

They were expected to aspire to the qualities of *virtus et honos* ('strength and honour'), qualities that became divine personifications and were worshipped in temples like those built by the general Pompey the Great in his theatre complex in Rome.


## Recruitment under the Republic

Back in the earliest days of the Republic, the financial costs of serving were borne largely by the men. Military service started out being at the soldiers' expense (pay was not introduced until the very end of the fifth century BC (see Chapter 4). Under the original system, military service was an obligation which all free Roman men were bound to fulfil, each according to his financial circumstances. It was a temporary arrangement because there was no standing army. With the war or campaign over, each man returned to his farm or business and attempted to resume his normal life. It was only possession of land in the first place that had made him eligible for service. There was little prospect of making a reliable career out of being a soldier, even though he might be called up again one day during his 20-year term of eligibility, which could be served between the ages of seventeen and forty-six. Luck, or misfortune, played a big part in a man's chances of becoming a lifetime soldier. Sometimes, if he was lucky and the army had been victorious, a soldier brought home the proceeds of his share of the booty. In 297 BC, during the Third Samnite War, shortly after the seizure of the city of Murgantia, the Roman general Decius encouraged his men to sell their shares of the booty to the traders who followed the army around and promised them there would be plenty more to come.[3]

In 171 BC a new Roman army was being recruited for a war in Macedonia. Some of the veteran centurions from previous campaigns were keen to sign up, until they discovered their former ranks were to be ignored. Furious at the thought of demotion, they appealed the decisions. Spurius Ligustinus, one of the veterans, stepped in to make a speech to try and calm things down. He is one of the earliest individual Roman soldiers (as opposed to officers) we know anything about in terms of his career. Ligustinus, who had first signed on as a soldier in 200 BC, proudly proclaimed his origins as a member of the Sabine people in Italy, his modest background on a small farm where he still lived, and his family of six sons and two daughters. He recounted a remarkable career which had seen him promoted to centurion in his third year as a soldier. He had fought in Macedonia, Spain, and against Antiochus III of the Seleucid Empire. He was a modest man and said 'it is for the military tribunes to decide what rank they think I deserve. I will take care that no one will exceed me in valour.' He chided the other old soldiers gently and reminded them that they should 'place themselves at the disposal of the senators and consuls and treat any position in which you are defending your country as an honourable one'. As a result the other centurions abandoned their appeal and signed up without further argument.[4]

## THE ARMY OF THE MID-SECOND CENTURY BC

In the middle of the second century BC a Greek historian and soldier called Polybius wrote a history of Rome. He wanted to explain why Rome had become so powerful by his time. Polybius was fascinated by the rise of Rome and was especially intrigued by how the Roman army worked under the Republic. He was also a friend of the Roman general Scipio Aemilianus, and was thus in an excellent position to pursue his project through the contacts he made. He travelled with Scipio during the Third Punic War and saw the destruction of Carthage in 146 BC at first hand. Polybius realized the army was the main reason Rome had come to dominate the Mediterranean area in such a short period of time.

When an army was raised a lottery was held which determined in which order of the 35 Roman tribes soldiers would be enrolled from.

The process of recruiting an army began when the Roman political system of voting by tribes elected the two annual consuls, the most senior magistrates. Next, 14 military tribunes were appointed and divided up among the four legions that were being formed, each consul commanding two legions. A laborious ritual followed in which batches of four men at a time were brought forward, and the officers of each legion chose one each in order of precedence. The word for a legion, *legio*, was derived from *leguntur*, 'are gathered', which referred to how soldiers were collected together when an army was raised.[5] Once a legion had 4,200 men it was considered complete, unless an emergency required it to be enlarged to 5,000. Each legion was allocated 300 cavalry, chosen from those who could fulfil a higher property qualification. The new conscripts were then obliged to swear an oath of obedience (see Oaths below in this chapter). The Roman army of this period was strengthened by ordering the magistrates in allied cities in Italy to supply designated numbers of men, chosen and organized in a similar way. These normally amounted to the same number of infantry as the Roman citizens, but three times as many cavalry.[6]

The new conscripts were sent home once they had been sworn in, but with orders to turn up on a day chosen for their legion to assemble. They were only exempt if they had to attend a funeral, were suffering from disease, had experienced an omen that could not be expiated, were obliged to attend an anniversary sacrifice, or had been attacked by a foreigner. Any other failure to turn up meant being branded a deserter.[7]

When the recruits reassembled they were organized into four classes, based on tiers of the property qualification and their ages. The oldest men were allocated to the 600 *triarii*, and those in the 'prime of life' to the 1,200 *principes*. Below them came the 1,200 *hastati*. The remaining 1,200, who were the youngest and on the bottom rung of the property qualification, were known as the *velites*. All except the velites elected twenty centurions, divided into ten centurions *priores* and ten centurions *posteriores*. Each centurion appointed his second, an *optio*. Each of the three senior classes was divided into ten companies known as maniples, with two centurions and two optiones each, divided into two centuries. The centurions then chose two standard-bearers, *signiferi*, for their maniples. These men had to be the bravest because they would have to lead by carrying the standards into battle and draw the troops on behind them. Finally, the velites were divided up among the maniples at 120 each. The 300 cavalry per legion were organized into the ten squadrons known as turmae. Three officers called *decuriones* and three optiones were selected from each, with one of the three decurions commanding the squadron.[8]

The four classes of infantry soldier were also differentiated by equipment. The velites carried a sword and javelins, and wore a plain helmet. The hastati were better equipped, carrying the so-called 'Spanish sword', two throwing spears and a long shield, and wearing a bronze helmet and leg greaves. The triarii and principes had largely the same equipment but carried thrusting spears instead of throwing spears.[9] The newly organized and equipped soldiers were then sent home again. They were given another date and place on which to reassemble, with no excuse acceptable for failing to turn up unless bad omens or unavoidable circumstances had prevailed; the normal practice was for the two consuls to choose different locations. The allied soldiers were also ordered to present themselves on the same occasion, so that they could be organized under the consuls' supervision.[10]

The picture Polybius painted of Rome's militia army reflected the enormous respect he had for the Romans, based on what he knew they had achieved in the First and Second Punic Wars. In reality the pristine and ordered distribution of troops with their equipment he described was probably rather more ragged. Absence, illness and other factors were bound to have made the numbers constantly variable, while the arms used were far more likely to have been based on what was available.

## A MANPOWER CRISIS

A Roman citizen cavalryman in the army described by Polybius had to serve overall for 10 years, whereas an infantryman had to serve for 16 years, both by the age of forty-six. Those below the property qualification threshold of 400 drachmas for army service had to serve in the naval arm. If there was a national emergency the infantry could be ordered to serve for 20 years, and no man could stand for political office until he had served at least 10 years.[11] It was therefore possible for a man to be taken away from his home to serve for two decades of his most productive life. That meant he could not farm or develop a trade, and nor was he likely to be able easily to father children at home.

Not long after Polybius wrote his account, the reliance on forcing huge numbers of free men into military service erupted into the greatest political crisis in the Republic to date. Tiberius Gracchus was a man of senatorial rank whose career proceeded on fairly conventional lines to begin with. He saw action in the Third Punic War with his brother-in-law Scipio Aemilianus, and in 137 BC, during the Numantine War in Spain (143–133 BC), served as *quaestor* to the consul Gaius Hostilius Mancinus, who was then commanding the army.

So many Roman citizens had been taken away for extended military service during Rome's wars that there was a crisis on the land. Greedy senators had been blithely helping themselves to property those men might otherwise have been farming, and had been creating estates operated by slaves, thereby ruining many peasant families. Tiberius Gracchus was not only worried that the increased number of slaves would lead to a rebellion, but also that so many free peasants had been dispossessed that it would be impossible for Rome to recruit enough men for its legions when they were needed. Everything that Rome had achieved might be lost. Neither was Tiberius the first man of his class to believe that a potential disaster was looming. But when a friend of Scipio Aemilianus called Gaius Laelius proposed measures in or around 140 BC to restrict the working of estates by slaves instead of free men, he was warned off and backed down, earning the ironic name *Sapiens* ('wise' or 'prudent').[12]

Tiberius Gracchus ignored the threat. His legal campaign for reform was to end in violent disorder in Rome, and in his murder. He waded into a quagmire of 'unimaginative conservatism and entrenched interests', his views of the Senate's arrogance perhaps compounded by his experience in the Numantine War.[13] War had made Rome rich, but the money had not benefited the ordinary masses anything like as much as it had the senatorial elite.

The unpredictable withdrawal of men from the land to serve in the army meant that it was difficult to operate farms efficiently. Unfortunately, the use of slaves on the land made more sense. The rich might have been greedy, but it was also unacceptable to compromise Rome's existence by threatening the routines on which agriculture depended. One of the reasons the senators had been able to appropriate the land in the first place was because so many men were absent and unable to operate their own farms.

Elected tribune of the plebs for 133 BC, Tiberius Gracchus tried to force through land reforms to rectify these problems. He was frustrated by senators who were determined not to give up the land they had appropriated. When he sought election for another term as tribune (to widespread senatorial outrage) he was murdered by a senatorial mob. A decade later his brother Gaius, also tribune of the plebs, tried to institute reforms (see Chapter 4). Elected for a second term without any protest in 121 BC, Gaius eventually committed suicide before he was himself destroyed by the Senate. Some subsequent tribunes of the plebs attempted to introduce further reforms, but with limited success.

## THE MARIAN REFORMS

Although lowering the property qualification was one way of increasing the numbers of eligible troops, the solution was eventually to be the creation of a professional standing army in which men could make a lifetime career. When in 107 BC Rome was fighting a war in Numidia, Rome's militia system found itself unable to service this conflict as well as others. The consul Gaius Marius, placed in charge of carrying on the Numidian War, decided to take volunteers instead of relying on the traditional methods of recruitment. Highly ambitious, and scornful of the privileged senatorial elite, he had already ingratiated himself with the troops by living as they did. Now he chose what Plutarch called 'poor and insignificant' men whom previous commanders had passed over.[14] These new recruits owed their chance to improve their lives to Marius and they responded accordingly. He also made some provision for retirement grants in the form of seized Gaulish land handed to his men.[15] Such endowments offered them a permanent route out of destitution. This transference of loyalty from the state to the person of the general was significant. Soldiers were always opportunists. Their loyalty was easily bought, but the buying and selling was done by generals. The result was the best part of a century of intermittent civil war.

Marius also did away with the idea of arming and equipping soldiers differently according to their relative wealth. His new recruits had no status anyway. Marius' troops went into battle with the same equipment and a sense of shared purpose, status and identity. They carried their own kit and were better able to look after themselves on the march and when they camped. Another significant difference from the old system of raising legions on a basis of need was replaced by Marius' idea of legions as permanent organizations with individual identities. This engendered fierce loyalty and a sustained sense of common purpose.

The Marian military reforms started a process in the last century of the Republic that led to ambitious Roman military leaders forming what amounted to personal armies. Marius himself determined to be given command of the war against Mithridates of Pontus in 88 BC. He sent tribunes to take over the army of his arch-rival, Lucius Cornelius Sulla, which was based at Nola. However, the type of loyalty Marius had encouraged backfired. Sulla's men were loyal to Sulla. They killed the tribunes and stripped and humiliated praetors sent out by the Senate to encourage them to back down. Sulla then marched into Rome with six legions. This spectacularly outrageous breach of precedent was one of the most infamous acts of the Republic.[16] Marius fled. An ugly civil war followed, which resulted in Sulla becoming dictator. This led to the age of the imperators, a period of instability characterized by powerful armies loyal to their generals, such as Pompey and Caesar. It eventually reached its climax in the triumvirate of Octavian, Antony and Lepidus, formed in 43 BC after the assassination of Caesar in order to pursue his killers. The three political associates became bitter rivals and squared up to each other, swaggering into Rome with their legions and praetorians. It was hardly surprising that the triumvirate collapsed as Antony and Octavian went to war with each other over control of the Roman world.

## The army of Augustus

In 31 BC Octavian won supreme power by defeating Antony at Actium, and established his regime.* Augustus, as Octavian was known from 27 BC, wanted an army of career soldiers whose loyalty to his person was synonymous with loyalty to the state. He reduced its size of around half a million by more than 50 per cent, paying off veterans from his legions and from Antony's, leaving an army in AD 5 of the approximate size described by Dio.[17] Augustus also made permanent arrangements for his legionary veterans by providing them with discharge grants and the opportunity to settle in new

---

* His legal name was now Gaius Julius Caesar, since from 44 BC he had been Caesar's adoptive son and heir. This can be confusing to a modern readership, so his traditional name of Octavian is normally used, as it is in this book. In fact his real birth name was Gaius Octavius.

colonies on conquered land. By taking responsibility for his veterans he managed to combine the soldiers' loyalty to him with loyalty to the state. Whether they were citizens or provincials in search of citizenship, soldiers could now embark on a career in Augustus' army and be confident that they would be taken care of when they had served their term.

## ENLISTMENT AND LEVIES

There were no Roman army press-gangs that we know of. Under normal circumstances there was no routine need during the time of the emperors to force young men into the army. The prospect of pay, a profession, food and board, retirement grants and the chance of excitement and prestige were incentives enough for men with no other future. Even so, the emperor Tiberius believed only the 'impoverished and vagrants' were likely to take up soldiering voluntarily.[18] Vegetius, writing several centuries later, was keen to point out that being a legionary was extremely hard work. Some potential recruits were put off, he claimed, by the severe discipline and the physical load, and – if they enlisted at all – preferred to sign on with auxiliary units instead because of the prospect of lighter duties.[19]

Any would-be soldier aged in theory between seventeen and forty-six had to prove his eligibility by passing an approval process called *probatio*, along with a medical examination. Physical fitness was obviously important. Vegetius commented on how it had always traditionally been the case that a cavalryman or soldier in the first cohort of a legion had 'to stand six or at least five [Roman] ten inches' (1.7–1.8 m) in height,[20] although he advised that by his time a shortage of suitable men meant such limits should be overlooked.

Intelligence was only a consideration insofar as a new recruit (*tiro*, a 'beginner') had the right mental attitude and the ability to understand his training.[21] Mental acuity was not necessarily regarded as a virtue. Soldiers with minds of their own could be dangerous men, especially if they were natural leaders. Spartacus, the celebrated leader of the slave revolt in 73 BC, had originally been a Roman auxiliary. His intellect was identified as having been a key factor in the danger he presented to Rome.[22] The same applied to Civilis, a Batavian tribal leader who had joined the Roman army as an auxiliary commander before leading a destructive revolt against the Romans on the Rhine.

Men like Spartacus and Civilis were exceptional. In contrast, during the Parthian campaign conducted in 216 by Caracalla, two soldiers started squabbling over a skin filled with wine, each claiming it as his booty. Unable to settle the row, they presented themselves to Caracalla and asked him to adjudicate. Dio said it was remarkable that they could be so disrespectful as to trouble the emperor with such a trifle. Caracalla told them to divide the wine equally between themselves. The men obliged by cutting the skin in two, in the expectation that they would each walk away with half the wine. The wine of course spilled out onto the ground and they lost the lot. It served them right for their stupidity, which was perhaps what Caracalla had in mind.[23]

Sometimes soldiers could be hoodwinked. Gordian III, who succeeded in 238 after a period of turmoil and several other brief reigns following the death of Severus Alexander in 235 (see the Maximinus story later in this chapter), was only about thirteen and as small as one might expect a boy of that age to be. To impress the soldiers he was allegedly carried around on the shoulders of a tall man, apparently successfully.[24]

Each new recruit was carefully scrutinized for any physical defects and also for features that suggested he was alert, well-toned, with an erect head, broad chest, muscular shoulders, strong arms, long fingers, narrow waist, and slim buttocks, all of which would indicate his personal qualities, 'just as one would examine a horse or dog', advised Vegetius.[25] The emperor Trajan, however, declared that a man born with only one testicle, or who had lost one, could still serve, though what had inspired such a decision is unknown.[26]

Distinguishing marks became the way a soldier was identified, not only so that he could not easily desert but also presumably to help pick him out should he die in battle. Gaius Longinus Priscus, aged twenty-two in AD 103 in the reign of Trajan (98–117) when he was sent to join a cohort in the Fayum, an oasis and extensive settlement in Egypt to the west of the Nile, was distinguished by having a 'scar on the left eyebrow'.[27]

Only Roman citizens could join the Praetorian Guard or the legions, although the latter were sometimes supplemented in emergencies by freedmen or even slaves. Provincials were able to join the auxiliary forces. Slaves were normally excluded, along with deserters, adulterers, exiles and those condemned to being savaged by wild beasts. In later dates being a Christian was also a disqualifying factor, since a Christian's loyalty to the state could not be guaranteed; this would change only when Christianity was legitimized.[28] Men whose civilian professions were dismissed as being more suitable for women, such as 'fishermen, fowlers, confectioners, and weavers', were not thought ideal.[29]

Disease, desertion and discharge (whether honourable or not) meant the demand for new soldiers never let up.[30] Sometimes pressing military requirements might oblige a general to go out actively looking for recruits. A levy (known as a *delectus*, 'selection') was a rare event and generally only followed a crisis or preceded a major war. In 275 BC, the consul Marcus Curius was forced by circumstances (the war with Pyrrhus) to call a sudden levy in Rome. When no one came forward, Curius proposed a lottery. All the names of the tribes of Rome were put in an urn. The first to be drawn was the Pollia tribe, but the first eligible young man of that tribe on the list did not come forward. Curius ordered that the man's property be sold at auction. The reluctant conscript was furious and protested to the tribunes. Curius decided not only to sell the man's property but also the man himself, on the basis that the Republic had 'no need of a man who did not know how to obey'.[31]

In 215 BC, during the Second Punic War, the huge losses to date had occasioned a recruitment crisis. The normal system had to be suspended. A levy was organized for all men over seventeen, and even some younger ones, to raise four legions and 1,000 cavalry. The numbers, however, had to be made up with the purchase of 8,000 slaves at public expense.[32] Valerius Maximus, referring to the same occasion, claimed that all the new recruits, 24,000 in all, were specially purchased slaves. 'Sometimes', he said, 'a noble spirit gives way to expediency', adding that even 6,000 debtors and convicts were recruited. They became known as the *Volones* and were remembered as having been volunteers.[33] In 63 BC a senator called Lucius Sergius Catalina led a plot to topple the consuls Cicero and Hybrida. In the midst of the emergency Cicero ordered the praetors to 'administer the oath of enlistment to the populace' as a means of preparing everyone in case soldiers had to be recruited.[34]

A rebellion in Pannonia in AD 6, during Augustus' reign, took place because most of the garrison had been sent to fight in a campaign in Germany led by the emperor's son-in-law Tiberius. After the Romans suffered a large number of defeats and heavy losses in the war to crush the revolt, Augustus sent out Germanicus, Tiberius' nephew, with a fresh army. It was made up not only of freeborn citizens but also of freedmen. Augustus had to pay the cost of their manumission and compensate their former owners for six months' keep.[35]

Three years later, in AD 9, an even more serious catastrophe occurred when the XVII, XVIII and XVIIII legions were lost in Germany. To make up the numbers – around 15,000 legionaries alone – Augustus had to resort to desperate measures. 'There were no citizens of military age left worth mentioning', said Dio, and those who were available did not want to sign up, which was hardly surprising given the appalling news. The auxiliaries had suffered heavy losses too. Augustus had no choice but to force men into the army. He made them draw lots: every fifth man who was under thirty-five had his property seized and his right to vote removed. Every tenth man over thirty-five suffered the same fate. Extreme though the measures were, they still did not produce enough men prepared to join up. Extraordinarily, Augustus started executing some of those who had ignored the call to arms, but ended up having to draw lots among men who had already completed military service to send back into the army. He also forced freedmen into the army, though in their case it is possible the opportunity represented more of a privilege and a step up.[36]

Augustus' levy showed how dangerous trying to force men into the army could be, especially if the men concerned were unsuitable and unwilling. In Syria in 58 the general Domitius Corbulo was preparing for a campaign in Armenia but found that Legio III Gallica and Legio VI Ferrata were relying far too heavily on veterans who were taking part in absolutely no military activity, and instead swanned around in the local towns on private business. Corbulo dismissed them but had to organize emergency levies in the provinces of Galatia and Cappadocia. How far this involved coercion is not clear but an element of pressure must have been applied, though the opportunity of regular pay and security of employment may have been enough for some. The levies nevertheless failed to raise enough recruits. Corbulo had to have another legion sent out, as well as auxiliary cavalry and infantry.[37]

In the Civil War of 69 Corsica was one of the Mediterranean islands kept on the side of the short-lived emperor Otho by a Roman fleet patrolling the nearby waters. However, the governor of Corsica, Picarius Decumus, loathed Otho and decided he would use the island's resources to support Otho's rival Vitellius who was trying to topple him and become emperor himself. Picarius had two senior fleet officers executed and organized a levy on the Corsican men – who were furious, having no interest in or knowledge of military discipline. Realizing the war and other Roman forces were a long way away, they decided to act. When Picarius was unattended in the baths, a group made their way in and killed him and his staff. They tried to curry favour with Otho by sending him Picarius' head, but events in Corsica went unnoticed in the middle of the tumult sweeping the Empire at the time.[38]

Later in 69, Vitellius, who was now emperor, faced the prospect of many of his army going over to his rival Vespasian. Vitellius made a reckless offer: anyone who signed up with him would be discharged after victory over Vespasian, and then receive all the grants and privileges a veteran would usually receive only after a full term of service.[39] The offer was meaningless: not long after, Vitellius was defeated and Vespasian became emperor.[40]

Regardless of the theoretical minimum requirements, in reality almost anyone might be signed up or kept in service. Hadrian had to order that no one be enlisted who was 'in military service younger than his age warranted' or so old that it was inhumane to keep him on.[41] During the reign of Marcus Aurelius a plague, possibly smallpox, ravaged the Empire and killed a huge number of people, including many troops. With the Empire's frontiers increasingly under assault it was essential more men be found. The emperor resorted to having slaves trained for military service, as well as Dalmatian bandits. Naming them after the Volones, the slaves who had volunteered to serve in the army in the Second Punic War, he called them the Volunteers, which suggests they were given the option of remaining slaves or agreeing to serve. The *Diomitae*, a form of military police similar to the Cohortes Urbanae but who worked in the Greek cities of the Eastern Empire, were incorporated into the Roman army, along with gladiators who obviously already had useful weapons experience. These were called *Obsequentes* (the Compliant).[42]

Caracalla later used a levy as a punishment. He arrived at Alexandria in the winter of 214–15 and proceeded to hurl abuse at the population because he had heard they had been mocking him. He issued an order that anyone with the necessary physical attributes should be enlisted, but subsequently ordered that they all be killed, copying the example set by the pharaoh Ptolemy Euergetes. It formed part of a larger massacre of Alexandrians instigated by Caracalla, who instructed the soldiers he had brought with him to do the killing.[43]

The great and mighty legions were made up of men who hailed

from opposite ends of the Empire and anywhere in between. Legio III Cyrenaica and Legio XXII Deiotariana spent at least some of the period 30 BC–AD 110 in Egypt. A remarkable document records the places 36 of the legionaries had come from. Seventeen alone were from Galatia (now central Turkey); others were from Syria, Cyprus, Gaul, Italy and even Egypt, among other places. Two are mysteriously labelled *castris*, 'in the camp'.[44] The only sensible interpretation is that the two were born in a fortress or its environs, presumably to the unofficial wives of soldiers, and had followed in their fathers' footsteps. Italy, Gaul and Spain, however, were the most common sources of legionaries in the earlier Empire. Between the reigns of Augustus and Caligula, evidence from inscriptions indicates that the majority came from Italy north of Rome and south of the Alps.[45]

For the most part the men were immensely strong and resilient, apart from when poor leadership allowed garrisons to fall into indolence. Among the bodies recovered on the beach at Herculaneum in excavations during the early 1980s was that of a man in his late thirties wearing a military belt and a gladius. He had been crushed by the pyroclastic flow when Vesuvius erupted in the late summer of 79, was not wearing armour and therefore must have been dressed in an everyday tunic. The man was also equipped with an adze and three chisels slung over his back. At some earlier date he had suffered a serious wound to his left thigh, but the recovery and healing to the bone indicated an individual who was in excellent condition and well nourished.[46] He cannot be identified as a soldier for certain, but it is most likely that he was, and if so he was probably a praetorian on duty in the region.[47] Finds of known Roman soldiers' corpses are extremely unusual, but one such was the skeleton of a soldier in his mid-twenties found at Viminacium; this showed that the mere wearing of armour and equipment had left pressure marks on a number of his bones, such as his shoulder blades and pelvis.[48]

Praetorians were often unusually tall and powerful, and were specially selected for that reason.[49] Vinnius Valens was one of the most famous individual praetorian soldiers, for the reason that he was a celebrated strongman. The legendary muscleman lived in the time of Augustus. He was said to have been able to hold wine carts in the air while they were unloaded, and also to have been capable of stopping wagons with one hand. Valens was commemorated with a tomb that was still standing in Pliny the Elder's time. Two centuries later a Thracian legionary called Maximinus, famous for his physical strength and height, so impressed the emperor Septimius Severus that he had Maximinus transferred to the Praetorian Guard. Maximinus did well in the Guard. In the year 235 he led a rebellion against the last emperor of the Severan dynasty, Severus Alexander, toppling and murdering him. Maximinus then ruled as emperor himself until he too was killed in 238 while on campaign in Italy against his rivals. It showed how far a man could go in the Roman army even if it did end in tears.[50]

Maximinus' short reign is sometimes seen as a turning point for the Roman army.[51] Prior to 235 soldiers were loyal to an emperor if they respected him and showed leadership, though this could be amplified if like Hadrian he had had a respectable military career and continued to try to live like a soldier. He was not expected to pull out his sword and join in the fray. From Maximinus on the emperor became a warrior-ruler, fighting alongside his men in battle and in some cases dying beside them too as Philip I and Trajan Decius did in 249 and 251 respectively. Any emperor deemed to have fallen short as a warrior-ruler, however well he had started out, was liable to be assassinated by the soldiers and replaced with perfunctory ease by someone at least temporarily perceived as more convincing in the role (and who also was able to promise generous pay and handouts).

## HONOURS

Being decorated after a battle or war was a major event for a soldier or even a whole unit. A range of crowns (*coronae*) and other awards was available, depending on the nature of the action. They included the triumphal, siege, civic, mural, camp and naval crowns. The triumphal gold crown was awarded to a commander who had been awarded a triumph. An ovation crown made of myrtle was available for the commander who had had an easy victory, fought a low-grade enemy (such as slaves) or fought a war that had not been officially declared. A general who relieved a siege was presented with a crown, made from grass grown in the besieged location, by those he had rescued. The oak-leaf civic crown was awarded to a citizen soldier who saved the life of another in battle. The first man to scale the walls of an enemy's stronghold won a gold mural crown, while the gold camp crown went to the first man to fight his way into an enemy camp. The gold naval crown went to the first soldier to board an enemy ship. There was also the incongruous notion of the olive crown presented to someone who had been awarded a triumph but who had not fought in a battle. These awards were a matter of enormous prestige but could be abused by commanders anxious to curry favour with their men. Cato made a point of accusing Fulvius Nobilior, consul in 189 BC, for awarding crowns on the slightest pretext.[52]

Lucius Antonius Quadratus of Legio XX was decorated twice during the reign of Augustus with bracelets and necklets, awards proudly commemorated on his tombstone with representations prominently displayed on either side of a legionary standard.[53] The stone was found at Brescia in northern Italy, where he had retired after military service. He had almost certainly served in Illyricum under the governor Marcus Valerius Messallinus, who in AD 7 had confronted a rebellion there with only half of Legio XX at his disposal. Despite that handicap, Messallinus managed to fend off the rebels, who allegedly numbered 20,000. For this he was awarded a triumph.[54]

Towards the end of the Jewish War in 71, Titus, eldest son of the emperor Vespasian, had the whole of his army in Judaea parade before him, and commended the soldiers for their bravery, exploits and obedience. He had already been provided with a list of the men who had acted the most bravely. The men concerned were promoted and awarded gold crowns, gold necklets, gold spears and a share of the spoils. They would brag about the occasion for the rest of their lives. The ceremony was followed by a thanksgiving sacrifice of oxen that were then distributed among the army as food.[55]

The men of a victorious auxiliary cohort or ala could find themselves elevated to Roman citizenship, while a legion might be awarded special titles. If on the other hand a unit fell short of what was expected, then ignominy was the normal outcome. After celebrating his army's success in Judaea in 71, most of Titus' men had cause to be proud. All, that is, except the soldiers of Legio XII, which he remembered had been defeated by a Jewish force at the Battle of Beth-horon and lost its aquila standard. The legion fought well thereafter, but was still punished by being removed from its comfortable garrison in Syria and stationed beyond the Euphrates. Conversely, Legio X Fretensis was repaid for its success by being allowed to garrison Jerusalem.[56]

The emperor Probus served as a soldier early in his adult life. He was said while holding the rank of tribune to have performed with exceptional bravery during a Sarmatian war, allegedly crossing the Danube to carry out his feats. The account of his life is not considered to be especially reliable, but the story of his military career probably has at least some truth in it. He was presented with four spears, two rampart crowns, one civic crown, four white banners, two gold bracelets, one golden torque, and a sacrificial saucer weighing five pounds (probably of gold). After this he was promoted to command of 'Legio III' (possibly Italica).[57]

## LEGIONARY CAREERS

Military tombstones from early in Britain's conquest period give us a good idea of the sorts of places from which some of the legions were drawn in the middle of the first century AD, as well as the careers of individual soldiers. Wroxeter and Lincoln were two fairly short-lived fortresses, established within a few years of the invasion in 43 and used for only twenty years or so. Gaius Mannius Secundus, a soldier of Legio XX, passed through Legio XIIII Gemina's base at Wroxeter while serving on the staff of the governor as a *beneficiarius*, but seems to have died while he was there. The word *beneficiarii* referred literally to those who were 'beneficiaries' of the extra privileges awarded to men on such duties. These could include guarding other visiting imperial officials, clerical tasks or even something as extreme as torturing prisoners.[58] Secundus' tombstone says he came from Pollenza in Piedmont, Italy. He was fifty-two when he died and had served 31 years, so he had evidently stayed on in the army long after he needed to.[59]

Marcus Petronius came from Vicenza in north-eastern Italy and joined Legio XIIII Gemina when he was twenty years old. He had served for 18 years, some of that time as a signifer, when he died at Wroxeter.[60] Gaius Saufeius served for 22 years in Legio VIIII, signing up when he was eighteen and dying in Lincoln aged forty. He came from a city called Heraclea Lyncestis in Macedonia (now Bulgaria). At twenty-three, Lucius Sempronius Flavinus was a little older when he signed on, but only managed seven years' service with Legio VIIII before expiring at thirty. He had come from Clunia in Hispania Tarraconensis (northern Spain). Titus Valerius Pudens was twenty-four when he was recruited from Szombathely, a Roman colony, in Pannonia (Hungary) but also died at Lincoln also aged only thirty, after serving just six years.[61]

As the unusual career of Quintus Vilanius Nepos illustrates there was no fixed pattern of promotion through the Roman army. His tombstone text says nothing about his earliest positions, the first of which he assumed when he was enlisted at eighteen. It starts with the fact that he was a centurion of Cohors XIII Urbana, a job which must normally have involved policing duties in Rome or other Italian cities. However, it seems that Vilanius Nepos was sent off to Dacia and Germany to fight in Domitian's wars, perhaps because the whole cohort was ordered there. He was decorated for his service in both conflicts, receiving 'collars and bracelets' for his performance in Dacia. He died aged fifty after 32 years' service.[62]

Petronius Fortunatus was eighty (or thereabouts) when a monument was erected in the third century AD to commemorate his remarkable career in the army, though ostensibly it was a memorial for his son. He had survived until a considerable age for the ancient world, and especially for someone who spent his life as a soldier. Fortunatus joined the Roman army as a legionary in the frontier province of Moesia Inferior (now parts of Bulgaria and Romania), signing up with Legio I Italica. Unlike many legionaries he moved from one legion to another once he had been promoted to centurion; he was decorated in the Parthian War and went on to serve in legions in parts of the Roman world as far removed as Syria and Britain. An army man through and through, he stayed a soldier for over 46 years. His son joined up too and served for six years before dying at thirty-five.[63]

Moving between the legions and the Praetorian Guard became common in later years, especially after Septimius Severus cashiered the praetorians on his accession in 193 and replaced them with his best legionaries. Aurelius Vincentius came from Thrace, joined Legio XI Claudia where he served for five years and then moved to Cohors III Praetoria for 11 years. He died, still serving, aged forty in Caesarea in Mauretania, where he had perhaps been sent on official business.[64]

Legionary centurions could find themselves promoted up and out of a legion to command an auxiliary unit. Marcus Censorius Cornelianus was born in Nîmes in Gallia Narbonensis, a part of the world that many legionaries came from. He rose to the rank of centurion in Legio X Fretensis during its long stay at Jerusalem in Judaea. He seems to have made a sudden jump from that job to the command of Cohors I Hispanorum milliaria at the coastal fort of Maryport in north-western Britain, a remarkable move across the Empire.[65] However, rather than being styled a prefect or tribune in the manner of a normal equestrian leader, including other attested officers of the unit at Maryport, he was called a *praepositus* or 'commander' (literally 'foremost position'). The post may well have been an emergency appointment following the death of an equestrian prefect, perhaps in battle.

This also seems to have happened at Dura-Europos in Syria after a Persian attack on the fort in 239 when the commanding officer was killed and a legionary praepositus was installed.[66] The centurion Flavius Betto was detached from Legio XX and placed in 'acting command' of Cohors VI Nerviorum on the Antonine Wall in Scotland at some point between *c.* 143 and 160. Betto's unusual name suggests he came from Gaul (or even Britain). Since the Nervians had been originally recruited in northern Gaul, Betto may have been picked to stand in as their commander because of his ethnicity and ability to speak their native language.[67]

It was not unusual to find brothers in the army. Gaius Canuleius fought for Caesar with Legio VII in Gaul. He survived the war, in which he was decorated, and became an *evocatus*, dying when he was thirty-five at Capua, probably his home town. His tombstone commemorates his brother Quintus, who died during the same war aged eighteen. It is easy to imagine the two brothers enlisting together and setting out on what they imagined would be a thrilling and profitable future. Gaius may have been one of those who landed in Britain during one of Caesar's two invasions (55 and 54 BC).[68]

Marcus Caelius was a primus pilus centurion killed in the disaster of AD 9 when three legions were wiped out. His brother Publius appears to have served with him but escaped and set up a cenotaph to Marcus at Xanten (see Chapter 7). Gnaeus Musius, an aquilifer with Legio XIIII Gemina in the early first century AD, came from Veleia in northern Italy and had signed on when he was seventeen, dying at Mainz (Mogontiacum) in Germany after 15 years' service when he was thirty-two. His brother Marcus Musius commissioned an elaborate tombstone showing Gnaeus in his armour and holding both his shield and his standard, standing between two columns beneath a decorative pediment.[69] Marcus tells us nothing about himself but he was clearly on hand when his brother died, making it likely he served in the same legion. In the mid to late first century, Quintus Sertorius Festus served in Legio XI Claudia, rising to the centurionate. His brother, Lucius Sertorius Firmus, was in the same legion, where he was a signifer and an aquilifer; he outlived Quintus, eventually marrying a woman called Domitia Prisca and erecting a tomb for his brother and their parents in Verona.[70]

## AUXILIARIES

Auxiliary units usually started life when they were recruited in specific areas from territories that had become allied to Rome or had been conquered. As such they were frequently given titles that recorded the region where the unit was first raised, such as Cohors IIII Gallorum, 'the Fourth Cohort of Gauls'. Some unit names were descriptive rather than ethnic such as the Ala I Contariorum ('the First Ala of Lancers'), and had clearly been raised to offer a specific fighting skill. Although the units were originally based in or near the place where they were raised, by the late first century AD they were far more likely to have been relocated to remote provinces in order to prevent them becoming the armed muscle behind home-grown resistance. Most auxiliaries were dispersed around frontier provinces, with Britain, Germany, Syria and Egypt amongst the most heavily garrisoned. Having been trained and organized as Roman army units, they could be lethal foes if they changed sides. In time, of course, these units needed new recruits, and while some were sourced in the original province it is obvious from tombstones and other records of auxiliary careers that over the longer term there was no fixed rule about the ethnic identity of the units they served with. The ethnicity of auxiliary units was often diluted by recruiting from wherever they were stationed, unless there were specific regional fighting skills involved. Even that did not preclude recruiting men from anywhere so long as they were up to the job.[71]

Understanding ethnicity in the Roman army can sometimes be

very challenging. Part of the problem is that in our time it is not only easy to confuse a Roman ethnic label with one of our own because they sometimes look or sound the same (for example, Syrians), but also to assume the Roman label is literal and clearly defined. In his description of the Roman army during the reign of Augustus, Cassius Dio cryptically explained that 'picked foreign horsemen ... were given the title *Batavians*, being named after the island in the Rhine called Batavia because the Batavians are superb horsemen'.[72] He seems to be saying that ethnic titles could be manufactured by the Roman army to create an artificial sense of pride based on popular reputations of excellence. In this case it sounds as if the best provincial cavalrymen were hired first and foremost for their skills, and were then gathered together into a cavalry unit with the honorific label *Batavorum*, 'of Batavians', even though it had no relevance to the men concerned.

Indus was a member of Nero's personal bodyguard of 'German' troops, known as the *Germani corporis custodes*. Indus died in Rome in service aged thirty-six and was described on his tombstone as being from 'the Batavian Nation'. The tombstone was erected by the 'Guild of Germans' and his 'brother Eumenes', a distinctly Greek name – in this instance 'brother' was being used in a professional sense. Batavians are among the most common attested members of this special imperial personal bodyguard, along with the German Ubii tribe.[73] The 'German' label then was really something of a stylistic convention that blurred individual ethnicity into a generic identity by which the unit had become known.

An obvious source of information is a soldier's personal name, assuming he retained it. Bodiccius, the standard-bearer with Cohors I Brittonum mentioned in Chapter 1, had a British-type name, so in his case it is easy to conclude that he was a Briton, and the same applies to his co-heir Albanus and the man they buried, Virssuccius.[74] This is a very rare attested instance of a 'community of Britons' in a British-named unit. Conversely, in 240–50 Ala I Brittannica recruited to its ranks around six Thracian men who served with others of Pannonian origin, although this was ostensibly a cavalry wing made up of Britons. One of these was the third-century trooper Marcus Ulpius Crescentinus, 'born in the nation of Pannonia Inferior', and who served in Ala Brittonum before rising eventually to serve in the emperor's mounted bodyguard, the singulares Augusti, and dying in Rome.[75] Likewise, Mucatralis was serving with Cohors II Flavia Brittonum in Moesia Inferior when he died during the reign of Trajan. He was a Thracian of the Bessi tribe.[76] There is no reason to assume that for the most part the position was different in any other auxiliary unit.

Just as Thracians and Pannonians are attested as serving in 'British' units, so we find non-Thracians serving in 'Thracian' units. Like so many auxiliary soldiers, Sextus Valerius Genialis emerged from provincial obscurity. He was a member of the Frisiavones tribe whose territory had been absorbed into Gallia Belgica. Genialis signed up at the age of twenty, perhaps burying his original name beneath a Roman one – which also shows that he acquired Roman citizenship at some point along the way (or claimed to have done so). He was forty when he died at Cirencester (Corinium) in Britain, probably in the late first century, having spent half his life in the army. By then he was serving as a cavalryman in the Ala Thracum.[77] Oclatius, a signifer with Ala Afrorum, belonged to a cavalry unit originally raised in Africa. By the time he died the unit was based in Neuss (Novaesium) in Germany, but his tombstone tells us he was from the Tunger region (modern Belgium).[78] In both cases, cavalry skills were clearly of greater importance than where the man came from.

Practical considerations meant that all sorts of pragmatic solutions had to be found when a unit needed more men. A soldier in Egypt, writing to his mother in the second century AD, described how a cavalry wing of Thracians and a cohort of Africans had recently left for Mauretania. He added that the African cohort was being brought up to strength by taking men from his own (unnamed) cohort, which was due to follow on. To add to the potential confusion to anyone reading the letter today, he used a colloquial term for the Thracian cavalrymen, calling them an 'Ala of Moors'.[79]

At some point during his reign in the late third century, Probus was said to have raised 16,000 recruits from barbarian tribes to serve in the border garrisons. He realized though that if he distributed them in the form of whole units it would be painfully obvious how dependent Rome had become on such men. Instead he ordered them to be divided up into groups of 50 or 60 and dispersed among existing units.[80] If the story is true, the new recruits' ethnicity was immediately and deliberately buried.

These examples are a reminder of how simplistic perceptions of the Roman army have sometimes been, with ethnic titles being taken at face value and often bolstered by the belief that archaeological finds of specific classes of regional artefacts such as brooches or pottery support this. The small auxiliary fort at Walldürn near Mainz in Germany is known to have been occupied in the 230s by an *exploratores* scout unit that supposedly included Britons. Yet, excavation there has produced 'no British-made objects'.[81] It is difficult to avoid the conclusion that some auxiliary units bore no more relation to their nominal nations 'of origin' than many modern European football teams. At any rate, there was clearly no fixed rule that applied either to the units themselves or to individual soldiers. Without specific written information about any individual soldier it is impossible to say where he had come from for certain.

Sometimes an auxiliary unit's specialist form of warfare makes it more likely that the individual soldiers came from the unit's original home. Monimus, son of Jerombalus, and his father had names that were compatible with Cohors I Ituraerorum's origins in Syria. The cohort's particular skill was archery; although this is unmentioned on Monimus' tombstone. He was, however, depicted on his memorial with arrows and his bow gripped in his left hand. He enlisted relatively late, at the age of twenty-nine, dying aged fifty in the early first century AD at Mainz, where his cohort was then based.[82] The use of Ituraean archers had been pioneered by Caesar.[83]

Apion, a young Egyptian in the second century AD, was attracted to the idea of a life in the Roman navy, perhaps because he had gained experience on the Nile or beyond the Delta in the eastern Mediterranean. The navy was the seaborne part of the Roman army, though pay was lower and length of service longer even than that of auxiliaries. Apion evidently fulfilled the criteria for eligibility, signed up and embarked on a dangerous journey to Italy, coming close to being shipwrecked en route. Having made it alive to the Roman fleet base at Misenum on the northern side of the Bay of Naples, Apion was allocated to a κεντυρία (*centuria*, century) called the Athenonica (by *centuria* he meant the company of a ship called the *Athenonica*) and promptly set about writing home. (The letter is in Greek, the everyday language in the Eastern Empire.) Apion thanked the Graeco-Roman Egyptian god Serapis for keeping him alive but his focus was mainly on how he had 'received from Caesar three gold coins for travelling expenses'. Three gold coins (*aurei*) were equivalent to 75 silver *denarii*, a quarter of a Roman legionary's annual pay in the second century. As a member of fleet troops Apion would be on a lower rate, so those three gold pieces amounted to something like four months' pay. Apion was evidently delighted by the cash, but he had something else to tell his father Epimachus. He had been given a new name on enrolment and would henceforth be known as Antonius Maximus. The name was neutral and typically Roman. It instantly obfuscated his origins but this was not something that would have bothered Apion.[84]

In around 120 Oronnous, another Egyptian recruit, joined an auxiliary cavalry unit and changed his name to Achillas (*sic*).* The Greek hero's name had become popular in Roman culture and literature as a synonym for power and martial skills.[85] Becoming 'Roman' was a major reason for joining the armed forces. According to the second-century Greek philosopher Aelius Aristides, who was probably exaggerating in his efforts to be as pro-Roman as possible, recruits willingly handed over their original identity in return for becoming Roman and were grateful for the opportunity:

---

* Using here a Greek variant spelling for Achilles.

[Soldiers] became reluctant for the rest of their lives to call themselves by their original ethnic identity ... the day they joined the army, they lost their original city, but from the same day became fellow citizens of your city and its defenders.[86]

Apion already knew the Roman army had systems that his world depended on. The letter was to be sent via military mail across the Mediterranean from Misenum to the fort manned by a cohort at Alexandria. From there it would be forwarded to his father at a village in the Fayum called Philadelphus. Another Egyptian who joined the same fleet in the second century AD wrote, also in Greek, to his mother Taesis in Karanis to reassure her that all was well. Apollinarius had arrived in Rome and been told he had to go to Misenum. He had not yet been told which centuria he had been allocated to because he would find out on arrival at the fleet base, but he decided to write to his mother before leaving: 'I beg you then, mother, look after yourself and do not worry about me, for I have come to a fine place.' He was almost bound to say that. Soldiers writing to their mothers at all times and places have always tried to sound optimistic. Apollinarius wanted to hear her news and about his brothers, and promised to write to her whenever he could. His letter was elegantly written on one side of a papyrus with the letters clearly and neatly formed, and was perhaps produced by a scribe to Apollinarius' dictation.[87]

Not every father was pleased at the thought of his son going off to serve in the army. Augustus cashiered a cavalry trooper because he had gone to the extreme of cutting off the thumbs of both his sons to ensure they could not serve.[88] Perhaps the trooper should have known about the case of Gaius Vettienus, who took the extreme measure of cutting off two fingers from his own left hand to avoid being called up to serve in the Social War of 91–88 BC. He was sentenced to having his property seized and to spend the rest of his life in chains in disgrace.[89] Four centuries later, under Diocletian and Maximianus, a soldier called Dizon brought a complaint that he had paid money to avoid military service, presumably unsuccessfully since he is described as a soldier in the law code of Justinian that records the case:

> *The Emperors Diocletian and Maximianus to the soldier Dizon*
> If it is established by indisputable evidence before a competent judge that you have paid a sum of money to the person of whom you complain, in order to avoid military service, you can recover it with his assistance. He [the judge], being mindful of public censure, after the money has been refunded, will not suffer the crime of extortion to remain unpunished.[90]

Many years later, under Theodosius I (379–95), all these methods of evading military service were specifically prohibited. Any man doing such a thing was to be 'branded' and obliged to endure military service 'as forced labour'.[91]

## DISQUALIFICATION

Even though the recruitment process meant a man had to demonstrate he was not a slave, some managed to get through, passing themselves off as free men in order to escape a life of servitude. In 93 Claudius Pacatus, an ex-centurion, was sent back to his master when it was found that he was really a slave. He had clearly proved his worth, but that was not enough to save him.* In the early second century AD Pliny the Younger was a senator serving as governor of the province of Bithynia and Pontus by the Black Sea. On the emperor Trajan's orders, an official called Sempronius Caelianus sent Pliny two slaves who had been masquerading as new army recruits. It was not clear either to Trajan or Pliny whether the slaves had turned up of their own accord to become soldiers, or whether they had been sent

---

* See Chapter 15.

as substitutes to stand in for free men who did not wish to serve. If the former, Trajan said, 'they will have to be executed'. Their crime would have been trying to pass themselves off as citizens. Once in the army they might well have remained undetected.[92] However, if they had been sent as substitutes then, Trajan said, it was those who had sent them who were guilty. Unfortunately, we do not know what happened next.

Sometimes even serving soldiers could be summoned to swear that they were entitled to be in the army. Titus Flavius Longus was an optio with Legio III Cyrenaica in Egypt in 92. Someone must have called his eligibility into question because he had to find several soldiers in the legion to act as his guarantors. Fronto, Longinus Celer and Herennius Fuscus 'declared on oath' that Flavius Longus was a freeborn Roman citizen 'and had the right of serving in a legion'.[93] Whether he – or they – were telling the truth or not is lost to history.

## Oaths and loyalty to the emperor

'The soldier loves he who wages war', said Ovid, meaning that was the basis of the soldier's loyalty and common interest.[94] In the days of the Republic, Roman soldiers swore an oath known as the *ius iurandum* (literally the 'sworn law') to the Roman state, an act which became compulsory after 216 BC.[95] An oath was regarded as 'inviolable and sacred' by the Romans.[96] Harking back to this tradition, under the Empire all soldiers had to swear an oath to the emperor, while in the Christian Empire in the fifth century the oath was also sworn to the Holy Trinity.[97] The vast majority of Roman soldiers never went to Rome, let alone saw the emperor in person, but they were constantly reminded of his name, likeness and power in the continuous cycle of religious ceremonies the units followed every year, by the statues that stood in the fort headquarters building, and by the images on the coins with which they were paid. The oath was unconditional. It meant they had to obey the emperor's orders, even if that meant the killing of families and children, and promise never to desert.[98] The traditional definition of a deserter was any soldier who in war went far enough away not to be able to hear the trumpets.[99] Soldiers could also be obliged to swear oaths to their commanding officers; in 200 the Praetorian Guard swore such an oath to Septimius Severus' praetorian prefect Plautianus 'by his Fortune' and offered 'prayers for his preservation'. The oaths did not save him. Plautianus was immensely powerful and described as virtually sharing imperial power.[100] He had high hopes for his family's future. His daughter was married to Severus' son Caracalla who hated Plautilla and her father. In 205 Caracalla had him murdered and exiled his wife, later killing her.[101]

On exceptional occasions the emperor might visit a military base. On 1 July 128 the emperor Hadrian visited Legio III Augusta at their brand-new fortress of Lambaesis in North Africa (in modern Algeria). He inspected the legion, watched its manoeuvres, which they had obviously been practising for him, and then addressed the men in a speech that was transcribed and recorded. He expressed his admiration for their competence, paying tribute to various different parts of the legion and its attached auxiliary units. To a cavalry unit he said, 'defences which others build in several days, you have finished in one. You have built a wall which requires a great deal of work ... you dug a ditch straight through hard and rough gravel and made it smooth by levelling it ... I can congratulate you.' He praised Ala I Pannoniorum with 'If anything had been wanting in your performance I would have noticed it, if anything had been obviously bad I should have mentioned it, but throughout the whole exercise you satisfied me consistently.'[102]

With such a dispersed army there was the possibility that legions and auxiliaries, or even the entire garrison of a region, would ally themselves instead with the ambitions of a legionary legate or a provincial governor. Caesar's men during his command in Gaul were fiercely loyal to him. Without their support he could not possibly have pursued his ambitions. When he crossed the Rubicon in 49 BC to confront the Senate and his erstwhile colleague – and now bitter enemy – Pompey, he was breaking the law that forbade a Roman general to bring an army into Italy. This occasioned a crisis in Italy and further divided Roman military forces, a situation in which the statesman, orator and lawyer Cicero found himself caught up. He was at a villa in the Naples area when a message arrived from three centurions serving with cohorts currently stationed in Pompeii who wanted him to lead resistance against Caesar.[103] Caesar's 'immense army' continued to follow him, resulting in a period of civil war that lasted on and off until Actium in 31 BC finally brought the conflict to an end, and with it the Republic.[104] Such affiliations could and did lead to civil war on several later occasions, for example in 68–9 and 193–7. By the third century, would-be emperors vied to outbid each other for army support and the period degenerated into generations of intermittent civil war.

Unit loyalty led to bitter rivalry. Legionaries looked down on the provincial auxiliaries, who themselves looked down on irregular units. The Praetorian Guard looked down on them all. Manlius Valerianus was a praetorian centurion based at Aquileia in Italy. 'I commanded a century in a praetorian cohort', he proudly claimed on his tombstone, 'not a barbarian legion', dismissing them as uncivilized.[105] The hatred was probably mutual, unless of course a legionary was promoted to the Guard and came to enjoy its extra privileges. Aurelius Saturninus was one such man. He served in the third century as a *tesserarius* in Legio II Italica before being elevated to the post of *eques* in Cohors VIII Praetoria in Rome, where he died at the age of twenty-eight.[106]

Sometimes a little spot of what seemed like harmless rivalry could erupt into outright violence with disastrous consequences. During the Civil War in April 69 the emperor Vitellius was presiding over a drunken dinner party at Ticinum (modern Pavia) in the aftermath of the defeat of his rival Otho at the First Battle of Bedriacum and Otho's subsequent suicide. The general disorderly mood spread to some of the soldiers in Legio V Alaudae and Gaulish auxiliary infantrymen. One legionary and a Gaul decided in a fit of high spirits to engage in a wrestling bout. During the fight the legionary was knocked down and his Gaulish opponent took the opportunity to mock him. In a trice 'the spectators took sides, the legionaries attacked the auxiliaries, and two cohorts were wiped out', said Tacitus. If he can be believed, that means 1,000 auxiliaries at least lay dead. At that moment a cloud of dust and the light reflecting off arms led to the rumour that Legio XIIII from Otho's defeated army was turning back to attack them. In fact it was the rest of Vitellius' army arriving. Either way a very dangerous situation ended. It remained a dreadful example of how volatile soldiers could be, and how tenuous their ability to cooperate with one another was.[107]

There is a curious footnote to the world of Roman military careers. Esteemed civilian magistrates in ordinary towns were sometimes given the title 'military tribune by popular acclaim'. It was added to their honours and included in inscriptions recording their careers. One such was Marcus Lucretius Decidianus, a senior magistrate at Pompeii in the Augustan period. He was also awarded the title 'engineering prefect'.[108] There is no suggestion that he was either a veteran or had ever served in the army in any capacity. No military units are named in the inscription on the base of his posthumous statue. Nor does the style of the titles match a military context – no military tribune would have been made so by 'popular acclaim'. Evidently the notion of military status was so ingrained in Roman culture the labels had become a suitable way of according a prominent man appropriate respect.

# THREE

# GLORIA EXERCITUS

## Making Soldiers

*As if [the Romans] had been born with weapons in hand, they never have a truce from military training, and never wait for emergencies to happen. Their peacetime exercises are no less tough than real warfare. Every soldier applies all his energy to training, as if he was on the battlefield.*

Josephus describes the way the Roman
army maintained permanent readiness[1]

Vegetius said that the Romans had conquered the world because of their military training and discipline.[2] Life as a Roman soldier began with training, but it also involved leadership, promotion, learning to use the arms and equipment, and the conditions of service. The priority was to create men with tough minds and tough bodies.[3] For the officers, drawn from the upper tiers of Roman society, there were different challenges. One of these was the fact that they could find themselves in positions of leadership without the slightest practical experience or knowledge of the job. Some were able to make the most of the trained men in their charge. Other commanders led their men into disasters, negating the training and preparation.

## TRAINING

The Latin word for army, *exercitus*, was the same as the word for 'disciplined' or 'trained'. Both were derived from the verb *exercere*, 'to keep busy' or 'to practise'. Training and exercise were thus fundamental to the Romans' concept of what an armed force was supposed to be.[4] They were also essential if morale, on an individual or unit basis, was to be maintained, especially in the chaos of battle.

Training meant learning everything from the use and care of weapons to the building of camps and forts and the bridging of rivers, as well of course as being instructed in tactics and battle formations.[5] 'Daily exercise', said Vegetius, was the only way to 'toughen them'.[6] New recruits were subjected to a minimum four months of basic training which was designed to find out which of the men were unable to cope.[7]

The great generals of the Republic had set the bar high. Scipio Aemilianus made sure his troops were subjected to a rigorous regime before ordering them to besiege the city of Numantia in Spain in 134–133 BC. Under his personal supervision, the men were made to build a new camp complete with fortifications from scratch every day and then dismantle it by sunset, as well as level the fortifications. The efforts paid off, but the siege lasted at least eight months and only ended when the Numantians started committing suicide. The rest set the city on fire and surrendered.[8]

Nothing had changed by the time of the emperors. 'He whom you do not want to panic in a crisis, train (*exerce*) him before the crisis', said Seneca, using a military analogy as general advice, in a letter written during the reign of Nero. Seneca described it as a given that Roman soldiers practised manoeuvres in peacetime, built fortifications against non-existent enemies, and wore themselves out with labour so they would be up to the challenge when a real war broke out.[9] A few years later Josephus described a rigorous and ongoing training regime which 'lacked none of the vigour of genuine warfare'. Each soldier drilled every day with the same enthusiasm he exhibited in real war, for the Romans were 'born to arms' and never waited around for circumstances that required them to be prepared for fighting, so he said. In other words, the men treated every day as one to be ready for war. As far as Josephus was concerned, this meant Roman soldiers were well able to withstand the sudden shock of going into battle, avoiding both confusion and fear, the result being inevitable victory. Their exercises were 'bloodless battles and their battles exercises with added blood'.[10] Army training was led by hard-nosed men like Titus Aurelius Decimus, a *campidoctor* ('camp instructor') of Legio VII Gemina who in 182 made a dedication in Tarragona in Spain to Mars Campestres (Mars conflated with the spirits of the parade ground).[11]

Discipline was another central feature of army life, but it was fragile as we have already seen. Centurions played a key part in the everyday disciplining of soldiers, but this could be counter-productive. When the mutiny led by Percennius (see Chapter 4) broke out among the Pannonian legions in AD 14 over pay and conditions some of the men were away from the base working on roads and bridges. News reached them of the insurrection. They promptly uprooted their flags, ransacked villages and a neighbouring town, their centurions powerless to stop them under a hail of insults and fists. Next the angry men turned on the praefectus castrorum Aufidienus Rufus. They pulled him out of his carriage, forcibly loaded him up with baggage and jeered at him while asking how much he enjoyed the experience. Tacitus, who always backed authority, was on Rufus' side because the experienced camp prefect had a long career of tough soldiering behind him, rising up all the way through the centurionate and believed in ruthlessly enforcing the same hard work on the men.[12]

Next the men returned to their camp and provoked more rebellious behaviour. During these disturbances one harsh disciplinarian of a centurion called Lucilius was killed. He had already earned himself the nickname *cedo alteram* ('bring me another!') in reference to his habit of breaking his vine rod symbol of office over the back of one ordinary soldier after another, and calling for a fresh stick to be brought. The VIII and XV legions were on the point of coming to blows over another centurion called Sirpicus for similar reasons. Only the intervention of Legio VIIII saved him.[13] A little later the same year a mutiny was stirred up among the Rhine legions (I, V, XX and XXI) over the way pay and conditions had been ignored. Many men bore the scars of beatings they had endured, and their first target was the centurions 'who had fuelled the soldiers' hatred for the longest'. They struck each centurion with 60 blows to match the number of centurions in a legion, killing some and severely injuring the rest, and then threw them into the rampart or into the Rhine.[14] Authority broke down completely for a short time and only Germanicus was able to calm the men down.

There are many other attested instances where leadership and training in the Roman army fell far short of what was needed. The outcomes were defeat, disaster, and dissent on a number of notorious occasions (see Chapter 7).

## LEADERSHIP

For many men of senatorial rank a period in the army was only part of a career that might ultimately lead to the consulship or to the most senior (and most lucrative) provincial governorships. For an equestrian, a posting as a junior officer in a legion could precede the command of an auxiliary unit and then perhaps any one of an increasingly prestigious range of procuratorial posts or prefectures, one of which was command of the Praetorian Guard. Appointment to senior positions such as military tribunates or legionary commands was almost entirely based on personality, patronage, family and contacts. The attractions of military leadership were sometimes similar to those of the ordinary soldiers, including the prospect of glory and the chance of profitable spoils that could either be sold or used for personal or professional favours. Quintus, Cicero's brother, campaigned in Gaul and Britain with Caesar. In late 54 Quintus promised his brother a gift of slaves acquired during the fighting, for which Cicero was very grateful.[15]

There was no such thing as a Roman military college or education, or anything remotely equivalent. Officers learned 'on the job', though they had the opportunity to read treatises on military command. These included *The General*, written by a philosopher called Onasander and dedicated to Quintus Veranius Nepos, a governor of Britain who arrived in the province in 57 but died within the year. There was a grand tradition in Roman military culture of recording and admiring the leadership skills of commanding officers, especially those of the Republic. Numerous instances appear throughout this book, but the following examples are from collections put together in the days of the emperors to provide guidance in military leadership.

Soldiers needed to be convinced that a battle was going to plan and was under the general's control. When a prefect in Sulla's army suddenly changed sides at the start of a battle, taking with him a cavalry unit, Sulla had the presence of mind to claim ingeniously that the turncoat had done so on his orders. Not only was a panic avoided, but the rest of his men were also given to believe that what had happened would work to their advantage. It was not the only time Sulla prevented panic breaking out in the ranks. An auxiliary unit under his command was sent off to fight but was badly defeated and effectively wiped out. Sulla was well aware that if news of the catastrophe spread in his army, the result would be chaos. His solution was to claim that the auxiliaries concerned were plotting to desert, so he had deliberately sent them into a dangerous spot. Sulla's soldiers were greatly encouraged by this display of leadership, and the disaster was neatly covered up.[16] When at the Battle of Munda in Spain in 45 BC against the army of his political rivals, Caesar's troops fell back, he promptly dismounted, ordered his horse to be removed from the battlefield and walked to the front line, where he stood alongside the troops as an ordinary infantryman to shame them into fighting beside him. The prospect of humiliation of course meant the men went back to the fray.[17]

These stories appeared in Frontinus' *Stratagems* in the late first century AD. Sextus Julius Frontinus was a senator who was consul three times and served in the late 70s as governor of the militarily challenging province of Britain, crushing tribes in Wales. He probably compiled his *Stratagems* after that experience, gathering the stories from other sources, many now lost. The result was effectively an instructional manual for officers, together with a final section about discipline (see below). It was all too easy, though, for young senators or equestrians to treat a few years in the army as an indulgence. Others were more applied, sometimes with the guidance of their patrons. Gnaeus Julius Agricola's career is known in some detail and demonstrates how a man of senatorial status could build up considerable military experience interspersed with other jobs. He was born in the Roman colony of Fréjus (Forum Julii) in Gaul in 40. His father was a senator but his grandfathers had been equestrians. Agricola's first post was as a military tribune, serving on the staff of the governor of Britain, Suetonius Paulinus, during the Boudican War. He moved on eventually to a quaestorship in Asia, then became tribune of the plebs in Rome, and in 69 took command of Legio XX in Britain, probably until about 73. Next, Vespasian made him governor of Gallia Aquitania in or around 74 and then governor of Britain in 77 or 78, succeeding Frontinus. He remained in charge of the province for an exceptional double term until *c.* 84. Tacitus, although as his son-in-law hardly an objective source, said Agricola had avoided the temptation of so many young men to turn soldiering into a pursuit of pleasures.[18]

The future emperor Hadrian first went into the army aged fifteen, but spent too much time hunting. His cousin and patron Trajan saw to it that he was recalled and thereafter oversaw the development of Hadrian's career. Trajan was himself a highly accomplished military leader 'who always marched on foot with the rank and file of his army', personally supervising where his troops were and what they were doing throughout a campaign.[19] When Hadrian was a little older he was made a tribune first of Legio II Adiutrix in Pannonia Inferior, and then of XXII Primigenia in Germania Superior. He was subsequently given the command of Legio I Minervia during Trajan's second Dacian war.[20] These experiences made him a stickler for military discipline when he became emperor.

Clodius Albinus, who as governor of Britain tried to become emperor in 193, had risen to that position after a stellar military career. He was helped in this by three of his kinsmen who had brought him to the attention of Marcus Aurelius and Commodus. As a result Albinus became the tribune of an auxiliary cavalry unit of Dalmatians before proceeding to lead Legio I Italica in Moesia Inferior and Legio IIII Flavia in Moesia Superior, and then moved on to other military commands.[21] He seems to have been a skilled general, but it is clear his family connections were essential to his being noticed and promoted. Marcus Aurelius was certainly impressed, writing to his prefects:

> Greetings. (Concerning) Albinus of the Ceionii family, a man of Africa but having not much of the African about him, son-in-law of Plautillus, I have given him command of two mounted cohorts. He is a disciplined man, sombre in life, dignified in his ways. I think he will be beneficial to camp affairs, certainly he will be no liability. I have ordered him a double allowance, simple military clothing suitable to his place, and quadruple pay. You should encourage him to show himself to the state, to have the reward that will be deserved.[22]

## Patronage and promotion

Precisely reflecting the way Roman society itself worked, patronage operated throughout the Roman army at every level. Not everyone approved: Agricola was said to have refused to appoint officers or ordinary soldiers to his staff during his time as governor of Britain on the basis of personal testimonies or just because he liked them.[23] But he was unusual. The senator Pliny the Younger was among those involved, as were most men of influence, in this complicated web of patronage. Pliny had obtained a military tribunate for Suetonius Tranquillus, the Roman historian and equestrian, by interceding on his behalf with a senator called Neratius Marcellus. Suetonius, however, decided to forgo the job before being formally appointed and asked Pliny to transfer it to a relative of his, Caesennius Silvanus. Far from being insulted, Pliny was 'equally pleased' by the thought that Silvanus would then owe his promotion also to Suetonius.

Some years later, when serving as governor of Bithynia and Pontus, Pliny asked an old friend of his, the veteran centurion Nymphidius Lupus, to join his staff. As a way of repaying the favour, Pliny wrote to Trajan to make a personal recommendation of Lupus' son, 'an honest, hard-working young man' of the same name who had already served as the commander of a cohort and been praised by senators. Pliny asked the emperor to consider him for promotion. He also wrote to Trajan about a man who had served in the army under the emperor and whose 'justice and humanity' made him worthy of recommendation.[24] Around thirty to forty years afterwards a young man called Publius Helvius Pertinax obtained the post of tribune in the cavalry thanks to the influence of a friend of his called Claudius Pompeianus. It set him on a career path that would lead him to becoming emperor in 193, although his reign lasted only a few weeks before he was murdered by praetorians.[25]

Vindolanda was an auxiliary infantry fort on Britain's northern frontier. Founded in the late first century Vindolanda remained in existence until the end of the Roman period, and went through multiple phases, serving for much of that time in the hinterland of Hadrian's Wall. The waterlogged levels of the early timber turf forts on the site have produced one of the most important collections of military documents from the whole Empire, mostly dating to *c.* 90–105. A fragment of a letter found there seems to be part of a similar recommendation of a 'good man' whose 'moral progress through liberal pursuits' and 'moderation' had made him clearly suitable for a position.[26] A rather more complete letter was written to Flavius Cerealis, prefect of Cohors VIIII Batavorum at Vindolanda, by someone called Karus. He endorsed a man called Brigionus who had asked Karus to do him the favour: 'Therefore I ask, lord, if you would be willing in what he has asked of you, to approve him to Annius Questorus, the centurion of the region at Luguvalium [Carlisle].' Karus said that if Cerealis obliged he would be in Cerealis' debt.[27] Such references were of course largely form-ulaic, as they are today.

A soldier could on the other hand do his best to pursue his own interests, as an ambitious young soldier from Egypt named Gaius Julius Apollinarius did in 107. In a letter home to his father, proba-bly written on his behalf by a scribe or friend, he explained how he had advanced himself one rung up the ladder of success in Legio III Cyrenaica by speaking to the commanding officer:

> I asked Claudius Severus, the consularis, to appoint me secre-tary on his staff. He said 'there is no vacancy. In the meantime I will appoint you *librarius* [secretary] of the legion with the pros-pect of promotion.' I went with this assignment therefore from the consularis of the legion to the cornicularius.

The legion was based in Bostra, Arabia Petraea, between 106 and 19 instead of its normal base in Egypt and so could be commanded by a man of senior senatorial rank, hence the reference to the *consularis*, instead of an equestrian (which was the case for legions stationed in Egypt).

Ordinary soldiers had the potential and the opportunity to build up specialized knowledge or skills, if they did not already have them. There was also a major incentive to be selected for special duties. Such men were called *immunes*, which meant that they were exempt

from the more tedious and mundane jobs of soldiering. Apollinarius'
new job as librarius was an *immunis* post and thus he was exempted
from the duties and fatigues that filled out an ordinary soldier's day.
This entitlement was enshrined in Roman law.[28] Immunes ranged
from surveyors to those who made equipment or armaments,
architects or lead workers; they could be responsible for preparing
the stone or wood for fort buildings, attending the sick or fulfilling
any one of dozens of different clerical duties, such as those keeping
a record of grain stocks, or serving as a beneficiarius on the staff of a
legionary legate or a provincial governor.[29] Apollinarius continued to
do well. By 119 he was still with Legio III Cyrenaica, but was serving
in military intelligence on detachment in Rome.[30]

Soldiers often belonged to associations known as *collegia*. These
were similar to trade associations or guilds in the civilian world and,
like them, were organized with officials, rules and protocols. The
collegia helped protect their members' interests and served to find
ways of increasing a soldier's income. In the third century, Legio III
Augusta's *librarii* at Lambaesis were members of such an association,
which included the senior clerical staff, the cornicularius Lucius
Aemilius Cattianus and the *actuarius* (registrar) Titus Flavius Surus.
The members paid into the collegium out of their wages but were
also interested in securing other sources of income for the organization. They collectively decided that when either the cornicularius
or the actuarius reached retirement age and was honourably discharged, the man promoted to fill either post should pay 1,000 denarii
to the veteran concerned. Similarly, anyone promoted to one of the
librarii posts was expected to pay the same sum for admission to the
collegium. A librarius who was honourably discharged could expect
800 denarii from the other librarii, and anyone who left the job for
unspecified reasons received 500 denarii. Over 40 clerks signed the
agreement, which was then formalized in an inscription. This was
only one such organization in one fortress. Associations for other
ranks existed at Lambaesis, and similar organizations with their own
individual arrangements also undoubtedly existed in other forts and

fortresses. It is doubtful whether any incoming librarii were exempt from the payments.[31]

Two of the most unusual jobs found in the Roman army were those of *speculator* and *frumentarius*. The Praetorian Guard included a body of 300 mounted speculatores ('scouts') who acted as the emperor's personal mounted bodyguard. The job seems to have evolved into a wider range of duties with individual soldiers in a legion being made into speculatores. There seems to have been an intelligence-gathering element to the job, which probably involved synthesizing incoming information and briefing the legionary legate or the governor of the province. This would explain why Valerius Celsus, speculator of Legio II Augusta, died in London, where several of his fellow legionary speculatores took care of his funeral.[32] There can be no question they were working for the governor of Britain at the time, probably in the late first or during the second century. Other duties of the speculatores may have included serving as executioners. At some point between 180 and 192 Gaius Julius Proculus, speculator of Legio XIIII Gemina, and Lucius Valerius Marcianus, a beneficiarius of the legionary legate, jointly honoured their patron, the senator Marcus Cassius Hortensius, at Ankara in the province of Galatia. This makes it likely Julius Proculus was at the time working for the legion rather than the provincial governor.[33]

Frumentarii also turn up in quite innocuous-looking legionary contexts: Publius Aelius Annianus, 'a soldier and frumentarius' of Legio X Fretensis, died in Athens.[34] A frumentarius was in theory in charge of organizing supplies of grain and fodder. The tasks obviously involved sourcing, interacting with traders and so on, and that inevitably meant picking up information that might be rather more useful than identifying available stocks of corn. The job thus turned into one associated with espionage. Hadrian used his frumentarii as personal spies to investigate the lives of his friends and his household so that he knew exactly what they were up to.[35] Publius Aelius Annianus, either Publius' father or another predecessor, must have become a citizen under Hadrian, so his role as a military frumentarius clearly postdates his acquisition of citizenship – though what he did on a daily basis is quite unknown. Elpinius Festianus was a frumentarius of Legio I Adiutrix and is specifically described on an inscription from Ephesus as working in a prison.[36] Another frumentarius, whose name is lost, served with Legio X Gemina in charge of 'armaments'.[37] Lucius Aemilius Flaccus, a frumentarius of Legio XX Valeria Victrix, died in Rome where Titus Sempronius Pudens, who seems to have been another frumentarius of the same legion, set up a tombstone to his 'best friend'.[38] Since the legion was based in Britain throughout the period, the two men must have been seconded like the ambitious young Gaius Julius Apollinarius to Rome on official duties, perhaps to do with spying or intelligence. The emperor Macrinus (217–18) used his frumentarii to spy on soldiers behaving in an immoral fashion.[39]

There was also the opportunity for promotion up through the ranks, each class being linked to a higher pay grade. Antonius Ammonianus' promotion through his cavalry unit was recorded on a papyrus from the reign of Gordian III in Egypt which lists a number of other men too:

Antonius Ammonianus [enrolled] in the consulship of Maximus and Urbanus [AD 234], promoted from sesquiplicarius of the Ala [name of unit missing] to decurion by Basileus, the most excellent man, prefect of Egypt, on the 9th of the Kalends of November, in the [consulship] of Atticus and Praetextatus [AD 242].[40]

A *sesquiplicarius* is thought to have been a rank that received one and a half times the pay of an ordinary member of the unit. Ammonianus had been promoted to decurion, apparently jumping the intermediate position of duplicarius (double pay), which now meant he was an officer commanding a 30-strong turma subdivision of his cavalry wing. Another man, Victorinus, was promoted that year to the same rank, but in his case he had formerly served as a cavalryman attached to one of the legions based in Egypt, demonstrating that promotion could result in a citizen legionary moving to an auxiliary unit.[41]

Aurelius Mucianus of Legio II Parthica, formed by Septimius Severus, was also skilled in spear throwing. Mucianus served for ten years, dying at the age of thirty, and was depicted on his tombstone, found at Apamea in Syria, holding four spears. He was described as having 'learning in spears', though whether he was an instructor or still in training is not clear; the text may be a modest claim to be still learning his craft.[42] Around the same time or later in the third century, Lucius Tullius Felix was serving with Legio III Augusta at Lambaesis, where he was training to become an aquilifer.[43]

The most successful soldiers rarely bothered to mention their earlier careers. Publius Anicius Maximus, in whose honour the city of Alexandria set up a monument in his home town of Antioch in Pisidia, in the province of Galatia, reached the dizzying heights of 'prefect of the army which is in Egypt'. Before that he had served with Legio II Augusta in Britain as praefectus castrorum during the war of conquest under Claudius (41–54), normally the most senior position a soldier who had risen up the ranks could reach in a legion. He had been decorated by Claudius for his exploits. Prior to his post with II Augusta, he had been primus pilus with Legio XII Fulminata, probably while it was based in Syria. Typically for a man in a senior role, the inscription that commemorated his career omitted details of his earlier and junior positions. Anicius Maximus had also served as prefect in Antioch in Pisidia, standing in for the Roman senator Gnaeus Domitius Ahenobarbus (father of the emperor Nero) who probably held some sort of honorary position there.[44]

One of the most successful soldiers in Augustus' army was said to have been Titus Marius from Urbino in Italy. He seems to have joined as an ordinary legionary but managed to work his way up 'to the highest military posts', though none of these are specified by our source, Valerius Maximus, and nor are any of the legions he served in. Despite his alleged achievements he is otherwise unknown. Apparently he became rich and promised he would leave all his money to Augustus – he even told Augustus as much personally the day before he died – claiming it was only thanks to the emperor that he had been so successful. He was lying. Augustus' name did not appear in his will.[45]

## DISCIPLINE AND PUNISHMENT

Soldiers faced the death penalty if they deserted, dodged duties or disobeyed orders. But there was no 'system' of punishment: it relied on a commanding officer's discretion. After the late first century an officer had the opportunity to turn to Book 4 of Frontinus' *Stratagems* for guidance, though there were certainly other manuals. Frontinus provided a series of examples in the form of anecdotes about how previous commanders had dealt with discipline, such as one he attributed to Marcus Cato who had told of how in earlier times soldiers caught stealing had their right hands cut off in front of their comrades. Aulus Gellius recorded how 'in antiquity' soldiers were disgraced by having a vein opened, but concluded this was probably carried out as a medical treatment for soldiers whose actions had been caused by mental illness. He wondered if it then became routine on the basis that any offending soldier was thought to be of unsound mind.[46]

As a result, soldiers were supposed to live in fear of what their commanding officer might decide to do to them.[47] However, while ordinary soldiers could be executed for failing to live up to expectations, generals and officers were often treated a great deal more leniently, even if their actions had had disastrous consequences.

Terentius Varro, the general responsible for the catastrophe at Cannae in 216 BC (see Chapter 7), not only escaped punishment or even execution but also found himself being thanked by a crowd grateful that 'he had not despaired of the state'.[48] Cornelius Scipio, a tribune of Legio II at the battle, was in its aftermath given no sanction and neither were the other three tribunes present at the time.

Instead Scipio was handed supreme command of the army on the spot, even though he was only nineteen.[49] It was an inspired promotion. He later led Rome to victory over the Carthaginians at Zama and became known as Scipio Africanus in honour of his achievement (see Chapter 9).

One of the simplest sanctions for the ordinary soldier was the withholding of pay, but there was no consistency about this. In 215 BC the Senate voted for a double tax to be levied in order to pay the army. However, the soldiers who had been so heavily defeated by Hannibal at Cannae the previous year were excluded, in dramatic contrast to how their commanding officers were sometimes treated, such as Varro, and allowed to escape censure.[50]

When Scipio Aemilianus reached Utica in Africa in 148 BC during the Third Punic War he was appalled at the standard of discipline the commanding officer Calpurnius Piso had maintained. The soldiers were layabouts and among their number were all sorts of camp followers and opportunists hoping for booty. Scipio accused the men of acting more like 'holiday-makers than a besieging army' because of their greed. He castigated them for putting money before duty. Camp followers were thrown out and could return only with his permission, if they brought 'plain soldiers' food' and at a time when their commercial transactions could be supervised. Instead of brutal punishment, though, when it came to the soldiers he opted simply for harsh words. They were told to follow his example. 'We must toil while the danger lasts. Spoils and luxury must be postponed to the proper time,' he said, promising great rewards for those who obeyed.[51]

Sometimes a soldier's punishment was public shaming. In 105 BC Marcus Aemilius Scaurus was confronted with the fact that his son was a member of a Roman cavalry unit that had been attacked and routed by Germanic Cimbrian tribesmen. After the survivors fled back to Rome, Scaurus sent a message to his son that he would rather come across his bones, having been killed in action, than see him in person guilty of such a disgrace. Confronted with the message, the younger Scaurus committed suicide by falling on his sword rather than suffer the shame (*pudor*) of his actions any longer.[52]

During the Social War of 91–88 BC, some of the men under Sulla's command were so enraged by an officer of praetorian status called Albinus that they stoned and clubbed him to death. Sulla did nothing to them except to put it about that he now expected them to fight even more bravely in an effort to atone for their appalling crime. He was more interested in keeping the soldiers on side to support him in his political ambitions.[53]

One form of punishment available for failing to stand and fight was decimation, which normally meant killing every tenth man. It was first attributed to Appius Claudius, consul in 471 BC during a war against a rival tribe called the Volsci when some of his soldiers disgraced themselves.[54] The sanction survived well into Roman military history. Crassus used decimation during the Spartacus War (see Chapter 5). In 34 BC Octavian was fighting in Illyricum and had surrounded the tribal stronghold of Promona; he placed a cohort outside the city gates, but when the enemy attacked them the Romans fled in terror. Octavian had to use the rest of his force to push back the rebels, who surrendered the next day. The cohort that had fled was told to draw lots so that every tenth man could be executed, as well as two centurions. After such a drastic punishment it seems almost petty that the rest of the offending cohort was ordered to eat only barley instead of wheat that summer.[55]

Decimation was a risky business. Determined to purge his army of any troublemakers or stirrers, Mark Antony was able to call on his tribunes to identify the culprits from the standard records kept of every individual soldier and his personality:

> From these he chose by lot a certain number according to military law, and he put to death not every tenth man, but a smaller number, thinking that he would thus quickly strike terror into them. But the others were turned to rage and hatred instead of fear by this act.[56]

Far from settling any disquiet, these actions provoked supporters of Octavian in Antony's army to go around handing out notes that encouraged others to change sides because of Antony's 'meanness and cruelty'. Antony tried to find out who was responsible, but no one would tell him. Eventually he had to back down, apologizing that what he had said was only military law and offering the men a donative of 100 drachmas each.[57]

During Domitius Corbulo's exacting campaign in Armenia in 58–9 he had to discipline auxiliaries for underperforming. Two auxiliary cavalry wings and three infantry cohorts had fallen back when faced with the enemy at a fortress called Initia. Corbulo told them they would now have to camp outside the Roman fortifications until they had proved themselves once again through successful raids and hard work.[58] On the whole it was a light punishment compared to what they could have been subjected to.

Avidius Cassius was a Roman general in the East who made an unsuccessful attempt to become emperor in 175. He had earned himself a reputation for ruthless discipline and a total intolerance of anyone usurping his authority. Punishments for his soldiers included cutting off the hands of deserters, or ordering their legs and hips to be broken, arguing that a living crippled criminal was a better example to the rest than one who had been executed. A unit of auxiliaries in his army killed 3,000 Sarmatians, who were camping by the Danube, on the suggestion of their centurions, without telling Cassius. Not even the tribunes knew. When the auxiliaries turned up with the booty to present to Cassius, he was furious. There was no reward for the centurions; instead they were arrested and crucified. The reason was simple. Cassius pointed out that they have might have been walking into an ambush and that would have damaged the Roman Empire's reputation, permanently destroying the barbarians' fear of Rome.[59] Marcus Aurelius was pleased with Cassius' approach, commenting that soldiers could only be kept in order 'by the ancient discipline'.

Cassius maintained a rigorous programme of discipline and

training, along with weekly inspections of equipment and shoes. Any soldier found in his uniform at the resort town of Daphne near Antioch was subjected to the punishment known as *discinctus*, meaning literally the deprivation of his sword belt.[60] In other words he was thrown out of the army. Given the punishments Cassius was known for, such men would have had good reason to be grateful (see another example of his harshness in Chapter 5).

Decimation was occasionally revived at much later dates. The emperor Macrinus (217–18) was particularly fond of decimating troops, especially when they had mutinied. He coined his own term for his method – centimation – arguing that he was thereby being especially merciful because he only killed one in a hundred soldiers, instead of one in ten or twenty. Macrinus also had a preference for punishing his soldiers in ways more normally reserved for slaves, such as crucifixion. A military tribune who had allowed a guard post on campaign to go unmanned was punished on Macrinus' orders by being tied up under a wagon and dragged along underneath for the rest of the march, whether he was still alive or not.

It was little surprise that Macrinus lasted only fourteen months as emperor. In one version of events, embittered soldiers with vivid memories of his cruel and humiliating punishments formed a conspiracy, and he was killed by a centurion called Marcianus. According to another account Macrinus ran away from a battle against soldiers who had defected to Elagabalus whom they made emperor, believing him to be the son of Caracalla, and was killed by his pursuers.[61]

There were also legal provisions for punishing individual soldiers in the normal course of the law. The father-and-son emperors Valerian and Gallienus (253–68) decreed that:

Our soldiers and centurions who have been convicted of military offences are only permitted to make wills disposing of their castrensian [camp or fort] property, and the remainder goes to the Treasury by the right of intestacy.[62]

In other words, a soldier found guilty of a military offence had the right to dispose through his will only of his personal belongings in the fort. Everything else he owned, perhaps property in his homeland, was forfeited to the state.

If the emperor learned about an errant soldier he could punish the man personally. The emperor Severus Alexander (222–35), the *Historia Augusta* reported, discovered that a soldier had treated an old woman badly. Since he was a wagon-maker by trade Alexander decided to throw the man out of the army and make him the old woman's slave so that he could build wagons and support her. This was part of the emperor's general strictness towards soldiers. He was also outraged to discover that legionaries in Antioch, where he had just arrived to embark on a campaign against Persia, were using their leisure time by spending it in the women's baths. The culprits were arrested and thrown into chains, resulting in a mutiny by their comrades. He was only able to calm the men down by pointing out that if they hurt him his successor would exact vengeance on them.[63]

## ARMS AND EQUIPMENT

Roman arms and equipment were admired by Rome's enemies from early on. The gladius sword became legendary, but it was only one component of a comprehensive range of equipment. At Cannae the brilliantly successful Carthaginian general Hannibal equipped his men with Roman armour and weapons captured in earlier battles. He won a major victory, allegedly killing at least 50,000 Roman and allied troops. The Carthaginian use of Roman weapons must have made the bitter pill even harder to swallow.[64] A shortage of Roman equipment was solved in 215 BC when a new army was put together in Rome. Spoils from previous successful battles were taken down from temples and porticoes where they had been displayed for distribution among the new soldiers.[65]

Serried ranks of designer Roman soldiers in gleaming, identical armour and pristine tunics, equipped with immaculate reproduction weaponry, form an image familiar from films and re-enactor events today. Josephus' description of Vespasian's army helps reinforce that impression with his comment that infantry and cavalry wore identical helmets and cuirasses.[66] The reality, as is quite clear from archaeological finds of helmets and other gear, is of a far more ad hoc and make-do-and-mend arrangement. Equipment was handed down, reused, customized and obtained from wherever necessary. Armour rarely survives, but there were several different types including mail armour, scale armour (*lorica squamata*), and the so-called *lorica segmentata* formed of overlapping concentric bands. Since stainless steel was not invented until modern times, preventing Roman armour from rusting meant regular oiling and polishing. Continual repairs and maintenance were necessary, whether the replacement of metal components like buckles and leather straps or of the pieces of armour itself. Properly maintained, though, Roman body armour could protect soldiers from everything except 'superficial wounds'.[67]

One bronze helmet dating to the first century AD, of the 'Coolus' type which fell out of use by the second century and was replaced with the 'Imperial Gallic' variety, did particularly good service. Lucius Dulcius, a soldier in a century commanded by the centurion Marcus Valerius Ursus, punched both their names into the neck guard, along with those of several other previous owners. In all at least seven soldiers used this one helmet at one time or another during its service life, each at least for long enough to consider it theirs.[68] It was finally thrown into the Thames at London, or into its tributary the Walbrook. The find spot makes it likely that the man who lost or discarded the helmet, or perhaps used it for a votive deposit, was serving in the governor of Britain's garrison. Another helmet, found at St Albans, belonged to at least three soldiers, including two called Papirius and Marcus.[69] Several more have been found, the names exhibiting varying degrees of legibility.

Similar arrangements applied to other equipment. Docilis was a soldier probably in Ala Gallorum Sebosiana at Carlisle in or around the year 100. On the instructions of his prefect Augurinus he drew up an inventory of lancers who were missing either their lance (*lancia*), a pair of smaller lances (*subarmales dua*), or what he called their 'regulation swords' (*gladia instituta*). Since cavalry were involved, the troopers must have carried their long sword (*spatha*), but Docilis clearly calls it a gladius. Evidently the word was applied quite casually to whatever the 'regulation' sword was for that type of soldier. We can only assume that, for example, Verecundus who had lost both his lance and his pair of small lances had done so in some sort of engagement.[70] Presumably the list was then used by Docilis to arrange for replacements drawn from stores, or which needed to be manufactured by the military *fabrica*. The emperor Hadrian, however, was specifically referred to in a late Roman biography of his life as wearing the spatha. It is quite possible though that by the time the biography was written in the fourth century, with a greater emphasis on cavalry in the Roman army, 'spatha' had started to become a more generic term for swords.[71]

A passing reference in the works of the rhetorician Quintilianus in the late first century AD gives us a better idea of the casual, even slang, words that soldiers must have used for swords as in the same way that today guns can be referred to as 'heaters', 'rods' or 'pieces'. *Gladius Hispaniensis* was abbreviated in everyday speech to *gladius* or *ensis* without any difference in meaning. According to Quintilianus words applied to a sword's characteristics were also utilized to refer to swords more generally, such as *ferrum* ('iron' or 'steel') and *mucro* ('point').[72] The future emperor Aurelian was an enthusiastic swordsman when in the army. As a tribune he was said to have earned himself the nickname *Aurelianus manu ad ferrum*, 'Aurelian-iron-in-hand'.[73] In the mid-second century AD Aulus Gellius itemized all the names of weapons he could remember. These did not include *gladius* but he did mention the *spatha*, *machaera* ('sword' or 'knife'), and also a small sword called the *lingula* from the Latin for a 'little tongue' which it was thought to resemble.[74]

In the early second century AD during the reign of Trajan an Egyptian soldier called Claudius Terentianus, apparently then in the Classis Alexandrina fleet, was about to be sent off in a detachment to Syria. He wrote home to his father in Karanis in the Fayum anxiously asking for important equipment to be sent to him which included a *gladius pugnatorius* or 'fighting sword' as well as a pickaxe, grappling iron, 'two of the best lances', and various items of clothing. Terentianus was particularly anxious that everything in his father's consignment be described in detail in a letter in case they were replaced with inferior substitutes en route. The letter's principal value is that it shows how soldiers might have to take individual responsibility for equipping themselves and that military-grade armaments were clearly available on the open market (see the contemporary letter from Hermoupolis Magna below, which itemizes the exact cost). However, as a soldier in the fleet Terentianus may have been dissatisfied with inferior arms available at his base in Alexandria and wanted to upgrade his equipment.[75]

Swords and their scabbards could be individually customized, especially if they were valued personal possessions. The decorated scabbard of the 'Fulham' sword, a gladius found in the Thames, bears a bronze plate depicting the legendary Wolf and Twins from Rome's foundation myth. The scabbard of the so-called 'Sword of Tiberius', recovered from the Rhine at Mainz, depicts in one scene Tiberius greeting Germanicus, while on what is thought to be the scabbard chape is what seems to be a depiction of the standards lost by Varus in AD 9, two of which Germanicus recovered.[76]

Aurelius Cervianus, who served with the Roman army in the mid-third century, probably as a commanding officer, somewhere on the Continent, was in possession of a bronze roundel that may have been a decorative feature attached to a shield. It was engraved with the names of Legio XX Valeria Victrix and Legio II Augusta (both based in Britain), Aurelius' name and the popular slogan *utere felix*, 'good luck to the user'. The roundel's find spot is unknown, but is assumed to have been somewhere on the Continent. The only context that would explain its presence there is the sending of detachments of both legions from Britain to the Continent in the third century during the reign of Gallienus.[77]

Soldiers also placed their names on other pieces of equipment, clearly bothered that someone else might help themselves to it. Victor, who served in a century commanded by a centurion called Verus, and seems also to have been based in London, punched his details into an iron spearhead that ended up in a well of first-century date.[78] At Newstead a soldier with the remarkable name of Compitalicius marked his name, and that of his centurion Barrus, on an axe in the late first century. His name was derived from the festival of Compitalia which celebrated the Lares spirits of the Highways and derived from the word *compita*, 'crossroads'. It came shortly after the festival of the Saturnalia which began on 17 December, on a date decided annually in Rome by a praetor, and continued for a number of days. Compitalicius had perhaps been born during the festival.[79] A particularly memorable example of named items is the collection of silvered-bronze horse trappings that belonged successively to cavalry officers Titus Capitonius and Verecundus around the year 50 at Xanten. One of them also punched on to a disc the words PLINIO PRAEF(ECTO) EQ(UITUM), '(under) Pliny, prefect of the cavalry'. This is the famous Pliny the Elder, whose *Natural History* is such an important source for the Roman world today and whose work is referred to several times in this book.[80]

Leather was a vital component in military clothing, but unsurprisingly it rarely survives unless found in ultra-dry or waterlogged deposits. Such are the fragments of leather doublets, worn under armour, found at Vindonissa.[81] Some leather goods or equipment were marked for the units that owned them, like a piece of tent(?) leather found in York incised with the words 'century of Sollius Julianus', who served with Legio VI Victrix there and on Hadrian's Wall.[82]

Woollen tunics, even if scraps survive, are impossible to attribute to an individual, but tombstones provide a better idea of clothing. Annaius Daverzus was a soldier who served with Cohors IIII Delmatarum at Bingerbrück in Germania Superior and died at the age of thirty-six after 15 years' service. His tombstone shows him dressed in a draped short-sleeved tunic reaching halfway down his thighs, with cloak, as well as a sword belt, spear and shield.[83] Oclatius, a signifer with Ala Afrorum at Neuss, is depicted similarly, armed with his sword and holding his spear-headed standard with an effigy of a lion's face.[84] In Egypt at a village called Philadelphia, in the year 138, the resident weavers were given an advance payment on the instructions of the prefect of Egypt, Avidius Heliodorus. He had ordered various articles of clothing to be made for the soldiers out of 'soft, fine, pure white wool'; they included four 'Syrian cloaks', as well as a tunic, and a blanket for use in a military hospital. These were destined for the garrison in Cappadocia.[85] There are countless examples of other items of quite trivial military equipment, and even personal possessions such as pottery vessels, that were laboriously punched or scratched with their owner's details. Januarius, a trumpeter (*bucinator*), took the trouble to inscribe his name on the base of a samian dish at Caernarvon, an auxiliary fort in Wales.[86]

Losing a weapon or piece of equipment, especially in battle, was a potentially humiliating and expensive disaster. In the days of the Republic men who lost a sword or shield would throw themselves on the enemy in the hope of recovering it, or accept death as an honourable alternative; anything rather than 'the inevitable disgrace and the humiliations they would face at home'.[87] There was also the question of the cost of equipment, for which there is little evidence. An anonymous letter found at Hermoupolis Magna in Egypt and dating to 19 June 115–17, in the reign of Trajan, refers to the cost and value of a number of commodities. These include an 'Italian sword' that cost 80 drachmas. It is named using a Greek version (σαμσειρα, 'samseira') of the Persian word 'šamšīr', coupled with the adjective 'Ἰταλικὴ' ('Italian'); the word is apparently unattested anywhere else, though it must mean a gladius or equivalent.[88] The writer says the price was 'below its value you will admit' and also refers to a cuirass that had been obtained for 360 drachmas. On the basis that at the time 80 drachmas were worth around 24 denarii (other estimates include 20 denarii), this makes the price equivalent to about one-fifteenth of a legionary's annual pay. But since the reference is to one incident in one place at one date, there is no means of working out whether it has any wider relevance, or whether that was what a soldier had to pay to buy such an item from the army through his unit's stores or on the commercial market.[89]

Junius Dubitatus was a legionary serving in a century commanded by the centurion Julius Magnus of Legio VIII Augusta, based at Strasbourg on the Rhine frontier probably during the reign of Hadrian. The legion is known to have been sent to Britain during Hadrian's reign, and this would explain why an elaborate bronze shield boss punched with Dubitatus' name, his legion and the name of his centurion, was found in the river Tyne at Newcastle.[90] It was decorated with the image of an eagle in the middle, surrounded by depictions of the Four Seasons, a bull, Mars and two legionary standards. The shield boss may have been a one-off personally commissioned by Dubitatus, because it bears no other names and seems to be especially elaborate. Dubitatus might usefully have borne in mind the words of Scipio Aemilianus, who in 134 BC noticed that one of his soldiers had an elaborately decorated shield. Scipio took this as a mark of cowardice, commenting scornfully that the soldier obviously put his trust more in the shield than his sword.[91] During Domitian's Dacian War of 86–8 one of his generals, Lucius Tettius Julianus, came up with the idea of ordering soldiers to write their own name and that of their centurion on their shields. The idea was that if any of them did anything particularly heroic, they would be more easily recognized.[92]

An episode in the Civil War of 68–9 described by Tacitus, and demonstrating how easily chaos could take over in the heat of the moment, has the advantage of telling us that the various parts of the army were normally supposed to look different. When in early

69 Otho ordered the arsenal in Rome to be opened in order to equip troops fighting for him against Galba, bedlam followed. The praetorians grabbed any piece of equipment, regardless of whether it was supposed to be for praetorians or legionaries or anyone else. Some even ended up with shields and helmets intended for auxiliaries. 'All was confusion', said Tacitus. [93]

Roman soldiers also had to carry tools needed for camp building and foraging, and provisions. That meant axes, saws, sickles, baskets, shovels, ropes and chains, with a total weight of about 60 Roman pounds (about 19.7 kg).[94] Hundreds of pack animals, such as mules, per legion could help, especially with carrying the tent and cooking equipment for each contubernium of eight men. Around 500 mules might be needed to carry the tents for one legion alone, animals that obviously needed feeding and watering.

In extreme circumstances the men could be ordered to carry only what they needed to stay alive. At the Battle of Thermopylae in 191 BC a Roman force under Manius Acilius Glabrio defeated Antiochus III. In the aftermath Glabrio's triumphant army struggled over a difficult mountain road, weighed down with baggage and booty, and thereby learned a terrible lesson. The road was so bad that the soldiers, laden with arms and equipment, struggled, and many fell over the edge.[95] In 133 BC at Numantia, Scipio Aemilianus told his men to carry several days' worth of rations to toughen them up, enabling them to endure long marches, but he smashed up any cooking equipment that he considered unnecessary on campaign. He also insisted that they go on foot rather than riding on mules, asking 'What can you expect in a war from a man who is not even able to walk?'[96] Pack animals moreover were slow, and required huge logistical support (see Chapter 6 for a discussion of the use of animals in the Roman army). The general Gaius Marius made his men carry their utensils and food themselves on forked poles, earning them the nickname 'The Marius mules'.[97] (Popular lore offered an alternative explanation of the name. Marius was said to have so praised a mule at the siege of Numantia for its health, obedience and strength that a 'Marius mule' became a facetious description of any soldier who behaved the same way.[98]) Three centuries later, Avidius Cassius ordered his men to carry for their sustenance only lard, vinegar and biscuit, on pain of severe punishment if caught with anything else. Basic food, hard work and exercise were felt to be effective ways of making soldiers ready for hardship in the future.[99]

## PERSONAL DISPLAY

Some Roman soldiers were given to extravagant personal display. The fortress of Legio II Augusta at Caerleon in Britain was one of the longest-established military bases in the Roman Empire. Excavation of the fortress baths and its drainage system has produced a remarkable assemblage of finds. Eighty-eight gemstones were recovered from the drain of the cold plunge bath (*frigidarium*); they had all originally been set in rings and had clearly fallen in while the soldiers bathed. They could be dated variously to the periods 75–110 and 160–230. They illustrate the sort of motifs the soldiers must have favoured on the rings they wore as they went about their daily business in the fortress. There were typically male and military figures such as Mars and Achilles, as well as more generic Roman imperial deities and personifications like Jupiter, Victory, and an eagle holding a trophy. Elsewhere in the fortress, gemstones depicting Hercules and even Alexander the Great have been found.[100]

A gold ring with a gemstone engraved with an image of a theatre mask was found at the fort of Housesteads on Hadrian's Wall. Gold could only be worn by equestrians or senators, so it must have belonged to the commanding officer of the auxiliary Cohors I Tungrorum stationed there.[101] The emperors also presented officers with ostentatious gold rings and brooches, especially at later dates. Diocletian was probably responsible for an officer's gold ring set with a gold aureus bearing his portrait and on the reverse the legend VIRTUS MILITUM, 'the valour of the soldiers'. Another, known from a number of examples, bears the inscription FIDEM CONSTANTINO, meaning '[I swear] loyalty to Constantine'.[102]

Other items recovered from the Caerleon baths included a gold pendant in the shape of a vase, a silver phallic pendant, and numerous bronze brooches used for fastening clothing, of a type always found on both Roman military and civilian sites. One of the most remarkable was found at Greatchesters (Aesica) on Hadrian's Wall. Known today as 'the Aesica Brooch', it was made of gilt bronze and is 4 in (10 cm) in height, decorated with swirling abstract Romano-Celtic designs. The most likely explanation for its presence is that its ostentatious size and decoration appealed to an auxiliary soldier from a frontier province stationed at the fort.

None of this love of display would have impressed the emperor Hadrian. He thought soldiers inclined to decorate themselves and their equipment were acting inappropriately. In Roman tradition 'luxury' was seen as a sign of effeminacy and moral degeneration. Hadrian spurned gold and other ornaments on his own sword belt, and was reluctant even to allow an ivory hilt on his sword.[103] By the third century, however, it had apparently become more common for officers to wear decorated belts, or so a story about Saloninus, son of the emperor Gallienus, would suggest. It was customary for soldiers to remove their sword belts when they attended a banquet. At one such banquet given by Gallienus, the light-fingered Saloninus stole all the bejewelled and gold-studded sword belts, while the soldiers dared not say a word. On a subsequent occasion, they kept their belts on; when asked why, they said they were 'wearing them for Saloninus'. Thereafter soldiers dining with the emperor supposedly kept their belts on. Gallienus' biographer disputed the story he had related, and said that in fact soldiers continued to dine unbelted with the emperor, but kept their belts on at other meals because of a tradition of eating in readiness to fight.[104]

Roman soldiers put up with the training and the harsh conditions because they expected to do well out of a military career, in the form not only of prestige and glory but also of pay, booty, and handouts and bequests from the emperor. Money, as we shall see, could play a decisive role in the management of soldiers; there were dangerous consequences for emperors and generals who failed to come up with hard cash. During the terrible days of the Second Punic War the Roman currency system was revised. The key coin was the new silver *denarius*, introduced *c.* 212–211 BC. It was struck to a standard weight, purity and size, which guaranteed its value.[105] The denarius was used by the Roman state to pay for the war, and soldiers were among its most important recipients. It became the staple coin of military pay for more than 400 years.

# FOUR

# GOLD AND SILVER

## Pay, Handouts and Bequests

*Ten asses a day was the assessment of body and soul: with that*
*they had to buy clothes, weapons and tents, bribe the bullying*
*centurion and purchase a respite from duty.*

The legionary Percennius whips up his aggrieved
comrades in Pannonia over pay and stoppages[1]

I n Rome's earliest days soldiers were unpaid, but in 421 BC plans
were laid for distributing public land and founding colonies so
that a tax could be levied on the occupants to pay the troops. It
is not clear whether the scheme was ever put into action.[2] In 406 BC
the Senate voted that soldiers be paid from the public treasury. A
few years later in 402 BC the Senate voted once more that those who
had volunteered to serve at the ten-year siege of the Etruscan city of
Veii beyond their normal obligations of service should receive payment as a reward.[3] There was, however, no regular system of paying
soldiers and it was not until the late third century BC that Rome
introduced a reliable silver coinage, in the form of the denarius. By
the mid-second century BC a cavalry trooper was paid one denarius

per day, an ordinary soldier one-third of that amount, and a centurion double what a soldier received.[4] However, there was no reliable system of payment, which was probably often in arrears if it came at all. Regularizing the army, as Augustus did a century later, meant that soldiers had fixed minimum terms of service and were also paid routinely, at least in theory. In 13 BC Augustus declared that praetorian troops would serve 12 years and legionaries 16, but he changed his mind in AD 5, increasing the periods to 16 and 20 years respectively.[5]

According to Tacitus, in AD 14, at the time of Augustus' death, a legionary received 10 copper asses (0.625 of a silver denarius) per day (equal to 228 denarii per annum).[6] He added that the Praetorian Guard did rather better and were paid two silver denarii per day (equal to 32 asses a day or 730 denarii per annum). The annual rate for a legionary was in fact 225 denarii per year, raised to 300 under Domitian (81–96). The Praetorian Guard's actual annual rate was 750 denarii. Tacitus had obviously rounded the day-rate figures for simplicity. Soldiers in the Roman army were paid three times a year, so their annual pay needed to be divisible into thirds, as 225, 300 and 750 denarii are. All these figures were subject to deductions for food, clothing and equipment. The cost of the army was already considerable, so much so that Domitian found it impossible to pay for the soldiers as well as for his public works. His money-saving plans to cut the size of the army had to be abandoned because this would have made the Empire's borders more susceptible, and he had to resort to more taxation and the confiscation of estates.[7] The army only became more expensive as pay increased over time, rising to 600 denarii per legionary under Septimius Severus over a century later. His son

Caracalla increased it to 900 denarii, a figure that was doubled by Maximinus I (235–8), all maintaining the requirement of being divisible by three.[8] The Praetorian Guard received proportionately far more, as they had from the outset. The exact figures are far less important than the trend. The rapid rate of increase in the early third century not only reflected inflation but also the increasingly cynical desire on the part of emperors to buy soldiers' loyalty, and the soldiers' greedy demands. In real terms the buying power under Maximinus was probably significantly less than in Domitian's time. However, it must always be remembered that routine deductions meant the individual soldier received only a proportion of his pay and even that was susceptible to delays. Little is known about military pay in the fourth century, in part because the coinage system of that period is itself far less well understood and also because it went through a series of changes and reforms with variable success.

Soldiers were unsurprisingly susceptible to the idea that they were being ripped off and ought to be paid more. Their loyalty proved increasingly fickle as the prospect of toppling one emperor in favour of another who made a better offer became their priority. Events in Pannonia and among the Rhine garrison in AD 14, when legionaries in both areas mutinied over pay and conditions, were ominous portents of later Roman history. Four legions on the Rhine – I, V, XX and XXI – erupted into mutiny after some of their number whipped them up. Their grievances included wanting more pay and for veterans to be discharged instead of being forced to stay on.[9] Although on this occasion the men were talked into backing down with promises of improved conditions, later generations of soldiers would come to realize how much power they really had.

Physical money appears to have been on hand in forts, at least so long as the state manufactured enough and was able to transport it to the soldiers' quarters. Little is known about this process, but during the first century until the reign of Nero silver and gold coins were minted in Lyon and Rome, sometimes simultaneously or at different times. During the Civil War of 68–9, coinage was minted where it was needed, for example by Galba in Spain and Vespasian in Antioch. Thereafter gold and silver was exclusively struck at Rome, apart from some occasional issues in Eastern cities. The only exceptions were the breakaway regimes, for example those of Postumus and Carausius in the later third century; both men produced coins for their armies in the main cities under their control, respectively Cologne and London.

Military pay seems normally to have been stored in the strongrooms that formed part of a fort's headquarters building. When the disaffected men of Legio V Alaudae and Legio XXI Rapax mutinied in AD 14 over pay and conditions (see Chapter 2), Germanicus made promises to give in to their demands, but they refused to go back to their winter quarters until they had been paid on the spot. The cash had to come out of chests containing the money for Germanicus and his entourage. Realizing how dangerous the situation was, Germanicus also paid off the II, XIII and XVI legions without them even asking.[10] As it happens, Legio V and Legio XXI remained bitter and Germanicus had to take military action against them.[11]

There were also sources of much larger sums of money for soldiers. Praetorians and legionaries received a gift in Augustus' will that was equal to one-third of their annual pay.[12] Later emperors sometimes followed suit in their wills. When Hadrian adopted Aelius as his successor in late 136 he paid out 75 million denarii 'to the soldiers'. If this was destined for the praetorians only, and it may well have been, this amounted to around 7,500 denarii each, a vast sum. When Aelius grew ill and died in early 138 Hadrian said the money had been wasted.[13] Gratuities were also paid out on an emperor's accession, or on the occasion of a marriage, such as when Antoninus Pius' daughter

Faustina the Younger married the future emperor Marcus Aurelius in 145. In 161, when Marcus Aurelius and Lucius Verus acceded after the death of Antoninus Pius, they promised 5,000 denarii to each member of the Praetorian Guard; this could have amounted to 50 million denarii for the praetorians alone.[14] In 202 Septimius Severus paid ten gold *aurei*, the equivalent of 250 denarii, to each praetorian to celebrate his ten years in power.[15] Spending power was bolstered by exempting soldiers from tax. In 58 Nero ordered that soldiers were to retain all their immunity from tax, other than on goods they were selling, which was illegal.[16]

On retirement a different ratio was used to calculate how much a soldier was due. Praetorians received 5,000 denarii, legionaries 3,000 denarii.[17] Augustus claimed that he had personally transferred 170 million sestertii (42.5 million denarii) out of his own personal estate to the 'military treasury' to pay retirement grants.[18] Such men were also eligible for land grants in military colonies (see Chapter 15). Later emperors and even empresses founded further colonies. For example, Agrippina the Younger, empress of Claudius, created the colony at Cologne (named Colonia Claudia Ara Agrippinensium as a result) where she had been born on the frontier during her father Germanicus' campaign in AD 15.[19]

Higher ranks did much better out of military service than ordinary soldiers, and sometimes there was a remarkable differential. By the reign of Augustus a centurion received 3,375 denarii per annum, a phenomenal fifteen times more than a legionary. Centurions in the first cohort of a legion were paid double that sum, and a primus pilus as the most senior centurion in the legion double again at 13,500 denarii. It was no wonder then that senior centurions could afford personal entourages including slaves and freedmen, the latter often taking charge of their former master's burial if he died in service. The pay differentials were maintained in later centuries, even though the actual amounts were increased.[20]

The promised accession handouts were not always paid. Didius Julianus, a short-lived emperor in 193, was toppled and killed by the praetorians for defaulting (see Chapter 11). Emperors who did not put paying soldiers first were liable to regret it. No wonder then that on his deathbed at York in Britain in 211 Septimius Severus advised his sons and successors, Caracalla and Geta, to 'be harmonious, enrich the soldiers and scorn everyone else'.[21]

Corruption was rife in the army, as it was everywhere else in the Roman world. In 69 it emerged that the already well-paid centurions of the Praetorian Guard had been receiving bribes from men keen to avoid doing any duties. As many as a quarter of the praetorians were loafing about, or away engaged on other jobs which paid them enough to buy off the centurions. The centurions were smart enough to make soldiers who came from better-off families do the worst jobs in order to maximize their income from bribes. The short-lived emperor Otho decided to do away with this abuse, but the reform meant he had to pay the centurions from state funds the money they would lose.[22] Hadrian finally banned military tribunes from receiving gifts from soldiers, which were presumably being offered in return for being left off the duty roster or in the hope of promotion.[23]

The pay of auxiliary soldiers is virtually unknown but was probably less than that of legionaries, and certainly no more. Auxiliary cavalry were probably the best paid and auxiliary infantry the worst. One theory is, or rather was, that auxiliary infantry for example received five-sixths of a legionary's pay, but this is completely unproven.[24] One unique piece of evidence concerns Clua, a Raetian cavalryman attached to an infantry unit of Raetians and based at Windisch (Vindonissa) in Germania Superior (in a part of the province now in Switzerland), at the fortress of Legio XIII Gemina where one of his pay receipts was found. Auxiliary units were often based with the legions alongside which they fought. In this case it was probably Cohors VII Raetorum equitata, which is known to have been at Windisch a few years later. The document refers to an advance of his September pay (the last third of the year), paid on 22 July 38, which was 75 denarii. Clua also drew out an additional 50 denarii from his existing account as if it was a bank.[25] Vegetius recorded how soldiers were obliged to hand over half their money so that it could be stored safely. Not only did that prevent the soldiers from wasting their cash, but it also discouraged them from deserting and thereby losing their savings.[26] It is not surprising that Clua as a cavalryman seems to have been on as much money as a legionary. But the document is unique, which shows how careful one would need to be in using that single example as a template for other auxiliary pay.

Just as any employee today, including the military, looks at the various stoppages and deductions on his or her pay slip with dismay, so did the average Roman soldier. In his case, deductions were mainly for food, clothing, arms and tents.[27] The practice went right back to the days of the Republic. Roman soldiers in the middle of the second century BC were given rations of barley and wheat, but the cost of the food, their clothes and even extra armaments, were all deducted from their pay. Oddly, soldiers of Rome's allies (generally various peoples in Italy who had sided with Rome) who were fighting in the same army received the same food rations for nothing.[28] The reforming tribune of the plebs in 123–122 BC, Gaius Gracchus, tried to institute an additional allowance for clothing. If this happened it seems to have fallen into abeyance. The grievances in AD 14 included the deductions for clothing and equipment.[29] There were also community deductions. Soldiers contributed to a burial fund. In a legion there were ten bags of money, one for each cohort's savings, and an eleventh bag contained money from every soldier to cover funeral expenses.[30]

The system of deductions continued throughout the army's history. In the year 81 Quintus Julius Proculus, a soldier from Damascus in Syria, was the subject of an annual pay and deductions summary. Julius Proculus was paid in the Egyptian denomination, being awarded 247½ drachmas three times a year out of 250 due (the difference may have been a currency conversion fee). This was the exact equivalent of the 75 denarii legionaries received elsewhere, though it has been suggested that Proculus and Gaius Valerius

Germanus, another soldier whose name appears on the same papyrus, were auxiliaries. At the end of the first four months he lost 182 drachmas in deductions. This paid for his food (80 drachmas), clothes (60), the annual Saturnalia feast (20), his boots (12), and bedding (10). He was left with 66 drachmas, but instead of this being handed over to him it was added to a credit balance which was by then 202 drachmas. By the end of the year his credit balance stood at 344 drachmas, but the third instalment that year had been wiped out by expenses, which on that occasion included 146 drachmas on clothes alone.[31] Julius Proculus was perhaps able to draw on his credit balance, as Clua did.

Just two years later, on the other side of the Empire in northern Britain on a remote and dangerous frontier, one legionary was in need of a loan. On 7 November 83 he turned to another legionary in the fort at Carlisle to help him out. It was the last winter of several years of campaigning on the northern frontier by the governor Gnaeus Julius Agricola. Quintus Cassius Secundus, a soldier of Legio XX, confirmed in writing that 'I owe Gaius Geminius Mansuetus, soldier of the same legion, 100 denarii which [I will repay . . .]'. The sum was considerable and was at least equivalent to four months' pay before deductions. The missing section must have cited the date by which the money was to be repaid and possibly a rate of interest. It is on one hand remarkable that Cassius Secundus was prepared to indebt himself to that extent, and on the other that Geminius Mansuetus had it to lend. In a world where banks as we understand them did not exist, private individuals with capital to spare might put it out to interest.

The transaction between Cassius Secundus and Geminius Mansuetus was recorded on a writing tablet found at Carlisle.[32] Such documents were potentially quite important if the debt was to be recalled. The emperor Gordian III made a ruling between 238 and 244 concerning a similar arrangement between two other soldiers, reminding them of the need for additional evidence to prove a debt should the original contract be destroyed:

*The Emperor Gordian [III] to the soldiers Priscus and Marcus*
Where the evidences of a debt have been consumed by fire, while it is unjust for debtors to refuse payment of the sums which they owe, still, too ready belief should not always be accorded to persons who complain of such an accident. Therefore, you should understand that where the instruments are missing, you ought to prove the truth of the statement in your petition by other evidence.[33]

Remarkably, a Cassius Secundus who was a veteran of Legio XX and died at the age of eighty was buried at the legion's base at Chester (Deva) sometime in the second century after an honourable discharge.[34] If this is the same man, then he clearly survived Agricola's war and lived on until his legion was sent north once more, this time to build Hadrian's Wall.

Hoarding coins was one way to keep personal cash safe. In the normal course of events a coin hoard might be added to, withdrawn from, and eventually removed. Some coin hoards were abandoned, either because the coins had become worthless from inflation or revaluation, or because their owners had forgotten where they had buried them, had died or had otherwise been prevented from recovering their money. These are the hoards that turn up today. Needless to say, coin hoards almost invariably tell us nothing specific about who owned them. They were most often stacked into bags or pots and buried without any identification, for the obvious reason that no one else was supposed to know they were there.

Occasionally the content of a coin hoard and the context in which it was found can give us an idea of who owned it, especially when the hoard seems to belong to a crucial date in Roman military history such as the invasion of Britain in 43.

Perhaps one of the men in that invading army was responsible for the hoard of gold coins buried at Bredgar near Sittingbourne in Kent and the river Medway, which the Roman forces crossed shortly after the invading force landed. The hoard consisted of 37 gold aurei ranging in date from the time of Julius Caesar to issues of Claudius made in 41–2.[35] The gold came to a huge sum of money, amounting to more than four years' pay before deductions for a legionary at that date (the annual pay for a legionary of 225 denarii was equal to nine such gold coins). The most likely owner therefore was a centurion or an officer because of the sheer quantity of gold involved. One recent suggestion is that it was a gift or bribe to an officer in the invading force from Claudius' freedman Narcissus who had been sent to encourage the reluctant troops to embark on the invasion. It is easy to imagine how soldiers advancing through northern Kent, all the while anticipating the battle by a river that did take place against two tribal leaders called Caratacus and Togodumnus in those first few days, might have taken care of their valuables by burying them first. Another hoard, this time found below the floor of a room in a barrack block in Legio II Augusta's base at Caerleon, consisted of 295 denarii and 2 sestertii deposited not before 177. It was probably the personal stash of one of the legionaries though in his case why he abandoned a sum of money equivalent to almost a year's pay is unknown.[36]

Sometimes soldiers kept money on their person. One member of the garrison at Birdoswald, a cavalry fort on Hadrian's Wall, was wearing an arm-purse made of bronze (pockets did not exist in the Roman world and such purses were a common means of carrying money about). This consisted of a small container with a band that fitted round the arm. Somehow the soldier had either dropped it or put it down for safekeeping, apparently while working on the earthen rampart that backed on to the fort's east wall. It turned out to have 28 silver denarii inside, which was about 9 per cent of a legionary's annual pay of 300 denarii at the time. The latest coin was dated to 119 in the reign of Hadrian, making it fairly likely that the purse was lost only a few years later and therefore probably when the fort was being built. The other 27 coins dated all the way back through the first century AD to the first century BC, the earliest having been minted in Rome in 125 BC. One of the denarii was struck by Mark Antony and had originally been paid out to a soldier in his army at the time of the

Battle of Actium in 31 BC. Mark Antony's coins were incidentally still in circulation half a century after the Birdoswald hoard; there were five in the Caerleon cache discussed above.[37] Further evidence that soldiers carried money around with them comes from coins dating to around 256–7, found at Dura-Europos on the bodies of soldiers killed in the mines when a tower collapsed and trapped them.[38]

Soldiers were, incidentally, not exempt from paying customs duties and seeing some of their money disappear that way too. Septimius Severus and Caracalla made a ruling around 198–211 concerning a soldier called Ingenuus, who clearly felt he should not have to pay the duties. He was ordered to do so, but was reassured that as a soldier he would not be liable to the normal penalties for defaulting on them.[39]

## COINS HONOURING THE ARMY

Some of the coins that passed through a Roman soldier's hands were important pieces of imperial propaganda, at the very least reminding them who the emperor was and who they were supposed to be loyal to. Rebel emperors took care to strike coins with their portraits on as fast as possible, primarily for that purpose. Promoting the army's achievements was an excellent way of trying to keep the soldiers on-side. Sometimes the message was blatant. In 41 the Praetorian Guard placed Claudius on the throne after Caligula was assassinated. Without an emperor the praetorians had faced saying goodbye to all their privileges. Claudius issued two types of gold and silver coins soon afterwards that made it clear he and the soldiers were in it together. On the reverse of one coin he was shown clasping the hand of a praetorian signifer, accompanied by the legend 'the praetorians received [by the emperor]'. On the other was a depiction of the Castra Praetoria with a legend that translates as 'the emperor received [by the praetorians]'.[40] The coins were almost certainly issued to the praetorians to reinforce their loyalty. As the coins worked their way into the circulating coin pool, the message was disseminated more widely.

Some emperors issued coins showing them on a podium and speaking to a group of soldiers, probably praetorians, with the legend AD LOCVT COH, 'Speech to the Cohorts'.[41] Nero struck a series of sestertii showing himself and a praetorian riding horses, equipped with spears, and accompanied by the legend DECVRSIO ('military manoeuvre').[42] However, this was more an artistic conceit than military propaganda, because the design was copied from coins of the Baktrian king Eukratides I (171–145 BC). Another image that emerged from time to time depicted a pair of clasped hands on the reverse of the coin, alongside the legend CONCORDIA EXERCITVVM, 'harmony of the armies'.[43] The hands represented the emperor and a symbolic soldier.

Coins commemorating individual legions and military units were more unusual. Mark Antony was an exception. In the immediate prelude to the Battle of Actium he commissioned the production of vast numbers of silver denarii, all with a more or less identical design that showed on one side a galley and his Roman titles and on the other a legionary standard, accompanied by a legend naming one or other of the legions in existence at the time. The coins had clearly been produced explicitly to pay his army, but because he used a lower standard of silver purity (about 80 per cent or slightly less) than normal for the period, around 90 per cent or better, Antony's denarii remained in circulation for a long time and are common finds today, like the one found at Birdoswald and mentioned above. They were not worth melting down or hoarding and were worth more as coin. For the same reason other denarii of a higher standard were preferred for hoarding. By the latter part of Nero's reign all denarii were being issued at 80 per cent fine, and apart from a brief restoration of 90 per cent fine coins under Domitian, 80 per cent became the norm once more under Trajan and from thereon began to drop steadily.[44] Only then did Antony's coins start to disappear from circulation. They had finally become worth more as a source of bullion, being as good as or better than the coins in production at the time. Septimius Severus also honoured legions that had supported him becoming emperor in 193 by striking silver denarii with reverses depicting legionary standards and naming each one.[45] These coins were almost certainly distributed to the men of each legion concerned. Gallienus did the same thing, honouring the Praetorian Guard and 17 legions, all of which had provided detachments for his field army, on coins struck at Milan in 260–1. The short-lived emperor Victorinus (269–71) of the secessionist Gallic Empire issued a number of gold aurei for some of the individual legions under the regime's control on the Rhine and in Britain as did Carausius, the rebel ruler in Britain from 286 to 293 though his legionary issues were in bronze (see Chapter 11).[46] One of the other last series to commemorate a specific unit were the tetradrachms made at Alexandria in 284 for the brother emperors Numerian and Carinus in honour of Legio II Traiana, and using the Greek legend **LEΓ B TPAI**. The coins used the Greek system of letters to indicate a number, in this case B for II. The legion had probably taken part in the Persian War fought by the two emperors.[47]

More often the Roman Empire produced special issues of coins that commemorated the emperor's victories. One of the largest issues in all Roman history was produced in 71 to celebrate Vespasian's victory in the Jewish War, in his name and that of Titus. These coins had the legend IVDAEA CAPTA, 'Judaea captured', with a male and female Jewish captive depicted next to a palm tree. A more neutral type bore legends meaning 'the victory of the emperor' and showed a figure of Victory inscribing a shield. Nonetheless, the implicit violence was clear.[48] But in the context of the era the intention was to demonstrate force, power and success – all suitable virtues for a Roman emperor and certainly ones to which his soldiers subscribed. Domitian issued a number of coins that advertised his success in Germania. Some went so far as to have no legend at all, and merely show the emperor on horseback riding down a victim. Others were more explicit, bearing legends like GERMANIA CAPTA and depictions of bound captives in emulation of his father Vespasian's Jewish War coins.[49] Commodus issued sestertii in 184 with the legend VICT(ORIAE) BRIT(ANNICAE) to commemorate a victory in Britain that year. Septimius Severus issued a far more extensive series of similar coins in all metals to commemorate his British campaign, and also his war in Parthia.[50]

These triumphant coins were part of the broader culture of the Roman world, in which the army was integral to the identity of the emperor and the state, reflecting an empire won and held on to by force of arms. In reality, though, most Roman soldiers spent a great deal of their time, like soldiers of all times and places, carrying out mundane tasks, kicking their heels in their bases or building the bases in the first place.

# FIVE

# A SOLDIER'S LIFE

## Garrisoning the Empire

*In a camp, well-chosen and entrenched, the soldiers lie safe
day and night within their works, even though they are in
the enemy's sight. The camp appears to resemble a fortified
city which the soldiers can build for their security wherever
they want.*

Vegetius describes a Roman army camp[1]

Roman soldiers spent vast amounts of time building temporary camps and permanent forts. Here they eked out their
lives anywhere from the Castra Praetoria on the outskirts of
Rome to forts in the freezing wilderness of northern Britain or the
suffocating heat of the Syrian desert. Forts were where soldiers were
trained and equipped, where they lived side by side with their fellows,
formed friendships and rivalries, fomented disputes and mutinies,
paraded their loyalty, and set out to war.

Each fort was different, yet almost all were built to the same basic
design. Thanks to archaeology, and to the accounts of Polybius,
Josephus, Pseudo-Hyginus and Vegetius, we know a remarkable

amount about the conventional forts of the imperial age. Scarcely any Roman military buildings survive in recognizable form, one of the few to do so being the headquarters building at the legionary fortress of Lambaesis. Modern reconstructions, including the fort at the Saalburg in Germany, and in Britain the west gate, commandant's house and barracks at South Shields and the baths at Wallsend, give us an idea of what some of these places once looked like.

## CAMPS AND FORTS

A legion or auxiliary unit on campaign might build a camp for a single night's stay. Known today as marching camps, the evidence of them that remains is usually mere marks in the ground visible only from the air and representing the defensive ditch and embankment that surrounded the compound.

Dismantling the camp was as important as putting it together in the first place. The process was similarly regimented and organized by trumpet calls. The tents and baggage were packed up and then the camp was set on fire, for the obvious reason that it would be madness to leave it available for the enemy. The soldiers then organized into marching order and left.[2] Marching camps like these are best known in Britain, where extensive and targeted aerial photography has identified large numbers in the north, though even here the only visible traces are the outlines.

Forts and fortresses of the first and second centuries AD tended to resemble each other in their main features, such as the headquarters building, the barracks and so on. In practice, though, no two Roman forts were the same, regardless of when or where they were built. If the soldier-surveyors who laid out a fort were working to a template, they did so quite independently of one another, adapting the basic design to the unit's needs and the lie of the land. A unit that needed a base for a season or two would probably build a fort out of earth and timber, using stone only for some parts of foundations and for structures like bread ovens that needed to be fire-resistant. These forts were quick and relatively easy to build as well as being durable, strong and effective – especially if the ever-practical Roman army took advantage of existing enemy fortifications. At Hod Hill in Dorset, in the south-west of Britain, a vexillation of around 700 legionaries and auxiliaries in the invading Claudian army converted the corner of a captured tribal hillfort shortly after 43 to serve as the northern and western defences of a new turf and timber fort. This Roman camp had simplified internal buildings, including basic barracks to suit a campaigning army in the field. It remained in use for around five years before being given up when the soldiers moved on.[3]

Only when a unit was permanently based somewhere would the fort be built of stone (assuming stone was available). Stone forts took longer to construct, but obviously were much more durable than those of earth and timber.

The archaeological evidence of forts found across the Roman Empire shows that throughout the period every single one was based on the theoretical principles described by a number of sources:

## THE POLYBIAN CAMP

In the mid-second century BC Polybius recounted with fascination how the Republican army of that date pitched camp, referring in this case to an encampment designed to house two legions: 'The tribunes take both the Romans and allies and pitch their camp, one simple plan of camp being adopted at all times and in all places.'[4] He was being over-simplistic, but the broad principle was correct. The fort was laid out starting with the general's tent, the *praetorium*, a word that denoted precedence over all others. Alongside were the tents of the tribunes. Beyond and around the officers' quarters, the zones housing the ordinary soldiers were established, respecting a road grid laid out based on right angles. The size of the fort was determined by the size or number of units involved, its dimensions being modified accordingly. The rampart was dug out all round at a distance of 200 Roman feet (59 m) from the tents and structures, half by Roman soldiers and half by allied troops. The theory was that each soldier knew what his individual job was; only those detailed to wait on the tribunes were exempt, others taking charge of pitching their tents and preparing the ground. By the time the camp was finished, said Polybius, 'the arrangement of the streets and the general plan gives it the appearance of a town'.[5] The camp was then prepared for the night which meant setting up watches and watchwords.

## Caesar's camp

The Roman army was relentlessly pragmatic, and if circumstances required the camp was adapted. During the Gallic Wars, Caesar was marching with the VII, VIII, VIIII and XI legions when he was confronted by an unusually large number of Gauls. Indeed, even he admitted to being 'surprised'. He therefore ordered the camp ramparts to be built to a height of 12 Roman feet (3.6 m), with a breastwork of the same size on top (normally a high rampart would mean a lower breastwork), a double ditch 15 Roman feet (4.4 m) wide with vertical sides, and three-storied turrets. The idea was to make it look as if he was scared and thus encourage the natives to attack, while also making it possible for fewer of his own troops to defend the camp while the others were out foraging.[6] There was another good reason for having such a deep ditch: in an earlier campaigning season the attacking Gauls had filled in the trench at another camp.[7]

## Josephus' camp

Josephus described how important it was to have the camp built. 'Whenever [the Romans] advance into enemy territory they hold back from any engagement until they have fortified the camp . . . only specialized units are involved in its construction.' He was at pains to point out how organized the process was, explaining how the men concerned with the building had all the necessary tools. The end result was a fort with a fully fortified wall around it, interspersed with four gates and with towers, and with a ditch six feet deep outside. Along the walls were mounted the artillery, which included machines for hurling spears and stones, and catapults, loaded and ready to face an attack. 'The inside of the camp is an exact street grid,' he said, and like Polybius compared the complex to an 'instant town'. Trumpets sounded the key moments in the day, like going to bed and getting up. In the morning the centurions went to the tribunes and the officers moved over to the headquarters to receive the day's orders from the commanding officer.[8]

## THE CAMP OF PSEUDO-HYGINUS

Hyginus was a Roman surveyor in the early second century AD to whom a treatise on constructing military camps used to be attributed.[9] The work is now thought to be by an anonymous author of the third century, sometimes called today Pseudo-Hyginus. The account describes details of how to construct a camp with suitable accommodation for legionaries, praetorians, auxiliary infantry and cavalry, as well as other specialized units such as scouts.

The camp described was designed to accommodate three legions and auxiliary units. Like the Polybian camp, it had a central administrative zone where the commanding officer and other officers were based. The rest of the area was given over to accommodation for troops, animals, prisoners and even booty. The overall layout was a classic playing-card shape based on proportions of 3:2.[10] There are several instances where we know forts large enough to hold several legions existed. Varus' legions, XVII, XVIII and XVIIII, were based at Xanten before their destruction in AD 9, though since the fortress technically had space for only two legions they may not all have been located there. Five years later in 14 in Pannonia VIII Augusta, VIIII Hispana and XV Apollinaris were all accommodated in one large fortress. More usually, a legion was based on its own in a single fortress, or even divided into vexillations and housed in separate forts. The emperor Domitian issued an order prohibiting two legions from being based in the same camp, for the specific reason that it was easier for a commanding officer to mount a rebellion if he had more than one legion to hand.[11]

## THE PERMANENT FORT

Perhaps the most remarkable fact about the permanent or semi-permanent forts built around the Roman Empire is that their basic layout was so similar, despite the vast range in size and detail (see the plan of Wallsend fort, p. xiii). A multi-legionary fortress, such as Xanten, could cover as much as 49 acres (20 ha), while an auxiliary infantry cohort could be housed in a fort just 2.5 acres (1 ha) in size. The tiny fort at Hesselbach in Germany housed only an irregular auxiliary infantry unit called a *numerus* but it still had a central headquarters building. Only the much smaller fortlets, intended for perhaps one or two dozen troops supervising a road, diverged from the norm. These usually had simple accommodation ranged round the inside of the ramparts, with a small courtyard in the middle. One of the few exceptions was the Castra Praetoria in Rome which had numerous barracks, and whose crenellated walls eventually reached 16 ft (5 m) in height, but had no external ditch. It also does not seem to have had the full range of internal buildings normally found in a fort, such as headquarters. The reason was probably that the emperors did not want the Praetorian Guard, or anyone else, to use it as a defensible stronghold against the regime.

Permanent or semi-permanent forts and fortresses were based on the overnight camp, but were much more formalized and developed. Even the short-lived legionary fortress of Inchtuthil in Scotland, built by Agricola's forces around 84, featured sophisticated facilities. Soldiers engaged on the building work were housed in a temporary compound nearby, while the officers occupied another temporary area. Earthen ramparts revetted with stone enclosed the fortress, which contained timber buildings. It had granaries (*horrea*), indicated by slot trench foundations that raised the floor above soil level to protect against damp and rodents, a workshop (*fabrica*) for manufacturing the metal components such as iron nails needed to assemble the buildings and arms and armour, as well as deal with repairs, and a hospital (*valetudinarium*). In the event, despite involving an estimated 2.7 million man hours to build, while work had started to add stone facings to the walls, the 53 acre (22 ha) fortress was given up and dismantled in about 87, leaving an almost textbook plan for archaeologists to excavate in recent times (see plan on p. xiv). Inchtuthil's only incongruous feature was a strangely small headquarters building (*principia*), normally the most imposing architectural structure in the fort; it also had no commanding officer's house (*praetorium*), though a space had been provided for it. The headquarters had presumably been erected quickly to provide an essential facility and focus for the rest of the layout; had the fortress remained in use longer it doubtless would have been replaced with a full-size version and a praetorium added. A stone bath suite for officers had already been begun in their temporary compound after it was no longer needed, but it too was demolished when the soldiers left. Inchtuthil's plan is one of the most complete and best known of all the Roman army's legionary fortresses (the other is Neuss), and according to one Roman historian 'is eloquent testimony to the sheer organizational capability of the Roman army'.[12]

More permanent forts and fortresses were even more complex and were consolidated in stone. In these cases a full-size headquarters consisted of a basilican assembly hall where the unit's commander could address his officers, and where unit records, standards and valuables were stored, together with a shrine. This looked out onto a courtyard surrounded on the three other sides by wings and a portico. At the heart of certain fortresses, like Caerleon and Neuss, were bath-houses, though at smaller forts these tended to be located at some distance outside the walls to save room and avoid the risk of fire. A praetorium provided some of the features and comforts of a traditional Roman townhouse where the commander and his family could escape from the noise, congestion and bustle of the fort. The praetorium was also where the commanding officer worked, and (unless there was a bath-house) it was usually the largest building within a fort, its size reflecting the commanding officer's importance. A hospital was sometimes included even in auxiliary forts.

Building work also involved repairing the forts and their infrastructure. However competent Roman military builders were, the sheer length of time involved in the occupation of some permanent forts over centuries meant that many structures fell down or into disrepair. At the small frontier auxiliary fort of Walldürn in Germania Superior, near Mainz, the baths had 'collapsed through old age' by 232 in the reign of Severus Alexander. A centurion called Titus Flavius Romanus of Legio XXII Primigenia was commanding officer at the time, his garrison being an irregular unit of exploratores scouts made up of Sturi and Britons. He took charge of the rebuilding, commemorating the fact on behalf of the unit on an altar dedicated to 'the Goddess Holy Fortuna', a deity often venerated in military baths.[13] At Lancaster in Britain a major repair project had to be organized at some point between 262 and 266 during the reign of the emperor Postumus, who had created the breakaway Gallic Empire of Britain, Germany and Gaul. The fort's baths and its associated exercise hall, here referred to as a basilica, needed completely rebuilding, probably because the fort had been unoccupied for a while, though the original structures might have been as much as 150 years old. The surviving inscription reads:

[For the Emperor Postumus] concerning the bath-house rebuilt and the basilica restored from the ground up, (having) collapsed through old age, for the troopers of the Ala Sebussiana

Postumiana, under our governor Octavius Sabinus, a man of senatorial rank, and under the charge of Flavius Ammausius, prefect of cavalry. Dedicated on 22 August in the consulship of Censor and Lepidus, both serving for the second time.[14]

According to an inscription found at the site, in 297 Flavius Martinus was a centurion in charge of an unknown cohort given the task of recommissioning the Hadrian's Wall fort at Birdoswald, which by then seems to have been virtually derelict. In the name of the Tetrarchy emperors Diocletian, Maximianus, Galerius and Constantius, and the governor Aurelius Arpagius, Flavius Martinus put his men to work. The task was a substantial one. The commandant's house was 'covered with earth and had fallen into ruin', and both the headquarters building and the baths needed restoration too. Fragments of a similar inscription found at Housesteads to the east along the wall suggest that Birdoswald was not the only major Wall fort that was unfit for purpose. These, virtually the last inscriptions known from the frontier, paint a picture of a Roman army trying to cope by patching up old and derelict military installations in a remote part of the Empire.[15] However, there was a rhetorical and formulaic element to these descriptions of dilapidation, and the decay may have been exaggerated accordingly.

By this time Roman military architecture had dramatically changed. In the middle to late third century new forts were being built with massive freestanding walls, projecting bastions and gate towers. The emphasis was on defensive capability and the same architecture is found in town walls of the period. Artillery on the bastions and gates could be directed at attackers trying to scale the walls. Diocletian's coastal base at Split in Croatia, built *c.* 305, was in reality a heavily fortified palatial compound with massive walls and projecting towers. Around half the palace's area accommodated his garrison. Some of the best examples of late forts are found in the deserts of Syria and Jordan and along the Channel coasts of Britain and Gaul. The latter, such as Richborough and Pevensey, were known as the forts of the Saxon Shore and seem to have been built to protect units and equipment charged with fending off seaborne raiders from Germany and Scandinavia.[16] These forts tend to be somewhat smaller than earlier ones, and little is known about their internal buildings, though that picture is starting to change thanks to modern geophysics. For example, a survey begun in 2012 at Brancaster on Britain's Norfolk coast revealed extensive traces of buried internal buildings including a conventional-looking headquarters, a granary and a horse-exercising compound.[17] Unfortunately the army had become less inclined to record its work in inscriptions, so we do not know very much about the dates and the units involved. The Notitia Dignitatum listed a cavalry unit of Dalmatians at the fort in the fourth century.[18]

## Architects, surveyors and engineers

The specialist military surveyors and engineers who oversaw the laying out and building of a fort were classified as immunes. Amandus was an *arc(h)itectus*, 'engineer', who made a dedication to the goddess of northern Britain, called Brigantia, at the fort of Birrens, north of Hadrian's Wall.[19] The fort had been originally established in the 140s but was destroyed and rebuilt *c.* 158, lasting until about 180, during which time it was occupied by an auxiliary infantry unit, Cohors II Tungrorum. Amandus tells us nothing about which unit he belonged to and it is possible he was sent on detachment from a legion to lay out the fort, if indeed he was responsible for that job rather than only passing through. Opponius Justus was an architectus with Legio XXII Primigenia who erected the tombstone of his friend Julius Paternus, another soldier of the legion, who had served a remarkable 33 years when he died at Bonn in Germania Inferior.[20] In one instance architecti are mentioned in association with *calcarii*, 'lime-burners', evidently soldiers whose specific job was to prepare the lime mortar on which so much Roman masonry construction depended. Architecti and calcarii from Legio I Minervia made a dedication to Jupiter Optimus Maximus and the Genius of the Vexillation in 188 at Iversheim in Germania Inferior.[21]

Attonius Quintianus was a *mensor*, 'surveyor' (but see the Glossary for other possible meanings), and an *evocatus*, which means he had signed up to serve again after discharge. He set up a dedication to the god Mars Condates at the auxiliary fort of Piercebridge, south of Hadrian's Wall.[22] Aurelius Castor, a mensor of Legio V Macedonica at Potaissa in Dacia, where the legion was based from *c.* 166 to 274, loyally set up an altar to Jupiter Optimus Maximus Capitolinus.[23] A variant on the job was an *agrimensor*, 'land surveyor', like Titus Claudius Tiberinus who served with Legio XI Claudia in Moesia Inferior *c.* 150–200.[24] Pudentus was a *mensor geometrae*, 'surveyor of geometry', with Legio XV Apollinaris in Trajan's time but seems to have died in North Africa at Sbiba (modern Tunisia), where he had perhaps been sent on detachment to help a civilian population.[25]

Permanent forts also needed men with maintenance skills. Gaius Caristicus Redemtus was a *plumbarius ordinarius*, 'lead-worker', in the Praetorian Guard at Rome, where he died aged forty having served 16 years.[26] Lead was the basis of all Roman hydraulic work because of its use in pipes and baths, though it was also used in roofing and for some weapons, such as sling-bolts. Lead-working was sufficiently important a skill for Caristicus to be classified as an *ordinarius*, a term that denoted equivalency of pay with that of a centurion.

## The vici and canabae

Once a legion or auxiliary unit was installed in a fort, informal towns known as *vici* grew up around even the remotest military establishments. *Vicus* was a word that could be applied to almost any small town-like settlement, though they were no usually no bigger than a modern village. Similar, but larger, settlements outside legionary fortresses were more likely to be known as *canabae*, 'the huts' or 'hovels'. Canabae and vici developed irregularly, straggling along the roads that passed by or radiated out from each fort with a haphazard collection of baths, latrines, shops, workshops, homes, temples (such as the shrine of Jupiter Dolichenus at Vindolanda), warehouses, buildings of indeterminate purpose and cemeteries. Activities ranged from the domestic lives of soldiers' families, the worship of local gods and the manufacture of sculpted monuments such as religious dedications or tombstones, to food storage and the drilling and repair of broken pottery – especially fine wares – with metal clamps. Most of the structures had basic rectangular plans, crammed up against one another with narrow street frontages. In the largest vici the area covered could easily exceed the size of the fort by some margin.

There was no clear division between the fort and the vicus, though some appear to have organized civic administration modelled on that of formally incorporated towns. Some vicus facilities may well have been built by the army for military purposes, not least because soldiers probably operated some of the businesses on a personal basis.

Few vici are known in detail because most archaeological attention is usually focused on the fort proper. However, modern geophysical surveying techniques are beginning to reveal extensive areas of settlement outside forts, for example at Birdoswald on Hadrian's Wall, and at Kastell Zugmantel on the military road between Wiesbaden and Mainz, occupied by a *cohors equitata*.[27] Zugmantel's vicus even included an amphitheatre, as well as a shrine of Jupiter Dolichenus. At Birdoswald the area of settlement was contained by Hadrian's Wall to the north and the sharp drop to the river Irthing to the south, but proliferated along the roads to the east and west of the fort. The wider archaeological evidence, and that of inscriptions, suggests the vici were mainly populated by a motley collection of military families, veterans, traders, middlemen, local people, and other opportunists who followed the army about, their own livelihoods depending on the soldiers' and veterans' spending power and personal needs. Forts were clearly only the focal points of what were often sprawling communities, emphasizing the enormous social and economic consequences of the presence of the Roman army.

Nor were all forts standalone establishments, distinct from the civilian settlements they attracted. Where the Roman army was concerned, there were always exceptions. At Dura-Europos there was a completely different type of relationship between the army and the civilian community. Here the town of Dura had already been in existence for centuries when the Romans took control in 165. The military camp of Cohors XX Palmyrenorum was developed within the north-western sector of Dura from 210 until 256–7, when the city was captured by the Sassanids and the Roman garrison destroyed. This 25 acre (10 ha) zone, which bore no resemblance to a conventional fort, possessed a number of specialized buildings of military character and function including barracks and a praetorium, distributed in what was effectively an enclave of the civilian town. An earlier temple in the military area served as the principia and was where the Feriale Duranum calendar of official religious ceremonies for Cohors XX to perform was found along with other papyri relating to the unit (see Chapter 16). Only a mud brick wall divided the military zone from the rest of the settlement, but a gate allowed movement.[28]

## Duties

The discovery of some remarkable documents in Egypt and Britain has produced some surprising revelations about how Roman army units operated in and from their bases. The Roman army was a thoroughly bureaucratic organization, records of anything and everything being constantly kept, but few survive today. Those found at Vindolanda and various locations in Egypt are therefore enormously important to our knowledge of the Roman army. They paint a different picture of life in a fort from the popular impression, encouraged by historians like Polybius, of hundreds of soldiers all engaged in the immediate business of war. A strength report found in Egypt, written on a papyrus dated to about 105 in the reign of Trajan,

describes how Cohors I Hispanorum Veterana, a part-mounted auxiliary infantry unit based at Stobi in Macedonia, had lost soldiers to
transfers, drowning, and death at the hands of bandits. Others had
been dispatched on various duties; some were 'in Gaul to obtain
clothing', which seems a remarkable distance to have to cover for
such a mundane purpose. Others were on service at the mines,
presumably supervising labour, or away trying to source horses.
A number of soldiers were on duties in Macedonia which involved
being 'across the river to protect the corn supply', guarding 'beasts
of burden', or on scouting missions.[29]

A number of exceptional waterlogged deposits found at the fort
at Vindolanda near Hadrian's Wall – or rather a succession of forts
dating from the late first century AD – have preserved documents
from around the same time as the Stobi strength report. They
include a similar report drawn up on 16 May, probably in the year
90. The details are better preserved and specify the whereabouts of
the 752 members of Cohors I Tungrorum, an auxiliary infantry unit
commanded by Julius Verecundus. It specifies that 46 men had been
seconded to be guards on the staff of the governor of Britannia, where
they would have served as *beneficiarii legati*.

While he was governor of Bithynia and Pontus under Trajan,
Pliny the Younger arranged for ten soldiers to accompany an imperial procurator called Virdius Gemellinus who had been sent out
to deal with the province's financial affairs. To Pliny's annoyance
Gemellinus' assistant, a freedman called Maximus, demanded
that he have six soldiers assigned to him too. Pliny decided to give
Maximus only two mounted soldiers in addition to those already
allocated to go with him on an expedition to collect corn, and asked
Trajan what he should do. Trajan wrote back saying that Maximus
could keep the two mounted soldiers and have two taken from
those allocated to Gemellinus.[30] Maximus was of course primarily
concerned about advertising his status. It was not the only time
this happened. The more soldiers attached to an official's staff,
the greater his prestige. Pliny was also informed by Gavius Bassus,

'prefect of the Pontic shore', that the ten beneficarii, two cavalry-men and a centurion assigned him from the provincial garrison by imperial order were insufficient for his needs. Bassus even wrote to complain to Trajan. However, Trajan wrote to Pliny and said it was not in the public interest to take soldiers off active service for such a purpose.[31]

According to the Vindolanda strength report, a remarkable 337 troops were on detachment in the nearby military base at Corbridge, effectively splitting the cohort in two, presumably for operational reasons. Several others were away elsewhere, including six in London, and thirty-one were sick. When all these were taken into account, only 265 members of the Tungrian cohort were fit and well and on hand to fight at Vindolanda, representing a little over one-third of the current paper strength.[32] This sort of information is extremely rare but is usually taken at face value today. The truth might have been rather different. All the absences on the report appear quite legiti-mate, and indeed the majority may have been. There was, however, a vested interest in making the absences appear above board, especially if some of the soldiers had bribed centurions to let them off certain tasks, a practice the Praetorians were very familiar with.[33] There is no reason to assume Vindolanda, or any other fort for that matter, was automatically immune to similar conduct. If so, the strength report might be partly fictitious.

There was a risk involved for soldiers sent away on state business: someone might help themselves to their property, which could include money, equipment, clothing or indeed anything he owned. In 224 a soldier called Flavius Aristodemus made a petition to Severus Alexander about this very matter. The emperor ruled that during the year following a soldier's absence, that soldier was legally entitled to claim back any of his goods that had been taken by someone else. Once the year was up, though, he could no longer 'interfere with the rights of the possessor'. In other words, after twelve months the property was lost.[34]

Legio III Cyrenaica and XXII Deiotariana, like all military units, maintained a roster of duties for their soldiers at any one time. These must have been drawn up on a regular basis, but virtually none have survived. One example, relating to the first ten days in October one year in the late first century AD, has however been found in Egypt, though it is not certain which legion it belongs to.[35] Marcus Domitius was away, looking after the granaries at Neapolis. Gaius Julius Valens had a much busier ten days ahead of him:

1 October:   training
2 October:   in a tower [on guard?]
3 October:   drainage
4 October:   boots
5 October:   at the armoury
6 October:   at the armoury[?]
7 October:   on guard duty at the headquarters building
8 October:   road patrol [this presumably means supervising civilian traffic]
9 October:   'in century'
10 October:  at the baths [probably on duty]

In this way countless numbers of Roman soldiers eked out most of their military careers. As all soldiers at all times and in all places know, warfare intervened only occasionally to break up the tedium. As one of the most highly trained and effective forces in antiquity, the Roman army also inevitably had an impact on the environment, one that is sometimes still visible today in the remains of their roads and quarries. In the year 100, for example, the IIII Flavia and VII Claudia legions were busily hacking out the living rock from mountains in Moesia Superior, and building supports, in order to construct a roadway along one side of the Danube. Given the physical effort needed, it was only appropriate that one of their dedications should be to Hercules.[36] The army's herculean efforts extended much further, to obtaining whatever was needed to keep the soldiers fed, housed and equipped.

# SIX

# LIVING OFF THE LAND

## The Roman Army and the Environment

*This war will support itself.*

Cato the Elder instructs his
men to take what they need[1]

The impact of the Roman army on the environment was gigantic, at least by the standards of the ancient world. The construction of forts and fortifications was only the start. Peacetime activities also included participation in significant mining and engineering projects, often involving soldiers in supervision and management (see Chapter 12).

The arrival of the Roman army in a frontier zone, especially a temperate area where trees were widely available, automatically resulted in colossal quantities being felled and prepared. Clearance must have been undertaken on a grand scale. Each turf and timber fort required vast amounts of wood for building and maintenance; felling and transporting it was a task so arduous that it helped provoke a mutiny in AD 14 amongst the forces in Germany.[2] The army also required wood for day-to-day heating, cooking and metalworking.

The military extraction of iron ore in the Weald of south-east Britain, for example, meant there was a constant and huge demand for charcoal for the smelting furnaces.

When Agricola's legionary fortress at Inchtuthil in Scotland was abandoned and cleared away *c.* 87, around 900,000 iron nails of various sizes that had been prepared for the fort were buried in a pit 12 ft (3.66 m) deep to prevent the local tribes from reusing them for weapons.[3] Pottery and glassware were smashed up and buried, but evidently melting down the nails and taking the iron away was thought too much trouble. They bear witness to the effort and logistics involved in mining, smelting and working the iron, as well as transporting it to the site either in pigs or as finished products. This kind of usage meant long-term management – the constant rotation of areas of woodland, making usable timber available within a reasonable distance rather than operating on a slash-and-burn basis, which was fine for conquest but hopeless for permanent garrisons. The legionary fortress of Caerleon may have required 370 acres (150 ha) of woodland to supply enough timber for the initial construction alone, and more for maintenance. A unique wooden writing tablet found at the fortress in a context dated to *c.* 75–85 seems to mention the collection of *materia*, 'building timber'.[4] A squared beam found in a Roman quay in London had a branded inscription of a cohort or ala of Thracians which had probably been responsible for felling it; the beam had perhaps been reused from a military building.[5] An inscription from the Rhineland, dated to 214, records that a vexillation of Legio XXII Primigenia was involved in the gathering of timber.[6] Although the Romans made some use of coal, they seem primarily to have relied on timber for furnaces. Estimates based on the small private bath-house at Welwyn suggest that a 58 acre (23 ha) area of managed woodland was needed to provide the fuel for just one small domestic facility. Based on floor area comparisons alone, that could mean that Caerleon's legionary baths needed over 13,800 acres (5,600 ha) of woodland to keep them running, unless coal was used instead – which in that case was a real possibility, because there were plentiful sources in the area.[7] Either way, a legionary fortress must have depended on vast supplies of local fuel. However, estimates are based on so many imponderables and unknowns that it is impossible to do more than conclude that the requirements must have been enormous and time-consuming, and probably also had a serious impact on the local environment. This may explain why Caerleon's baths had fallen into disuse by *c.* 230, long before the rest of the fortress.

When forts or frontiers were consolidated in stone, the Roman army became involved in quarrying on a similarly grand scale to the gathering of timber. Quarrying not only produced stone for building but also limestone which could be burned to create concrete and mortar; both were used in vast quantities in Roman stone construction, especially in the major buildings of a permanent fortress. Legio I Minervia had a lime-kiln depot at Oversheim, about 18.5 miles (30 km) from the fortress at Bonn. Caerleon's fortress baths were built using lime obtained by using furnaces fuelled by local coal to burn limestone quarried in the area.

Some of the best-attested quarries are in the vicinity of Hadrian's Wall. They include the so-called 'Written Rock of Gelt', where a number of soldiers took time out to inscribe their names on the quarry face, probably as a quality control measure. One announces:

A detachment of Legio II Augusta. The working face of Apr[ilis?] under the charge of the optio Agricola.[8]

This, and perhaps some of the other inscriptions, then belong to a rebuilding of the wall under Septimius Severus, by which time it was almost eighty years old. One of the reasons was that the Wall had not initially been well built. Although it was dressed with facing stones, the core was made of rubble and mortar. As a result of water ingress and frost, both common problems in the area, some of the facing stones had fallen off and the core had started breaking up. The survival of some of the quarry faces and cuttings made by the frontier garrisons still show how laborious and intensive the building and repair work was, as well as the visible impact on the landscape. In one of the upland sections of the wall at a place now known as Limestone Corner, the rock proved too much even for the normally indefatigable Roman army. They used water and wedges to split apart vast chunks of rock and then lifted some of the blocks out. Others proved more resistant and the work was abandoned. Today some of the blocks still lie in what was supposed to be the wall's forward ditch, and the wedge holes can be seen.[9]

The Hadrian's Wall quarries were exploited for military installations. Apart from settlements outside the Wall forts, there were few civilian contexts in the area in which masonry was needed. In more heavily settled regions, it is not usually possible to know whether soldiers working quarries were doing so to fulfil the army's own infrastructure needs or to meet the demands of cities or civil engineering projects (either of which soldiers could have been working on). The soldiers of Legio IIII Scythica, quarrying at Arulis near Belkis on the Euphrates, usefully recorded their presence for us not only by inscribing their names but also making dedications to Jupiter Optimus Maximus and Silvanus, for whom some created carved niches. One soldier carved 'Aurelius Carus to Silvanus'.[10] Two standard-bearers of Legio IIII, Julius Aretinus and Julius Severus, and a trumpeter called Rabilius Beliabus, banded together to make a dedication to Jupiter Optimus Maximus Silvanus the Preserver.[11] In Egypt an inscription recorded 'Annius Rufus, centurion of Legio XV Apollinaris, commander of the marble works at Mons Claudianus under Trajan, the Best Emperor', referring to a state-controlled quarry that supplied marble and granite for imperial projects.[12] Auxiliaries also found themselves assigned to the quarries in Egypt. Ammonios, a member of Cohors II Thracum, had worked in the quarries and died in service around 143.[13]

The manufacture of ceramic products was almost invariably carried out on-site, or within a few miles at a specialized military works depot. Legio X Fretensis had such a depot dedicated to the production of tiles and pottery at Bin Ya'nei (Jerusalem).[14] Far too fragile and cumbersome to be moved in quantity any distance, tiles were needed in huge numbers to roof timber and masonry buildings in fortresses and forts, and for heating systems. It was also illegal to carry a weight in excess of the equivalent of about fifty roof tiles in a wagon.[15] The resources needed (clay, temper, water and fuel) were generally available in most locations. Legio XX's manufacturing depot at Holt, about 7.5 miles (12 km) from the fortress at Chester, included kilns, workshops, baths and barracks.

Since tiles survive well, even if broken, they are particularly good evidence for how the Roman army made sure its possessions were clearly labelled. Wooden dies inscribed with the abbreviated name of the legion or unit were made and stamped into many of the tiles found at the sites of Roman military bases across the Empire; for example 'LEG I ITAL' for Legio I Italica, based at Novae in Moesia Inferior by the Danube. Such stamps are anonymous and merely name the unit. This was not enough for 'Julius Aventinus, a soldier in Cohors I Sunicorum', an auxiliary unit known to have been in Britain between 122 and 198–209. Aventinus wrote his name elegantly into a wet tile that had not yet dried and been fired above the stamp LEGXXVV, for Legio XX Valeria Victrix. The tile was found at the legion's depot at Holt. It would seem that at least part of the auxiliary cohort had been detailed either to work at the depot for the legion or to collect a consignment of tiles for a military building somewhere in the region.[16]

## HORSES AND OTHER ANIMALS

Each legion had 120 cavalry, according to Josephus, writing about the Jewish War in 67. Whether this number was invariably the case is unknown because this is the only reference we have to a legion's cavalry contingent, the *equites legionis*, in the first to third centuries.[17] Not enough is known about legionary fortress plans to conclude whether provision was always made for this many mounted troops. Auxiliary infantry cohorts with a mounted component (a *cohors equitata*) had either 128 or 256 cavalry, depending on whether the unit was quingenaria or milliaria in size. There were also the auxiliary cavalry regiments (*alae*) with 512 or 768 troopers. The result was that as much as one third or more of the whole auxiliary force was mounted, which in could have meant 75,000 or more in Hadrian's reign, let alone horses for the servants of the individual troopers, pack animals, and making good losses from war, disease and accidents.[18] Another estimate suggests around 110,000 horses and 60,000 mules were needed for the whole army including the fleets. Accommodation in cavalry forts included the 'stable-barracks', where men and horses shared the same building. Horses and other animals also had to be constantly replaced, adding to the logistical complications and manpower needed to manage them.[19]

Obviously therefore, horses were a very important part of the Roman army and were needed in huge quantities. When, during his war against Mithridates, Sulla laid siege in 87 BC to Mithridates' puppet ruler Aristion in Athens and to the port of Piraeus, he needed 10,000 pairs of mules to help operate his siege engines; all the animals obviously had to be fed and watered. The figure is obviously rounded but gives an indication of the scale.[20]

The mounted soldiers who fought with these horses and took care of them had to be highly trained, skilled and experienced. Not surprisingly, such careers tended to involve specialization. Marcus Ulpius Crescentinus, born in the frontier province of Pannonia Inferior, served for 26 years in the Ala Brittonum, then the Ala Praetoria, and rose to serve in the emperor's equites singulares Augusti before his death at Rome.[21] Gaius Cominius Commianus must have been especially talented. He was recruited at the age of sixteen as a trooper in the Ala Brittonum but died aged only twenty at Budapest (Aquincum) in Pannonia Inferior.[22]

The value of an exceptional horse to the army, and the prestige of riding it, was recorded in a story attributed to the reign of Aurelian.

During his rule, it was said, a horse had been captured from the Alani tribe or some other enemy which could run 100 miles a day for as many as ten consecutive days. The animal was not especially large or handsome, but the soldiers assumed that Probus, then serving as Aurelian's general, would help himself to it. Probus allegedly dismissed such a notion on the basis that a horse which could travel so far was better suited to someone interested in retreat. He ordered lots to be drawn so that the horse could be allocated to one of his men. By some strange coincidence there were four other soldiers called Probus in the army, and although the general's name had not been thrown into the urn it was the name Probus that kept being drawn out. The soldiers insisted that he take the horse.[23] The story is an intriguing one but unlikely to be true, at least for the most part. The biography of Probus was not written down until well into the fourth century and belongs to the unreliable Historia Augusta series.

Horses were the most prestigious and the most militarily essential, but far from the only animals to be found on military sites. On the evidence of bone remains, cattle dominated by as much as two-thirds the total number of animals at a fort or fortress, followed by pigs, with sheep or goats the least common. These animals must on the whole have been kept in the vicinity and managed for slaughter and consumption. There were other bones from animals like deer, hare and wild boar, which were hunted for sport and food (see Chapter 13).[24]

## WATER, FOOD AND SUPPLIES

The most essential resource of all was water. Building a fort next to a river might seem to solve the problem of maintaining a convenient water supply. In fact the colossal effort involved in moving quantities of water uphill made it a hopelessly unrealistic solution beyond the provision of small amounts for drinking. Housesteads fort on Hadrian's Wall relied in part on the local heavy rainfall and a system of tanks that caught the run-off, channelled a constant flow through the fort latrines and disgorged the waste into the vicus outside. Aqueducts were the only reliable means of supplying water, which involved locating a source of water at a greater height and bringing it into the fort or vicus through a combination of raised channels, open leats and buried pipes. Where aqueducts could not be built, or would take too much time, the only alternative was to rely on wells which had to be dug throughout the fort and beyond, such as at the Saalburg fort in Germany.

A cavalry regiment's requirements were even greater than those of an infantry cohort. An aqueduct was built for (and probably by) Ala II Asturum at Chesters on Hadrian's Wall between 180 and 184.[25] Since the fort had been commissioned almost sixty years previously, the works must have supplemented existing arrangements or replaced an older system. The water supplied the fort, its vicus settlement, and the substantial military baths and accompanying latrines, located down the slope from the fort towards the river Tyne into which all the waste was channelled. In 216 at Chester-le-Street an unknown cavalry regiment built an aqueduct to supply a bath-house built at the same time,[26] while at South Shields an aqueduct was built for Cohors V Gallorum in 222.[27] Military expertise in this area meant the army was also used to construct aqueducts that served civilian communities, for example in Judaea (see Chapter 12).

After water came food, but there was a tradition of admiring commanders who encouraged reliance on basic meals. It was all part of the Roman tradition of venerating tough, self-disciplined men of the old school. When travelling round the Empire, Hadrian followed the example of Scipio Aemilianus and restricted himself to 'camp food', made up of bacon, cheese and vinegar.[28] Onasander, in his manual of advice for a commander, said it was good practice for a general to live off the enemy's land to avoid causing damage to his friends and allies.[29] In peacetime food was a major consideration. In war food could occasion a crisis. In 216 BC the Roman garrison at Casilinum (near Capua) in Campania, Italy, was under siege by Hannibal. The

Roman troops' position became desperate as starvation set in. They started by using hot water to soften the leather on their shields so they could eat it. They also consumed rats or any other animals they came across, as well as digging out any plants they could find. The Carthaginians ploughed up the grassy soil round the town to destroy any plants. The Romans responded by sowing turnips, the idea being to show the enemy that they were confident of surviving the siege indefinitely. This enraged Hannibal who thought he was now going to have to wait while the turnips grew. Eventually Hannibal, a man who normally never agreed to terms, had to accept the soldiers offering a ransom for their freedom. More than half the garrison survived. A story emerged recounting how one of the garrison found a rat (or a mouse) and sold it to another man for 200 denarii, only to die of starvation himself while the other man lived.[30]

The Second Punic War provided the Romans with the opportunity to learn the importance of integrating supply chains into a campaign. In 212 BC Hannibal, who was then ravaging Italy, enticed Capua into withdrawing from its alliance with Rome. The city was promptly besieged by a Roman army supported by a grain depot at Casilinum to the north, a fortified river route to the west, and the newly fortified port at Puteoli.[31] In 205 BC, when Scipio put together an army and fleet to sail from Sicily to North Africa to defeat Carthage, the logistics were placed under the charge of a praetor called Marcus Pomponius. 'Food for 45 days, of which enough for 15 days was cooked, was put on board.' A remarkable 45 days' worth of water for the men and the cattle was also loaded onto the ships. Livy was unable to find out for certain how many men were in that army, discovering estimates that ranged from 10,000 infantry and 200 cavalry to 35,000 in total.[32] If we take the lowest estimate, then based on the monthly corn allowance of 64 lb (29 kg; see below), for 45 days each man would have needed around 97 lb (44 kg) of corn. That equates to 441 tons (448 metric tonnes) of corn alone for an army of 10,200.

Of course, one solution was to make sure the army lived off the land as much as possible as Onasander recommended. The notoriously parsimonious and ruthless Cato the Elder led a campaign into the Iberian Peninsula in 195 BC. Realizing it was the time of year when harvested grain was on local farmers' threshing floors, he told the contractors who would normally have sourced and bought grain on the wider market to go home. 'This war will support itself,' Cato said. His decision next to 'burn and lay waste' the enemy's fields might have been a rash one, but the gamble paid off.[33] Cato had spotted an opportunity, but it was not one that could be relied on. In 171 BC, during the Third Macedonian War, fighting seems to have been suspended in the late summer so that the armies could gather corn. Perseus, the Macedonian king, had his men threshing in the fields, while the Romans threshed in their camp.[34]

Keeping any army properly supplied and fed also relied on order. In 110 BC, during the Jugurthine War in Africa, the lazy and ill-disciplined soldiers of the army commanded by Postumius Albinus took to selling their grain rations and bought bread on a daily basis rather than make it themselves. In between times they and the motley crew of camp followers they had accumulated spent their time robbing local people, and helping themselves to cattle and slaves. These they sold to the traders who tagged along in their wake in exchange for luxuries like imported wine. The new general, Caecilius Metellus, had to sharpen up discipline quickly and force the soldiers to live off official supplies.[35]

Foraging in enemy territory was essential but always dangerous, especially if the men were separated and laden down with what they had collected; it was essential to send a foraging party out with other troops whose sole job was to guard them and maintain vigilance. Pompey had been dispatched with an army to defeat Sertorius, the rebel governor of Spain. But when the legion concerned was out foraging, Sertorius saw his chance to strike. The whole legion was 'cut to pieces', along with all its baggage animals.[36] In 56 BC, during Caesar's campaign against the Gaulish Veneti tribe, some of his men were captured after being sent out to forage for grain. Caesar sent envoys to negotiate for the men's release but the Veneti imprisoned them too, in the hope that they could be used to bargain for a release of hostages. Caesar refused and continued the war.[37] In 60 in Armenia, Corbulo's army had to ward off starvation by killing its horses and pack animals until it reached cultivated land and could steal crops.[38]

Even in Cato's time, the Roman army had been large, widely dispersed and often on campaign for years. In the long run it would depend on a sophisticated and reliable food supply chain. By the time of the emperors, with a standing army largely settled in fortresses and forts, the organization of food supplies reached a level unmatched until modern times. The army remained dependent on middlemen of various sorts who sourced and supplied food both for units as a whole and for individual soldiers. Of course the army could also grow its own food. Many documents found at Vindolanda are perfunctory lists of goods and supplies that were supplied to the fort at the end of the first century AD. Because the archive is unique in the history and archaeology of the Roman army, there is consequently no means of knowing how representative it is, at least in detail. But what must have been typical of the Roman army was the sheer quantity and range of commodities, the records of cost, the logistical arrangements and the numbers of people involved. A single line in Tacitus describes how the whole military frontier zone along the Rhine had 'Roman itinerants and traders scattered all over the countryside'.[39] Polybius mentions the inclusion of a market in the Republican fortress layout, while at Lambaesis two standard-bearers of Legio III Augusta, Sabinius Ingenuus and Aurelius Sedatus, were the agents in charge of a marketplace attached to the fortress where traders sold goods to the legion or the legion sold some of its own produce.[40] These references, and the Vindolanda documents, paint a picture of Roman forts as hubs in a thriving and ceaseless network of trade manned by countless individuals for whom the Roman army was the basis of their existence and livelihood.

One Vindolanda document refers to a saddle, something that might be expected at a fort, priced at 12 denarii. It also lists expensive lengths of scarlet, purple and greenish-yellow curtain. The purple curtain, measuring 11.5 (feet) (3.4 m) was priced at twelve times the cost of the saddle. Clearly luxury products, the curtains must have been destined for the residences of the officers and their wives whom we know to have been living there (see Chapter 14).[41] However, thanks to damage and decay many of the documents consist now only of fragments with tantalizing references. One laboriously lists the poultry consumed on various dates, including an occasion described as '17 January for the dinner of Brocchus'. Aelius Brocchus was the commanding officer at another fort in the area – which one is unknown – and was friends with Flavius Cerealis, who commanded Cohors VIIII Batavorum at Vindolanda. The two also hunted together (see Chapter 13).[42]

One Vindolanda letter, from a cornicularius named Severus to Candidus, a slave of a prefect called Flavius Genialis, refers to preparations for the midwinter Saturnalia festival. The two men seem to have been on good terms, probably because they had to work together on various arrangements involving the fort's administration and supplies. Severus wrote:

> Severus to his Candidus. Greetings. Saturnalia expenses. I ask, brother, you settle [them] at four or six [asses], and radishes at no less than ½ a denarius. Farewell brother. To Candidus, slave of Genialis, from Severus, cornicularius.[43]

One of the most famous of the writing tablets found at Vindolanda is a letter from Octavius to Candidus. Octavius was involved in some way with the trading of commodities, though it is not clear whether he was acquiring these in the capacity of a military official or whether he was a private trader hoping to sell them on to the army. His business seems to have been obtaining goods for the fort at Vindolanda. He wrote the letter while dealing with other incoming correspondence, to which he refers. Candidus may be the aforementioned slave of Genialis, an optio at the fort of this name, or may be another

Candidus altogether. Octavius' main concern was that he was sent money to pay for commodities he had acquired, such as '5,000 modii of ears of corn'. At about 15.4 pints (8.73 litres) per *modius*, that meant he had taken possession of 43,650 litres of corn ears. The weight of corn depends on its moisture content. Depending on the moisture level, each litre of corn might have weighed about 1.7 lb (0.789 kg), but this is very approximate. This means Octavius had bought 55,323 kg of corn, or over 54 tons. In Polybius' time soldiers received 'two-thirds of an Attic medimnus' monthly, approximately equivalent to 37 litres or about 64 lb (29 kg), totalling 767 lb (348 kg) annually. Assuming this was still valid, this crude calculation means that Octavius had enough corn for almost 160 men for a year. Or, to put it another way, it was almost exactly the right amount to feed a cohort of 480 men for one-third of the year. Since army pay and supplies were computed on the basis of three *stipendia* annually, this is surely no coincidence and may have been the purpose of the order.

Octavius had paid out a deposit of 300 denarii to secure the corn and needed at least another 500 to make sure the deposit was not forfeited. Among his other concerns was a consignment of hides which were still at the fort and military settlement of Catterick, on the road south to the legionary fortress of York (Eboracum). Octavius wanted to collect them and asked Candidus to authorize their hand-over, explaining that he would have been to get them already but had been reluctant to send his wagons down 'while the roads are bad'. This is not a paraphrase – the Latin (*dum viae male sunt*) says exactly that. Octavius was also bothered by the fact that 8½ denarii owed to him by a man called Tertius had not been paid. He was even more annoyed by a 'messmate of Frontinus' who had turned up asking for hides and promising to pay cash. They had arranged that he would come on 15 January, but he never showed up.[44]

The corn Octavius was so troubled about would eventually have been stored in a granary. These were some of the most distinctive structures in a fort. They also had to be the most robust. Settling grain causes enormous pressure on a granary's walls, creating heat and a fire risk. Masonry military granaries had conspicuous buttresses down either side unless they were built in a pair, in which case the adjacent walls of the two buildings supported each other. Granaries also had raised floors, suspended on piers or rows of parallel supporting walls, to maximize ventilation and provide some protection against rodent activity. At Wallsend, at the easternmost point of Hadrian's Wall, the two fort granaries sat in the central zone of the fort between the headquarters building and the hospital (see the plan on p. xiii). At South Shields, not far from Wallsend but closer to the mouth of the river Tyne, the fort was enlarged to accommodate an exceptional number of granaries so that it could act as a supply base for Septimius Severus' campaign in Caledonia. As a result as much as two-thirds of the fort was given over to granaries. The rest was occupied only by a small number of barrack blocks and the headquarters buildings, all other conventional fort structures being done away with.

Other documents found at Vindolanda refer to all sorts of goods, including *cervesa* ('beer'), wine, *muria* ('brine', but usually translated as 'fish sauce'), barley and pork fat.[45] The availability of these was totally dependent on the province of Britain's infrastructure. Vindolanda, like all other forts on the northern frontier, was linked to a network of roads and rivers leading to the sea that gave it access to goods available across Britain and beyond. Some of the commodities which were transported up to Vindolanda proudly bore their manufacturers' name stamps and trademarks, such as the leather shoes made by Lucius Aebutius Thales, son of Titus, stamped with his name, vine-leaves and cornucopiae.[46] It is possible he worked on the northern frontier but London or York is probably more likely, servicing the military frontier market through middlemen. Finds at Vindolanda include a small lead mirror frame manufactured by Quintus Licinius Tutinus of Arles in Gallia Narbonensis.[47] There is every reason to believe that other frontier forts across the Empire were equally well served and supplied.

Clearly there was a level of unit administration and bureaucracy in organizing the supply and transport of consignments of goods. Other evidence points to sub-divisions of the cohort and even individuals all taking care of their needs. The century of Africanus (its centurion), based at Vindolanda, had its own quernstone, which must have been used to prepare the flour for the century's bread.[48] Quernstones are common finds on military sites, and indeed on all Roman sites. At Haltwhistle Burn, close to Hadrian's Wall, a Roman military mill was powered by a channel cut across a bend in the river, the outflow driving a waterwheel which turned the millstones. Similar mills were built into bridge abutments on the Wall itself at Chesters and Willowford, and must have been operated by army personnel to help feed the garrison.[49] Ovens are also found, usually near the defences, and located at a safe distance from the barracks and other buildings because of the fire risk.

At Newstead, during Agricola's governorship, the tribune Attius Secundus had his own amphora, which had come originally from southern Gaul and contained three modii of unknown contents.[50] Gaulish samian ware is especially common on Roman military sites and soldiers were sometimes keen to identify their own bowls, dishes and other vessels. In the vicus outside the fort at Papcastle (Cumbria in north-west Britain), Senecio marked his name on the base of a samian bowl which seems then to have been passed on successively to a Cato and a Tertius.[51]

Soldiers could help themselves from the stores of the local communities where they were based. This was not supposed to happen, but it certainly did. The locals were unlikely to be able to fight back. Sometimes the Roman authorities showed some consideration. According to Frontinus, the emperor Domitian ordered in 83, during his German war, that compensation be paid for any crops planted by the Cubii tribe which were on land enclosed by the new Roman fortifications. It was in this way, he said, that the emperor's justice 'won everyone's allegiance'.[52] However, since Frontinus wrote this down the following year, while Domitian, gradually emerging as a jealous and paranoid individual, was still emperor, it is fairly obvious he was treading carefully and probably exaggerated the emperor's generosity. The general Avidius Cassius was a ruthless disciplinarian and had zero tolerance for troops who took advantage of civilians. He ordered any soldiers caught stealing by force from provincials to be crucified immediately (see Chapter 2 for other instances of his brutal discipline).[53]

Even Rome was not immune. Maximinus I (235–8) allowed his men to steal from the land surrounding the city.[54] Civilian settlements of every type throughout the Empire were susceptible to Roman troops helping themselves. The village of Blagoevgrad (Skaptopara in Thrace) found itself ravaged by soldiers stationed in two camps nearby. Their depredations were ruining the village, and outraging the inhabitants, who felt they had dutifully paid their taxes and should not be subjected to any more costs. Of particular appeal to the soldiers were the village's hot springs. A fair was held in the town, as was a market, the latter taking place several times annually with a special tax-free fortnight in October. Fortunately the villagers had a means of solving their problem, or so they hoped. One of them, a soldier called Aurelius Pyrrus, owned land there but also served in the Roman army in a praetorian cohort in Rome. With his help, the villagers filed a petition to the emperor, Gordian III, on 16 December 238. Gordian instructed them to take their case to the provincial governor; this Pyrrus did, delivering a speech on the village's behalf. Unfortunately, we do not know what happened next.[55] But Gordian could hardly be blamed for delegating the case. In December 238 he was a month shy of his fourteenth birthday.

For all the training, organization and garrison building, and the effort poured into building infrastructure and sourcing food, the Roman army was supposed to be a fighting machine. Its reputation came down in the end to the ability to fight and win victories. Paradoxically, like armies throughout history, success often derived from the painful lessons caused by defeat, whether thanks to a lack of morale and training, poor leadership or bad luck. Tacitus was at pains to point out how intractable a foe the Germans, for example, had continued to be.[56] Rome's history was littered with the tales of terrible defeats. But in the end the Roman army demonstrated that what mattered was not what went right, but how it coped with disaster.

# SEVEN

# IGNOMINY AND DEFEAT
## The Roman Army's Darkest Days

*Augustus was so alarmed that for several consecutive months he did not cut his beard or hair, and sometimes he bashed his head in the corridors, crying out, 'Quinctilius Varus – bring the legions back!'*

Augustus reacts to the catastrophic news
that three legions under the command of
Varus had been wiped out by the Germans[1]

The Second Punic War nearly saw Hannibal and his Carthaginians wipe out the Roman forces. The legendary defeats at Trasimene (217 BC) and Cannae (216 BC) haunted Roman memory. But they were not the only times Rome came near to complete destruction at the hands of an enemy on the battlefield.

Although there are innumerable tales of Roman military bravery, like that of Caesar's centurions Lucius Vorenus and Titus Pullo (see Chapter 9), there are instances where Roman troops fell short of what was expected of them in the heat of battle. Some did not want to fight, particularly if they knew that the enemy was especially challenging. Others fought valiantly, but if they were defeated and lived to tell the tale they were treated by their countrymen as shameful failures. The stories in this chapter recount some of Rome's darkest days, and depict some of its most humiliated soldiers.

## HISTORICAL DEFEATS: TRASIMENE AND CANNAE

The great Republican general Scipio Africanus had no time for ill-prepared commanding officers. One of his sayings was that bleating 'in war the words "I had not expected [that]"' disgraced whoever said them. Scipio firmly believed that any military campaign should only be conducted after exhaustive preparations and planning had been undertaken. One of the reasons he said this was because of some of the disasters under his predecessors' leadership earlier in the Second Punic War. He added that one should only ever fight a battle with an enemy if an opportunity had arisen, or out of necessity.[2]

A series of major defeats in the Second Punic War nearly destroyed Rome, yet the city's ability to learn from catastrophes and come back fighting was absolutely fundamental to the development of the Roman army, its military skills and leadership. In 217 BC the Carthaginian general Hannibal was loose in Italy, roaming and laying waste at will, with the firm intention of provoking the Romans into action. He succeeded. The consul Gaius Flaminius was so outraged he ignored advice to wait for his fellow consul, Servilius Geminus, to arrive with his army. Flaminius completely misunderstood the depth of the threat as he marched along in the vicinity of Lake Trasimene early on the morning of 21 June. Believing Hannibal was some way off, he was horrified to find his army had been stopped by Hannibal's African and Spanish troops. Hannibal had stationed them there with the express intention of blocking Flaminius' men and trapping them between the lake and the mountains.

Flaminius had rushed straight into a trap. Livy claimed that Flaminius displayed admirable coolness under the circumstances and tried to rally the soldiers by telling them only their 'brave exertions' could save them. It made no difference to the outcome.[3] The Romans saw the Africans and Spaniards before they realized there was an ambush waiting for them. They were also caught out by the morning mist they were marching through. Hannibal had sent his other troops up on to the high ground overlooking the road. Stationed above the mist, they could send visual signals to each other to coordinate the attack. The Romans had no idea where Hannibal's men were until they heard the shouting but could not see anything; in the 'din and confusion' the centurions and tribunes were unable to issue any orders.[4]

The Romans were attacked on all sides. They were still in marching formation and had had no chance to reorganize into battle order. According to Polybius, Flaminius was in a state of 'the utmost dismay and dejection' and had exhibited 'a total lack of judgement'. When an Insubrian horseman called Ducarius recognized Flaminius as the general who had attacked Insubrian territory in 223 BC, he charged forward and killed him.[5]

The fighting was not even over. Troops at the rear of the Roman column were pushed into the lake, where Hannibal's cavalry killed them or they drowned. Only 6,000 Roman troops managed to escape and fight their way to higher ground, where they could see how disastrous the battle had been. They tried to hold out in a nearby village but were encircled by Hannibal and captured.[6] Critically, the whole scenario had prevented the Roman army from operating in disciplined battle formation. 'It was no ordered battle', said Livy. Every man fought for himself in a frenzy that lasted three hours, apparently ignoring even an earthquake which coincided with the mayhem.

It was a 'disaster memorable as few others have been in Roman history', and it was impossible to pretend to the Roman people that it had been anything else. In fact it was almost the first time the general population had ever been told about a defeat.[7] In the end a mere 10,000 Romans managed to struggle back to Rome in dribs and drabs. Meanwhile, 4,000 Roman cavalry under the command of Gaius Centenius which diverted to help out at Trasimene rode right into Hannibal's hands and were lost too, half being killed and the other half captured.

To a beleaguered Rome the situation seemed catastrophic, but the senators kept their cool and spent days discussing what to do.[8] The critical lessons were not to let a commanding officer act unilaterally, and not to allow an army to become trapped. The Roman army's greatest skill was organization and discipline. Both had broken down at Trasimene, and one of the reasons was its reliance on recently raised and untrained recruits.[9] The results were disastrous losses and Hannibal's freedom to continue to roam Italy at will.

The solution, it was decided, was to confront Hannibal in over-whelming numbers and to make sure the men were fully trained and confident. Aemilius Paullus and Terentius Varro were elected as the consuls for 216 BC and placed in charge of the military preparations, Paullus dealing with recruitment and training. When Hannibal seized the town of Cannae, helping himself to Roman stores there, the scene was set for a major showdown. The Romans organized themselves into an exceptionally large army. Eight legions were formed, double the normal number. Each legion was also enlarged to 5,000 with 300 cavalry attached, adding a further 2,400 men to the 40,000-odd infantry. Another 40,000 allied infantry were raised, along with around 5,000 allied cavalry. The Roman army of 216 BC now numbered nearly 90,000 men, an extraordinary number for the time.[10]

Unfortunately, the two consuls put their egos first and disagreed about what to do. Camped 5 miles (8 km) from the Carthaginians, Paullus wanted to avoid fighting there because the land was open and it would give Hannibal's cavalry the opportunity to attack with impunity.[11] That was what Hannibal himself wanted, said Livy, because 'in a cavalry action ... he was invincible'.[12] Varro, who was much less experienced, did not concur. It was, noted Polybius, 'the most dangerous situation which could happen'. The normal practice was for the consuls to take overall command on alternate days, which would not matter if they were cooperating. But they were not, and they adopted different tactics in an enthusiastic display of incompetence. Varro took advantage of his prerogative the next day to order the men to break camp and advance. Hannibal, who must have been stunned at his good fortune, capitalized on the unexpected opportunity and attacked, but the Romans managed to fight back and hold off the Carthaginians until night fell. The following day Paullus took over and told the army to pitch camp by the river Aufidus, but decided to split his force, ordering one-third to guard a ford over a mile away on the opposite bank so it could protect foraging parties.[13]

What followed was a series of inconclusive encounters, the news leaving the people in Rome half mad with nerves. On 2 August Varro once more took overall command. He led the army over the river without even bothering to tell Paullus and placed them in battle formation, egged on by the troops who were desperate to fight. The main body in the middle was the infantry under the command of Geminus Servilius, with Roman cavalry on the right and allied cavalry on the left. Hannibal brought his 40,000 infantry and 10,000 cavalry out to face the Romans in a similar configuration. The only significant difference was that he ordered his centre forward to create a 'crescent-shaped bulge' so that they could start the battle.[14] He had also ensured that the Romans were facing into the sun and the dust, whipped up by the wind that was common there.[15]

The fighting only became truly vicious when the cavalry met each other on the Roman right, beside the river. The area was so small that they dismounted in the congestion and started fighting hand to hand. The Roman cavalry gave way, many being killed; Varro was one of those who fled, along with 70 surviving Roman cavalry, adding ignominy to his stupidity in the lead-up to the battle. The Roman infantry made much better headway against Hannibal's Celts, but they had inevitably advanced into a trap. It was precisely what Hannibal had planned all along. His African troops, arranged on either side of the Celts, enveloped the Romans. Meanwhile the allied cavalry on the Roman left fell back; Hannibal's brother Hasdrubal was able to leave them and support the Africans in the heart of the infantry battle.

In the melee Paullus tried to rally the troops but lost his horse. A tribune called Gnaeus Lentulus rode up and offered his own horse, he said, to the 'only man without guilt in the disaster of this day'. It would save him and allow him to avoid the shame of a consul's death. Paullus declined, telling him to go to Rome and warn the Senate to fortify the city, and leave him to die among his soldiers. At that moment Paullus was killed in a rain of enemy missiles and Lentulus escaped. Gradually the Carthaginians whittled down the Roman army by working inwards from the outside, systematically killing the Romans as they went.[16]

Hannibal captured 10,000 Roman infantry, all of whom had been left by Paullus in the camp as a reserve. About 3,000 more fled from the battlefield, as did 300 allied cavalry. Polybius was appalled at how Varro managed to escape too, commenting that his flight from the battlefield only matched the disaster of his tenure of office. The rest, numbering allegedly about 70,000, were believed dead. In contrast, Hannibal lost about 5,700 men in total – less than 10 per cent of the Roman losses. These at least were the figures recorded by Polybius. Livy suggested something more like 50,000. Like all such details they were rounded and probably exaggerated to some degree. Nonetheless, it is clear the battle was a catastrophe for Rome and a relatively cheap victory for Hannibal. Livy described the Roman army as having been 'annihilated in a massacre', the terrible news being passed from house to house in the city.[17] The lesson was a hard one for Rome. The decisive factor had been Hannibal's cavalry, proving that it was better to have fewer infantry than the enemy and more cavalry.

A few days later another Roman army was wiped out by Celts in Gallia Cisalpina. The only thing that could be said in the Romans' favour was that when the news of these hair-raising disasters reached

Rome, the Senate held its nerve. As Polybius noted, 'through the special virtues and their ability to keep their heads', the Romans were eventually able to fight back. Their ultimate victory in a war they were to win a decade later would leave them 'masters of the whole world'.[18]

The events of 216 BC remained in Roman consciousness long afterwards. Cannae was remembered as an occasion 'when the survival of the Republic hung on the loyalty of our allies'.[19] By imperial times a story had grown up that Varro had suffered such a terrible defeat at Trasimene because he had insulted Juno. As aedile in charge of the games he had installed an exceptionally good-looking young male actor in Jupiter's carriage to hold the necessary religious accessories. It was a clear allusion to Jupiter's cup bearer Ganymede, with whom Jupiter had fallen in love. Therefore, it was alleged, Juno had been infuriated by Varro's insensitivity and had made sure he was defeated. When the tale was recalled in later years sacrifices had to be made to make amends for Varro's contempt for Juno's sensibilities. The point is not whether the story was true or not, or even whether the Romans believed it, but rather the notion that battles or campaigns were susceptible to the whims of deities in response to the actions of men.[20]

The eventual Roman victory in the Second Punic War was won at astronomical cost. Trasimene and Cannae reinforced the Roman sense of destiny that they had prevailed and overcome enormous adversity. The nature of the evidence we have tells us little or nothing about the individual soldiers involved. They are absent from the record. What we do know is that in the longer run the families they left behind were often dispossessed of their land by the greedy Roman aristocracy. Even the survivors all too often came home to find their peasant farms had been stolen and absorbed into vast senatorial estates run by slaves, leading to a political crisis in 133 BC (see Chapter 2). It was the ensuing crisis in the Roman Republic that began slowly to pave the way for the age of the emperors and the establishment of a professional standing army.

The surviving Roman forces from Cannae who made their way

back to Rome were not thought worthy of compassion. In Roman eyes they had failed ignominiously. It would have been far better had they died in battle. Instead they were punished by being sent to Sicily and forced to live as if they were still on campaign for as long as the Second Punic War lasted.[21] In 212 BC the remnants of the army defeated at Cannae appealed to the general commanding in Sicily, Marcus Claudius Marcellus, to take their concerns to the Senate. In their letter they found an oblique way of implying that the fault for the defeat clearly lay with their leaders and reminded him that the officers had proceeded to pursue normal careers, some even becoming provincial governors. They drew attention to the lenient treatment meted out to other defeated Roman armies and begged for the chance to overturn their disgrace. Marcellus agreed to send the letter on to the Senate, but the appeal was rejected.[22]

In 205 BC Scipio, who was assembling a new army in Sicily to mount an invasion of Africa to destroy Carthage, took a different view. The soldiers themselves believed that under Scipio they could redeem themselves:

Those who were left of the soldiers who had fought at Cannae felt convinced that under Scipio, and no other general, they would be enabled, by exerting themselves in the cause of the state, to put an end to their ignominious service. Scipio was far from feeling contempt for that description of soldiers, inasmuch as he knew that the defeat sustained at Cannae was not attributable to their cowardice, and that there were no soldiers in the Roman army who had served so long, or were so experienced not only in the various kinds of battles, but in assaulting towns also. The legions which had fought at Cannae were the V and VI. After declaring that he would take these with him into Africa, he inspected them man by man. Leaving those whom he considered unfit for service, he substituted for them those whom he had brought from Sicily, filling up those legions so that each might contain 6,200 infantry and 300 cavalry.[23]

The Cannae survivors had finally exonerated themselves, but it had been a long time coming.

## Humiliation – the Spartacus War

The slave revolt led by Spartacus in 73 BC presented Rome's army with an unprecedented challenge, and it came about because of the grievances of a man who had once fought for Rome. The Thracian Spartacus had reputedly once been a soldier in the Roman army – glossed over in the famous motion picture about his life. Having somehow been enslaved, he was eventually sold for gladiatorial training at the school run by the *lanista* Lentulus Batiatus at Capua.[24] Spartacus' military training as well as his intelligence stood him in good stead when, according to Appian, he decided to lead a rebellion by overcoming the training school guards. Plutarch's version has 200 gladiators planning to escape, with 78 managing to do so and subsequently electing Spartacus as leader. (Given the chaos there must have been at the time, it is hardly surprising the accounts do not tally precisely, not only at this point but also throughout.) Spartacus and his band of rebels headed for the slopes of Vesuvius, having been joined as they went by runaway slaves and even some free peasants. Any loot or booty gathered along the way was shared out equally, an egalitarian gesture which soon encouraged many others to join them.

A slave rebellion was something that terrified the Romans. Two major revolts had erupted in Sicily in 135 BC and 104 BC and had proved extremely dangerous and difficult to crush. There were so many slaves that there was a real risk they would realize that by working together they could easily overcome their masters. Despite that, no one in Rome appreciated quite how dangerous Spartacus' rebellion was. Arrogance and complacency set the Romans on the route to another military disaster. A couple of scratch forces rather than proper armies were thrown together and sent after the rebels, who beat them off easily. Spartacus' army was now thought to number 70,000, and thanks to his experience he was able to oversee the manufacture of weapons and military training. Only then did the Senate dispatch a proper consular army of two legions.

To begin with they successfully defeated a force of 30,000 fighting under another gladiator leader called Crixus, who was among 20,000 men killed. Spartacus decided to lead his slave army north through Italy but was cut off by one of the consuls, while the other blocked his retreat. Since the consular army cannot have numbered more than about 10–12,000 men, however, splitting them was a bad move. Spartacus attacked the two halves of the Roman army in separate engagements and defeated both with a force that had now grown to 120,000, the consuls fleeing in separate directions. Plutarch adds that Spartacus headed on towards Gaul, defeating Cassius, the governor of Gallia Cisalpina, and an army of 10,000. Spartacus ordered the execution of 300 Roman prisoners as a sacrifice in honour of Crixus. Next he marched on Rome, killing all his remaining prisoners and his pack animals so that he could advance as fast as possible. By now the consuls had regrouped and together tried to block Spartacus. He defeated them again.

The fact that a trained Roman army was being overcome time after time by a slave contingent led by a former slave who had once fought for the Romans himself was a devastating and terrifying humiliation for the Roman military. Spartacus was also smart enough to understand that his force, for all its size, lacked the equipment and training to take Rome. Instead he captured the city of Thurii, where he only allowed merchants to bring in iron and brass so his men could manufacture more arms. Another victory over a Roman army followed.[25]

Spartacus' rebellion had lasted three years by the time it finally came to an end. In 71 BC Marcus Licinius Crassus stood for one of the praetorships that year. No one else would stand for fear of the immense task ahead, but Crassus was extremely wealthy and even more ambitious. Appian said that, determined to win at all costs, Crassus raised an army of six legions, to which he added the consular army of two legions. However, he reduced the number of men in two consular legions by decimation as a punishment for their failure to defeat Spartacus in the preceding two years. According to another story that circulated (which Appian quotes), he was in fact defeated himself by Spartacus to begin with and had to decimate his whole army, killing around 4,000 men. In Plutarch's version Crassus sent Memmius, one of his legates, with two of the legions to fight Spartacus. Memmius was defeated and some of the survivors fled. Crassus regrouped them but took the 500 most cowardly and decimated them.[26] Whatever the truth, it worked. He killed two-thirds of a group of 10,000 of Spartacus' army who were encamped some way from the main force. With that victory under his belt, Crassus led his men to rout the rest.

Spartacus' army fled south to the sea, hoping to cross to Sicily, but was trapped by Crassus. Spartacus tried to avoid joining battle after he lost 6,000 men when trying to break out, and instead spent his time trying to harass Crassus' army. Back in Rome, the thought that the siege might last any longer led the Senate to invite Pompey and his army, which had recently returned to Italy from the war in Spain against the rebel Quintus Sertorius, to head south and help defeat Spartacus.* That infuriated Crassus, who believed Pompey would end up with all the glory and was desperate to fight Spartacus before his rival arrived. Spartacus tried to negotiate terms but Crassus refused. Spartacus then broke out and headed for Brindisi in the hope of making his escape by sea. He gave up when he heard that Lucullus, proconsul of Macedonia, had arrived back in Brindisi after defeating Mithridates of Persia. Spartacus now turned to face Crassus in battle and was defeated. He was killed

---

* Quintus Sertorius was an enemy of Sulla who ended up seizing power in Rome's Spanish provinces, prosecuting a successful resistance against Roman armies until his assassination in 72 BC.

while making his way towards Crassus, killing two centurions as he did so, as were all the rest of his army except 6,000 men whom Crassus crucified along the Via Appia from Capua to Rome. The Spartacus War had put Rome in the most extraordinarily precarious position. Spartacus' success came close to undermining the military prestige Rome had won over several centuries, and made a mockery of the Roman military system Polybius had so admired only eighty years earlier.[27]

## THE DIGNITY OF THE STANDARDS — CARRHAE, 53 BC

In 53 BC, almost twenty years after he brought the Spartacus War to an end, Marcus Licinius Crassus and a huge Roman army set out to the east to confront the Parthians under their commander Surena, Orodes II's general. Crassus was a political associate of Caesar and Pompey, and at the time the three men were united in an unofficial alliance known today as the 'First Triumvirate'. Crassus was moreover the richest man in Rome.

The battle that followed at Carrhae (near present-day Harran, eastern Turkey) was a catastrophe. Most of the Roman force was annihilated. Crassus was killed, as was his son. Even worse, the Roman standards were lost. All the auspices had been bad from the outset. For some reason, Crassus was handed a black cloak instead of the white or purple one a general would normally wear to battle. The soldiers were in a depressed and silent mood instead of being eager to fight. The legionary eagle standards proved to be a problem too. A centurion struggled to pull one of them up out of the ground, while another swivelled round to face the wrong direction. Crassus 'made light' of such bad omens, said Plutarch. Valerius Maximus put it all down to the inevitable consequence of human vanity in the face of heaven.[28]

To begin with, Crassus spread his men out on a wide front to stop the Parthians surrounding them. Soon he changed his mind and ordered them to rearrange themselves into a hollow square with 12 cohorts and a cavalry wing on each side. The idea was they would be able to face an attack from any side. Keen to bring the Parthians to battle, he refused to make camp by the river Balissus and allow his parched troops to quench their thirst. They had to eat and drink where they stood before moving on. When they did advance, Crassus marched them into a trap, even though his army was much bigger than Surena's. Surena had ordered his men to hide their armour behind their clothing and had concealed his main army behind his advance force. With a graphic sense of theatre he then ordered the Parthian soldiers to work up a terrifying and disorienting din using drums and bells, before exposing their armour – an impression amplified by their painted faces and the way they wore their hair over their foreheads in the style of the Scythians to make themselves look as frightening as possible.

The Parthian force surrounded Crassus' hollow square. He ordered an advance, but it collapsed almost instantly under a hail of arrows. The Romans could not escape. When Crassus realized the Parthians had brought up a supply chain of camels laden with fresh arrows, any hopes that they were about to run out of ammunition evaporated. Crassus told his son, Publius Licinius Crassus, to lead an attack as the Parthians withdrew, before the opportunity disappeared for good. The younger Crassus took eight infantry cohorts, 500 archers and 1,300 cavalry. He raced right into a trap. The Parthians stopped, turned, used their cavalry to kick up dust and embarked on a massacre. Roman infantry troops died or suffered in agony trying to pull out the barbed arrows, which ripped open their wounds even wider. Others had arrows through their feet and hands, pinning them to the ground or fixing their hands to their shields. Young Crassus pressed on, only for his Gaulish cavalry to fall foul of the Parthians' longer spears and the heat. They fell back, taking over a patch of higher ground to give them a defensive advantage, but only exposed themselves to another hail of arrows. The wounded young Crassus turned down a chance to escape and ordered his shield-bearer to kill him. The Parthians then turned on his father.

Before Crassus could advance to relieve his son the Parthians returned to face him, started on their drums again and held up his son's decapitated head. They mocked Crassus for his cowardice compared to his son's valour. Roman morale plunged further, a speech from Crassus failing to lift their spirits. The first day of the battle ended with the Parthians once more surrounding the Romans and killing many with arrows and spears. Crassus spent the night in despair, so his officers decided to organize a general retreat into the city of Carrhae, abandoning many of their sick and wounded. In the morning the Parthians started the day by killing 4,000 of those left behind. In addition, four cohorts that had become separated from the main Roman column were wiped out after getting lost.

With the remaining Romans stuck in Carrhae, Surena offered a truce if they would leave the region. It was agreed that he and Crassus would hold a conference, but the next day the Parthians told the besieged Romans that if they wanted a truce they would have to hand Crassus over. The meeting, when it took place, was a disaster. Surena offered Crassus a horse but one of his men pulled the animal's reins while Crassus was seated on it. His officers tried to restrain the horse but violence broke out and Crassus was killed, allegedly by a Parthian named Pomaxathres. That sparked an eruption of killing which started with the deaths of some of Crassus' party and ended with the slaughter of 20,000 Roman troops and the imprisonment of 10,000 more, and the Roman standards were lost. Led by Cassius Longinus, one of Crassus' officers, only 10,000 soldiers made their way back to Syria (some of them must have recorded their experiences in accounts later read by Appian and Plutarch).[29] A particularly distasteful account of the aftermath was related by Cassius Dio. He said there was a story that the Parthians had poured molten gold into Crassus' mouth in mockery of his riches and the way he had pitied those who could not afford to bankroll a whole legion from their personal wealth.[30]

The disaster was an object lesson in what could happen when soldiers, demoralized and far from home in extremely arduous conditions, were confronted with an imaginative and resourceful enemy. Carrhae was a body-blow because a superior Roman force had been destroyed with comparative ease. It was not until the reign of Augustus, when Parthia conceded control of Armenia following a campaign led by the emperor's stepson Tiberius, that the standards were returned. Tiberius received the standards on 12 May 20 BC, after Tigranes II was restored to the Armenian throne; he also recovered the standards lost by Mark Antony when Orodes II's son Phraates defeated his army in 36 BC.[31] With their return, some Roman dignity had been restored and the shadow of the defeat laid to rest. Over four centuries later, the tale of Crassus' defeat and how he had been 'annihilated' was well remembered in Roman lore.[32]

## AGRIPPA IN SPAIN

In 19 BC Marcus Agrippa, Augustus' general, right-hand man and son-in-law, went out to Spain to deal with a rising among the Cantabri. Enslaved after their defeat in a war earlier in Augustus' reign, the Cantabri had killed their masters, returned to their tribal homelands in northern Spain, and whipped up a rebellion that involved making plans to attack the Roman garrisons. 'But he had some trouble with his soldiers', said Cassius Dio of Agrippa. Many were too old and worn out by the continuous wars of recent years; evidently the recruitment of new, younger soldiers had fallen short of requirements.

Once Agrippa had weighed in, telling them off, encouraging them and trying to inject some optimism, eventually they agreed to obey his orders. Agrippa might have thought he had sorted out the problems, but he was soon to discover he had not. The Cantabri proved to be an intractable foe. They had 'gained practical experience' of fighting and were highly motivated by the thought that if they were captured they were bound to be killed. Strabo described how the Cantabrians had a 'ferocity and insensibility' to suffering that meant when they were captured and crucified they sang 'their paean of victory'.[33] No wonder then that a large number of Roman soldiers were lost in the fighting, and Agrippa had to punish numerous other men 'because they kept being defeated'. He even stripped Legio I of its title 'Augusta' out of disgust at its failure in the war. Only then did he manage to turn round the fighting, killing many Cantabrian fighters and capturing others.

Agrippa had won a victory, but he turned down the triumph Augustus offered him, perhaps because it had been such a close-run thing. Legio I later redeemed itself in Germany, possibly in the aftermath of the disaster of AD 9 (see below), becoming known as Legio I Germanica.[34] Meanwhile Legio IIII Macedonica and Legio X Gemina had to be stationed in Spain to control the area for several generations to come.

## VARUS LOSES THREE LEGIONS

In AD 9 Publius Quinctilius Varus was in his third year as governor of Germania, a region in which the Romans still only held certain districts. Confident that the province was at peace, Varus was managing a programme of urban colonization designed to encourage the locals into Roman ways of life.[35] The Germans had been left in no doubt about Roman military prestige. A magnificent twice life-size bronze equestrian statue, undoubtedly of Augustus, was erected at Waldegrimes, a civilian settlement. The bronze horse's head survives, decorated with a figure of Mars on the bridle.[36]

In fact Varus was being willingly lulled into a false sense of security by the German tribes, who helpfully pretended to be acquiescent and peaceful. Two tribal leaders, Arminius and Segimerus, posed as his friends and were allowed to share his tent, even while they plotted against him; so trusted was Arminius that he was made a

Roman citizen and an equestrian. As far as the Roman historian Velleius Paterculus was concerned, 'fate took control of Varus' plans and blindfolded the eyes of his mind'. Consequently, the foolhardy governor 'did not keep his legions together'. Instead he gave in to various requests from widely dispersed settlements for the dispatch of soldiers to act as police by guarding locations and supply traffic, and catching thieves. When the Germans instigated an uprising, so that Varus would have to set out to suppress it, Arminius and Segimerus excused themselves, claiming they were heading off to put together units of allied auxiliary soldiers to help out. It was, of course, a lie. They killed the Roman soldiers sent by Varus and then waited for the governor and his army to arrive.[37]

Varus blithely led a Roman army of three legions, XVII, XVIII and XVIIII, from Xanten to Mainz right into the Teutoburg Forest, a dangerous place for soldiers who preferred to fight pitched battles in the open. The force of more than 15,000, together with its baggage train, which included women and children and other followers, swanned into a trap 'hemmed in by forests and marshes'.[38] The soldiers had been mixed up with the wagon train, an unforgivable lack of planning, so it was impossible to get into any defensive formation when the Germans attacked.[39] When the bewildered Romans counterattacked they found themselves crashing into each other and into the trees. Foul weather hampered their progress as they desperately tried to cut down the trees and build the roads the army so depended on for movement and control of the terrain, while floundering in the mud during constant attacks from the tribesmen.

A bad situation became a great deal worse over the several days it took the disaster to unfold. More torrential rain and a gale on the fourth day meant the Romans became weighed down with equipment that they could not use; their hands were soaked and they could not throw javelins or fire arrows, or even hold their shields. It was a decisive moment. Varus' surviving men were unable to escape or even effectively defend themselves. They were now besieged in the middle of a forest. More Germans had arrived, 'largely in the hope of plunder'. The Romans who had survived thus far were starting to run out of food. They decided to show their few German prisoners that they had excellent stocks of food and could hold out. Next, having cut the prisoners' hands off, they set them free so they could go back to Arminius' men and tell them there was no chance of starving the Romans into surrendering.[40]

In reality the position was hopeless. One of the camp prefects, Lucius Eggius, surrendered and gave himself up to be tortured to death. A legionary legate, Numonius Vala, fled but was captured and killed. Varus had the presence of mind to fall on his sword, as did his other surviving officers. It was an honourable way out and had the useful bonus of avoiding an agonizing death at the hands of the enemy. The eagle standards were seized by the Germans from the spot where Varus died. When the remaining soldiers realized what had happened, some committed suicide and others gave up, throwing down their weapons and waiting to be killed. Escape was impossible. They were slaughtered to a man 'like cattle'. The German tribes took delight in mutilating Varus' body before decapitating it and sending the head to Tiberius, the stepson and heir of Augustus and currently in command in the region.[41] When the news reached Augustus in Rome he was devastated and incredulous.*

During his campaign in 15, six years later, Germanicus came across the grisly remains of Varus' camp. He found the remains of the Roman soldiers' bodies 'cut to pieces, in a ditch and on the plain nearby bleached bones either in piles or scattered', no doubt where wild animals had carried them off to feast on. There was worse to come. Dying in battle was one thing, but those captured had suffered far worse fates. 'Skulls were impaled on the trunks of trees. In the groves of trees nearby were barbarian altars where the tribunes and the most senior centurions had been slaughtered.' Even the animals had been butchered. Germanicus found the remains of horses that had been hacked to pieces. The few survivors who escaped to tell the

---

* See the quote at the start of this chapter.

tale had reported how Arminius had made a speech from a tribunal on the spot. They pointed out the gibbets and torture pits his men had specially prepared for the massacre.

Germanicus' men set about the grim task of gathering up the bones and burying them properly.[42] They had already recaptured the eagle standard of Legio XVIIII when they attacked the Bructeri tribe en route to the site of the disaster. One of the other standards would have to wait until the next year when an informant told Germanicus it was buried in a grove and had only a light guard; the third did not turn up until 41, when Publius Gabinius recovered it after defeating the Chauci.[43]

Miraculously, the tombstone of one of Varus' men, found at Xanten, has survived. Marcus Caelius, a centurion in Legio XVIII, died in the horror in the Teutoburg Forest along with thousands of others. Of course, in reality his body had never been recovered; if it had been, it was jumbled up among the piles of mutilated and bleached bones found by Germanicus, and could never have been identified. The tombstone was thus really a cenotaph, set up in commemoration by his brother Publius. It also depicted and named Marcus' freedmen, who must have accompanied their former master and died alongside him. Publius added a note, more in vain than in hope, that their bones could be interred there too if they were ever found and identified.[44] As for the equestrian statue of Augustus erected at Waldegrimes, it was hacked to pieces by tribesmen and the head thrown down a well, ironically ensuring its survival.

## How Tacfarinas humiliated Rome

Tacfarinas was a Numidian Berber leader who had once fought for the Roman army as an auxiliary. He turned out, like Spartacus, to be one of many men trained in Roman military skills who subsequently became major threats. In 17, during the reign of Tiberius, he deserted. Using his Roman military training he formed a band of rebels in North Africa, organizing them into 'detachments and squadrons'.

He used Roman discipline and gave his men Roman-type weapons, though that same year he was routed by a Roman army led by the governor of Africa, Furius Camillus.[45]

Three years later Tacfarinas reappeared, though his plans to fight with trained men seem to have gone awry. He started with a guerrilla war, burning villages, but then 'blockaded a Roman cohort' in its fort. This must have been an auxiliary infantry unit in an outpost. Its commander was an officer called Decrius, a brave and experienced man who regarded being besieged as shameful. He drew his men up outside the fort so they could fight a battle in the open when Tacfarinas attacked. Unfortunately the cohort crumbled at the first attack, leaving Decrius to race around under a hail of projectiles chasing after the men who were running away and cursing his standard-bearers for standing by while that happened. Decrius was wounded in the chest and hit in one eye but he fought on singlehandedly until he was killed, completely forsaken by his men.[46]

The engagement showed how a Roman force, albeit an auxiliary one, could give up on the spot when confronted with an unexpected and frightening foe. The fallout was devastating for the cohort. The new proconsular governor of Africa, Lucius Apronius, was disgusted at the dishonour the men had done the Roman army and ordered a decimation of the unit. 'In a rare deed for that time and of ancient memory he chose by lot every tenth man of the disgraced cohort, and executed them by cudgel.' In other words the men were beaten to death by their comrades. The brutal punishment seems to have motivated other troops. Shortly afterwards a force of 500 veterans routed Tacfarinas' army, and a soldier called Rufus Helvius who saved a Roman citizen was decorated for his bravery.[47]

## THE IGNOMINY OF LEGIO XII FULMINATA

In 62, following several years of war in Armenia (see Chapter 8), violence started up again when the Parthians under Vologaeses attacked. Lucius Caesennius Paetus, the governor of Cappadocia, had Legio V and Legio XII at his disposal but had been rather too lax in authorizing leave applications from soldiers, with the result that the legions were not up to strength. In the Battle of Rhandeia that followed the Romans were severely beaten, and Paetus had to send a desperate request to Corbulo in Syria for help. The legions fled, even allegedly being forced to go through the humiliation of walking under the yoke.[48] Corbulo was forced to withdraw the demoralized Legio XII and send it to Syria, its best men lost 'and the rest terrified'.[49]

Just four years after that humiliating defeat, Legio XII found itself once more confronted with a challenge, and an opportunity to recover its dignity. During the Jewish War then being prosecuted by Vespasian, a general at the time, and his son Titus, the legion was involved in an another ignominious defeat. Tension, which had existed for centuries between the Greek inhabitants of Alexandria and the Jewish colonists, had erupted into violence. The Jews of Alexandria had been given their own quarter in the city, along with various legal privileges that allowed them to pursue their way of life without harassment. A particularly ugly incident took place when the Jews tried to join in a public assembly discussing a proposed embassy to Nero. Three Jews were captured and taken off by the Greeks to be burned alive. The Jewish population rose up in out-rage and attacked the Greeks. The Roman governor of Alexandria, Tiberius Alexander, tried to ease the tension with negotiation but his plan failed.[50]

Unfortunately, Tiberius Alexander then decided to order Legio III Cyrenaica and Legio XXII Deiotariana, both based in Alexandria at the time, along with 2,000 additional troops recently arrived from Libya, to attack the Jewish colony. According to Josephus, 50,000 (a figure unlikely to be accurate) lay dead before Alexander called off his men, but the Greek mob continued the violence against the Jews until they were forced to back off.[51]

Cestius was governor of Syria between 63 and 67.* When the news from Alexandria reached him in Antioch he decided he would have to intervene. He took a substantial army with him: Legio XII Fulminata, 2,000 soldiers each from the three other legions based in Syria, six auxiliary infantry cohorts (about 3,000 men), and four cavalry wings (at least 2,000 troopers), alongside another 11,000 men supplied by client kings in the region, as well as soldiers from the legions stationed in Judaea (VI Ferrata and X Fretensis). With an army numbering at least 27,000, Cestius headed south into Judaea to punish the Jews. He started by sacking Acre (Akko in modern Israel) before dividing up his army so that he could widen his impact. This started badly. When Cestius left Acre he installed 2,000 men there, but they were killed by a Jewish attack. Cestius was not swayed from his plans. One of his detachments sacked Joppa successfully, and another ravaged territory near Caesarea.[52]

At that point Cestius detached Legio XII Fulminata with its commander Caesennius Gallus and sent him into Galilee. All went well for the Romans to begin with. Cities in the region offered no resistance, while any rebels had melted away into a mountainous region near the city of Sepphoris. Caesennius Gallus decided to attack the rebels even though it ought to have been obvious that they were in an advantageous position. They managed to kill 200 Romans before Gallus' men positioned themselves on even higher land. After killing 2,000 rebels in return, Gallus returned to Cestius in Caesarea convinced that Galilee was under control. Cestius believed he could now advance his army towards Jerusalem.[53]

In Jerusalem, the Jews abandoned their celebrations of the Feast of the Tabernacle and attacked Cestius' army. A near-disaster occurred when the Roman front line was broken, being averted only when some cavalry and infantry managed to regroup and plug the gap. The

---

* His name was Cestius Gallus but in order to avoid confusion with Caesennius Gallus, legate commanding XII Fulminata, the Gallus is omitted here.

Jews lost a trivial number of men compared to the several hundred Romans who died, managing to escape back to the hills with some of the Roman baggage animals. Cestius had to take refuge in his camp for three days while the Jews watched for any sign of further attacks or movements. Attempts at diplomatic negotiations to resolve the impasse failed when the most extreme rebels refused to cooperate and attacked the Roman ambassadors.[54]

Cestius ordered a full-scale attack on Jerusalem, for which he spent three days preparing. On the fourth day his men fought their way towards the city, but Cestius made a crucial error. On the advice of some of his officers, Cestius did not tell his men to attack the Upper City, instead ordering them to pitch camp outside its walls. The assault, when it came, was protracted and dangerous. The Romans organized a *testudo* (tortoise) formation which would allow them to undermine the walls without being battered by stones thrown down on them. The most extreme rebels began abandoning the city, believing all was lost. That meant the more moderate citizens were able to start planning to surrender and save themselves. Thinking his assault was going nowhere, Cestius told his men to stand down, with the effect that the rebels suddenly realized they had a chance of victory after all. During the lull they burst out of the Upper City and attacked the Romans. Cestius was chased back to his camp pursued by the Jewish rebels, who attacked and killed numerous men. Among the casualties was the legate commanding Legio VI Ferrata. Cestius decided he would have to withdraw.[55]

The retreat needed to be conducted as fast as possible. Cestius knew his men were in a tight corner. Except those that carried ammunition or pulled the artillery machines, the baggage animals were all killed. The route meant having to pass down through a narrow ravine which trapped the Romans beneath an ambush by the Jews, who hurled down missiles. It was impossible for the Romans to defend themselves and they were only saved when night fell and they could make their way to Beth-horon.

Here Cestius came up with a plan. While 400 men were posted

to make it look as if his whole army was still at Beth-horon, the rest crept out under cover of darkness and relocated 3½ miles (5 km) away. When the sun rose the Jews realized they had been tricked. They broke into Beth-horon and killed the 400 men left behind. Next they set off after Cestius, who was still on the move and accelerating as he ordered his siege and artillery equipment to be abandoned. The Jews failed to catch him but helped themselves to the free gifts, which they would subsequently use against the Romans.[56]

Cestius' campaign and the humiliating midnight retreat from Beth-horon had been a shameful affair. Legio XII was singled out for humiliation as the principal legion involved. Josephus reported that 5,300 infantry had been lost, along with 480 cavalry. The losses were huge for a Roman army and seem to have been the largest ever incurred in a province that was supposed to have been under Roman control. More alarming still was the relatively small number of enemy casualties. No doubt bearing in mind the defeat the legion had suffered in 62 under Caesennius Paetus at Rhandeia, Titus ordered Legio XII to leave Syria permanently when the Jewish war ended in 73. It had to abandon its base at Raphanaeae and was relocated far to the east at Melitene by the Euphrates, on the border between Armenia and Cappadocia. There it stayed until it disappeared from history after the late fourth century.[57]

There was a gruesome aftermath to Cestius' experiences. In 67 Vespasian ordered an attack on the town of Gabara. His soldiers, filled with loathing for the Jewish people after what had happened to Cestius, massacred all the inhabitants, regardless of their age. Gabara was burned to the ground, as were the nearby settlements, their inhabitants being sold into slavery if they had not already fled.[58]

## SELF-INTEREST IN THE JEWISH WAR

In 70, during the Jewish War, the general and future emperor Titus threw a legionary cavalryman out of the army. He was one of two soldiers captured alive during the assault on the royal palace and temple in Jerusalem. The other was an infantryman whom the Jews killed immediately by cutting his throat and dragging his body through the city. The cavalryman faced the same fate as his fellow captive but claimed he had information which would help the Jews survive. Brought before the leader Simon bar Giora, it turned out he had nothing to say, so he was promptly handed over to a Jewish commander called Ardalas for punishment. Ardalas had the cavalryman bound and blindfolded and took him out where the Romans could see him, so that he could be publicly executed as an example. However, as the executioner was drawing his sword, the cavalryman broke away and dashed back to the Roman ranks. This put Titus in a difficult position. He could not execute the man himself, yet the trooper had disgraced himself by being captured alive in the first place. Titus therefore cashiered him. His arms were taken away and he was expelled from his legion, a fate generally considered worse than death.[59] His treatment was officially known as *missio ignominiosa*, 'dishonourable discharge', a sanction that could be applied to any soldier for committing a misdemeanour of similar significance.

During the same assault the Jews decided to use fire as a weapon against the Romans, who were building ramps up to the building. They lured Roman soldiers up to the west colonnade by pretending to withdraw, but had packed it with wood and pitch. The soldiers clambered up without an order being given, only to find themselves engulfed in flames. Many were killed, or jumped and broke their limbs. A few managed to dodge the fire and held out against the Jews, who picked them off one by one. The last man standing, a soldier called Longus, did as a Roman soldier was supposed to do under the circumstances. Another soldier, Artorius, turned out to be rather less heroic:

> At the last a young man among them, whose name was Longus, became a decoration to this sad affair. While every one of them

that died was worthy of a memorial, this man appeared to deserve it beyond all the rest. Now the Jews admired this man for his courage but were unable to kill him. They persuaded him to come down to them, promising him his life. From the other side his brother Cornelius persuaded him not to tarnish his own glory, or that of the Roman army. Longus followed this last advice. Lifting up his sword so both armies could see he killed himself. There was one called Artorius among those surrounded with the fire, who escaped by an underhand trick. For when he had with a loud voice called to him Lucius, one of his mess-mate fellow soldiers, and said to him, 'I will make you heir of all my possessions if you will come up and rescue me.' Lucius came running up to receive him readily. Artorius then threw himself down on Lucius, and saved his own life but his weight crushed Lucius on the stone pavement, who died immediately. This disaster demoralized the Romans for a while.[60]

## CONQUERING CALEDONIA

Around 150 years later, the emperor Septimius Severus embarked on a vast campaign into Caledonia in northern Britain. The land beyond Hadrian's Wall had proved an intractable issue ever since the earliest days of the province. Despite a short-lived attempt to build a new frontier (the Antonine Wall) beyond Hadrian's Wall, the Romans seem to have resigned themselves to Hadrian's frontier as marking the end of Britannia, apart from a hinterland beyond which some outpost forts were built. Severus, however, had other ideas. Severus was keen to toughen up his indolent and swaggering sons Caracalla and Geta and give them proper experience of a military campaign. He, his sons, his empress Julia Domna and a vast army, set out for Britain in 208.

Severus took care to do everything he could to ensure the campaign's success. He ordered the construction of supply forts and preparation of logistics and communications to service the war. As the contemporary historians Cassius Dio and Herodian reported, the plans soon started to go wrong. The Severan army in Britain 'experienced countless hardships in cutting down the forests, levelling the heights, filling the swamps, and bridging rivers', said Dio. Even worse, Severus could do nothing to bring the Caledonians to battle. They declined to play by the same rules, avoiding a confrontation and never organizing themselves into battle formation. Instead, the Caledonians left sheep and cattle out to attract the Romans and encourage them to advance ever further into the unknown. These were not enemy tactics the Roman army could cope with. The soldiers became trapped in the swamps, despite their efforts to use pontoons to provide a firmer footing. Rome was not the first superpower to be sucked into a war with an enemy that knew the land and was able to melt into the shadows, and nor would it be the last.

The Caledonians swam in the swamps virtually naked, carrying a small shield, a sword and a spear, and spurning the use of armour which they knew would prevent them from making headway through the marshes. 'The enemy finds it easy to escape and hide in the woods and marshes', said Herodian. Bogged down in every sense and disoriented by the mist which hung over the marshland, the Roman soldiers were easy meat for the Caledonians, who were of course in their element. The idea was that the Romans would end up killing their own incapacitated and injured so they could beat a hastier retreat. According to Cassius Dio, 50,000 Romans died in the guerrilla war, an implausibly high figure since it equates to ten legions, but it is clear the losses must still have been enormous.[61] In reality, Severus' soldiers faced an enemy that harried and persecuted the Roman columns as they marched into the highlands of what we call Scotland.

The war went from bad to worse. In 210 Severus was infuriated by yet another tribal revolt. Dio said that he not only ordered total annihilation but had the wit to do so by quoting (and adapting) Agamemnon's words to Menelaos in Homer's *Iliad*:

'Let no one escape sheer destruction at our hands,
Not even the babe in its mother's womb; if it be male,
Let it nevertheless not escape sheer destruction.'[62]

It was an optimistic instruction. The reality for the average legionary or auxiliary was a terrifying ordeal that made the war last longer than had ever been planned. The hostilities continued inconclusively until early 211, when Severus died of illness and old age at York. One of the unlucky men who participated in the war was Gaius Cesennius Senecio, a centurion serving with Cohors II Praetoria. According to his tombstone in Rome, his body (which presumably means his cremated ashes) was brought all the way back from Britain for burial.[63]

## THE MYSTERY OF LEGIO VIIII HISPANA

Some legions had very mixed stories, experiencing successes and disasters. A few disappeared completely, their fates shrouded in mystery. The strange case of Legio VIIII Hispana remains an enigma. Its history dated back at least to Augustus' time, when it served in Spain. Next it was relocated to Aquileia, at the head of the Adriatic Sea in north-eastern Italy, and then, by the time of Augustus' death in 14, to Pannonia, where it was garrisoned in one fortress with two other legions (VIII Augusta and XV Apollinaris). All three were involved that year in a dangerous mutiny about pay and conditions.[64] The legion remained in Pannonia for a generation, apart from being sent in 20–22 with Legio III Augusta to fight in Africa against the rebellion of Tacfarinas in Numidia. In 42 Aulus Plautius was governor of Pannonia; he seems to have brought the legion with him when he invaded Britain the next year. After its drubbing in the Boudican Revolt when it lost around 40 per cent at least of its manpower (see Chapter 8), it built two legionary fortresses, first at Lincoln and then at York. During that time, between *c.* 78 and 84 the legion played a major part in the governor Agricola's invasion of northern Britain; it was nearly destroyed when the Caledonians attacked it in the middle of the night,[65] until Agricola sent cavalry and infantry to save the day, which they did in the nick of time as the sun rose. Legio VIIII Hispana fought well and recovered its dignity. The legion went on to participate in the victory over the Caledonian tribes at Mons Graupius in 84 and was last attested in York on a major building inscription in around 107, during the reign of Trajan.[66]

Crucially, however, Legio VIIII Hispana was absent from the building of Hadrian's Wall 15 years later and seems to have been replaced by Legio VI Victrix, which Hadrian brought with him during his visit in about 119. One of the VIIII's commanding officers around this time, Titus Aninius Sextus Florentinus, went on to govern Gallia Narbonensis and Arabia, where he is known to have been in post in 127.[67] By 130 Florentinus was dead, buried at Petra in Arabia, where his tomb inscription records most of the major stages of his career.[68] Working backwards, he appears to have commanded VIIII Hispana around 120–2, although unfortunately we do not know where the legion was that year.

The mystery of what happened to VIIII Hispana endures to this day. It may have been destroyed in fighting in Britain at that period, going down in a blaze of heroic glory, but Florentinus' career suggests the legion was still in one piece around the time Hadrian arrived in Britain, and could even have lasted to take part in the Jewish revolt that broke out under Hadrian in 132–5. It may therefore have been transferred from Britain and lost somewhere else. Alternatively, given its slightly chequered career, it may have been cashiered after failing to live up to expectations. Either way, it was never heard of again.

Oddly, Legio XXII Deiotariana disappeared from the records around the same time. Last attested in 119, it is completely absent from any later papyri or documents in Egypt, where it had previously been recorded on numerous occasions. Like VIIII Hispana it may have been destroyed in Judaea, but no unequivocal evidence has yet been found for its presence there either.

*

The stories covered in this chapter show how poor leadership, bad or inappropriate planning, reckless ambition, the selfishness of an ordinary soldier or sheer bad luck could compromise the mighty Roman army. There are many other examples of campaigns and battles that went badly for the Romans. Fortunately for the Romans, resilience and resourcefulness on the part of many soldiers, and the timely presence of some of the greatest military leaders in history, more often than not brought astonishing success – sometimes in the aftermath of occasions when it must have seemed that all was lost.

# EIGHT

# I CAME, I SAW,
# I CONQUERED

## The Roman War Machine Victorious

*The legions . . . dashed forward in wedge-shaped formation.*
*The auxiliaries charged in the same way, and the cavalry with*
*extended spears broke through what was powerful and in the*
*way. The rest took flight, though escape was difficult . . .*

The heavily outnumbered Roman army
defeats the Boudican hordes in 61[1]

Despite the tales of epic defeats, the greatest prospect for many Roman soldiers was the chance to go on campaign, especially if that meant a war of conquest, with all the chances of glory and booty that might bring. It was also the most terrifying. This chapter traces some of Rome's most remarkable warriors in republican and imperial times: artillery experts, those who committed acts of remarkable bravery in the heat of battle or who lived to tell the tale and dine off their heroic acts for the rest of their lives. These were the men who helped define Rome's greatest military successes and slay the demons of past defeats. They also showed what superb training, discipline and well-maintained morale could achieve.

As Polybius described it, the Roman order of battle was almost impossible to break through. The Roman soldier could fight in it individually or collectively, with the result that a formation of troops could turn to offer a front in any direction. The individual soldier's confidence was strengthened by the quality of his weapons. The result was, he said, that in battle the Romans were 'very hard to beat'.[2]

Josephus was staggered by the Roman war machine in action during the Jewish War, fascinated by the way the Romans never laid down their arms yet always thought and planned before they acted. As a huge admirer of the Romans, like Polybius he painted a very compelling and biased picture of an invincible force. He saw Vespasian, the future emperor, set out on campaign to invade Galilee and described how the legions went to war. The auxiliaries attached to the legions were sent out ahead to scout for ambushes and fight off any enemy attacks. Behind them came the legionaries, with a detail of ten men from every century carrying the unit's equipment. Road engineers followed to take care of levelling the surface, straightening out bends and clearing trees. Behind them came the officers' baggage train, guarded by Vespasian's cavalry and his personal escort. The legion's cavalry was next, followed by any artillery, the officers and their personal bodyguards, the standards and the legionaries' personal servants and slaves, who brought their masters' effects. At the back came the mercenaries who had joined that campaign, and finally a rear-guard to protect the rear of the column.[3] The Roman army had reached this arrangement after centuries of experience that had also involved terrible defeats and lessons.

The great achievements were rarely commemorated at the site of battles or campaigns themselves, although to do so was not unique. Actium, unusually, had a monument at the location of the conflict. Trajan erected a memorial at Adamklissi (Tropaeum Traiani, 'the Trophy of Trajan') in Dacia in honour of his victory there in

107–8,[4] while fragments of an inscription found in Jarrow church in Northumberland in Britain evidently once belonged to a huge monument built under Hadrian's rule to commemorate the 'dispersal [of the barbarians]' and the construction of his Wall by 'the Army of the Province' of Britain.[5] But more often Roman military successes were honoured with triumphal parades and monuments in Rome, the latter usually in the form of an arch, like those of Augustus, Claudius, Titus, Septimius Severus and Constantine I, or the columns of Trajan and Marcus Aurelius. Another stood at the port of Richborough in Britain, serving as a gateway to the province and commemorating the completion of its conquest in *c.* 85 under Domitian during the governorship of Agricola.* There were many more in provincial cities throughout the Empire. Victories and conquest were a matter of Roman national prestige and the emperor's standing with the mob was of the highest importance. Few ordinary people were ever likely to travel to the sites of former battles, so there was little point in going to great lengths to build monuments there.

## Zama

No Roman general ever went to war without thinking about his celebrated forebears. In 202 BC, when Publius Cornelius Scipio was still only thirty-four years old, the fate of Rome hung in the balance. The Second Punic War had been dragging on since 218 BC. Scipio had carried a vast army across from Sicily to North Africa in 204 BC and had been slowly wearing the Carthaginians down ever since. The following year, a major defeat had cost the Carthaginians dear when Scipio attacked two of their camps near Utica. It was said that 40,000

---

* By the late third century the Richborough arch was serving as a lookout tower for seaborne pirates before being demolished when the site was turned into a large fortified coastal compound, one of a number on the coasts of Britain and Gaul.

Roman silver denarii. The silver denarius was the staple Roman bullion coin and used
to pay soldiers. Left to right: 1. First denarius issue, Roma with (rev) the Dioscuri, Second Punic
War 211–206 BC; 2. Vespasian (deified) by Titus, Victory with shield and Jewish captive, 80; 3. Trajan,
military trophy, 103–12; 4. Marcus Aurelius, Victory inscribing shield for the Parthian War, 166.
All struck at Rome. Diameter c. 18–19 mm.

Marcus Favonius Facilis, centurion of Legio
XX. Facilis is shown bare-headed as was
normal on tombstones. Colchester, Britain.
About 43–7.

Legio XV Apollinaris re-enactors. Roman legionary re-enactors of Legio XV Apollinaris on display at Pram in Austria. A centurion leads. In reality, equipment and armour probably varied much more among soldiers in a single unit than this image suggests. Photo: Matthias Kabel.

Auxiliary infantryman. A re-enactor portrays an auxiliary infantryman of Cohors V Gallorum.

Auxiliary cavalryman. Auxiliary trooper re-enactor photographed near the outpost auxiliary fort of Drumlanrig, Scotland. He wears mail armour, carries a shield and wears a spatha sword. Note the horse's decorative harness.

Auxiliary cavalry horse trapping. Auxiliary cavalrymen were particularly fond of elaborate display. This silver plaque is from the equipage of a cavalry horse in a unit commanded by Pliny the Elder. Found at Xanten. Mid-first century.

Auxiliary trooper's parade helmet. Auxiliary cavalrymen sometimes took part in battle demonstrations, re-enacting episodes from myth or Roman history. This example was found at Ribchester and was owned by a soldier with the Spanish name Caravius.

Legionary fortress of Legio XX, Chester. This model shows the fortress in its completed state together with its canabae civilian settlement and extramural amphitheatre. The canabae was probably much more densely packed than shown here.

Porta Praetoria (main gate) at the reconstructed auxiliary infantry fort at Saalburg, Hesse, Germany. The fort was built in the late first century but rebuilt later in stone. The modern reconstruction dates from the early twentieth century. Photo: Ekem.

Part of Legio III Augusta's principia (headquarters) at the legionary fortress, Lambaesis, Algeria. The structure is a good example of the army's ability to produce sophisticated masonry structures, even in remote places. Photo: Zinou2Go.

Luxor (Thebes), Egypt. During the fourth century AD a Roman fort was built in and around the temple courts, by then mostly over fifteen centuries old. A chapel dedicated to the imperial cult of the Tetrarchy emperors (Diocletian, Maximianus, Galerius, and Constantius I) was adapted out of a chapel originally used for the worship of the Egyptian goddess Mut. A fresco survives showing Roman officers from the garrison honouring the emperors in a way all garrisons had been doing since the reign of Augustus.

The reconstructed praetorium (commandant's house), South Shields. Although the original building was of late date (*c.* 300), this reconstruction gives a good idea of the sort of houses commanding officers occupied in forts from earlier times. The open courtyard and ambulatory recalled Mediterranean style houses, even on Britain's northern frontier.

Amphitheatre of Legio II Augusta, Caerleon, Wales. Large enough to seat the whole legion, the amphitheatre was constructed around 90 to provide a venue for parades, displays of military skills, and entertainments such as gladiatorial bouts. It is one of few such military buildings visible today.

Arch of Titus, Rome. Marble relief depicting Titus at the triumph in Rome in 71, described by Josephus. Titus is in a quadriga with Victory, and accompanied by soldiers bearing spears. Constructed in 80.

Legio I Italica. A stamped tile, probably from a *pila* in a heated room, manufactured by the legion and marked as its property. From Olpia Oescus, a base built by Trajan on the Danube, prior to his Dacian campaign *c.* 106. Diameter 180 mm.

Classis Britannica. Stamped *tegula* tile from the roof of the fleet baths at Beauport Park in the iron-working region in the Weald of Kent and Sussex, Britain. Second century.

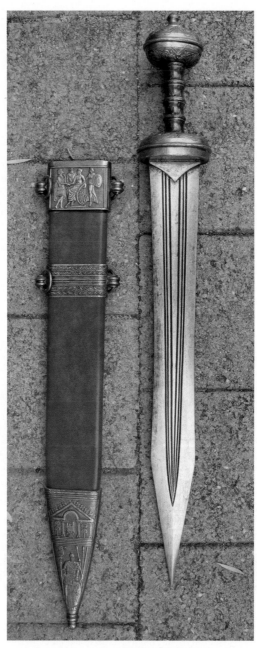

Modern replica of the *gladius* (short sword).

men, taken completely by surprise and unarmed, had been killed and 5,000 captured, as well as six elephants. Scipio celebrated the victory by dedicating the captured arms to Vulcan and then ordering them burned.[6] Polybius painted the picture of confusion, shouting, fear and raging fire caused by the assault and judged it to be 'the most spectacular and daring' of Scipio's attacks.[7]

The war, which Scipio had been ordered to bring to an end, was at this stage still far from over. During a storm shortly afterwards, a Carthaginian naval attack came close to wiping out his fleet. Sixty transports were seized by the Carthaginians and towed away.[8] A little while later three Carthaginian triremes attacked a quinquereme carrying Roman envoys. Although the envoys were rescued, a large number of Roman troops on the quinquereme were killed. This renewed Roman determination to finish the Carthaginians off.[9] When talks between Scipio and Hannibal broke down, fighting was inevitable. The stakes could not have been higher. Both Rome and Carthage were fighting for survival.[10]

The battle opened with a Carthaginian charge, heavily reliant on Hannibal's 80 elephants. This turned out to be a mistake. The animals were badly rattled by the noise of the Carthaginian trumpets, panicked and turned back to run into Hannibal's Numidian cavalry. Some of the frightened elephants reached Roman lines, causing serious casualties before being forced off the battlefield by Roman javelins. Gaius Laelius, Scipio's cavalry commander, took advantage of the opportunity to charge the Carthaginian cavalry and drive them into a retreat.[11] Only then did the battle descend into close combat as the rival infantry forces advanced towards each other. Thanks to Roman discipline and organization, their infantry formations held and were backed up by their comrades, despite a vicious assault by Hannibal's mercenaries. But the Carthaginian troops failed to support the mercenaries, who turned on the Carthaginians themselves. Only then did the Carthaginians start to show their mettle, fighting both mercenaries and Romans simultaneously, but the Romans managed to stand fast. Some of the Carthaginians fled from the battle, prevented by Hannibal from taking refuge with his veterans.[12]

Thus far the battle's confusion and the Carthaginians' problems had been largely self-inflicted. The Romans had done well but had not yet managed to take control. Scipio was furthermore prevented from attacking because of the sheer number of corpses and the quantity of debris and abandoned weapons in the way. He had the wounded carried off before ordering his men to reorganize themselves into formation by treading their way over the dead bodies. It was effectively a second battle.[13] Once they were in battle order they were able to advance on the Carthaginian infantry. The fighting proceeded inconclusively at first, since both sides were evenly matched; the attrition was only broken when the Roman cavalry returned from chasing away the Numidian horse and attacked Hannibal's men from the rear. Many were killed as they fought, others as they tried to escape. It was a decisive moment. The Carthaginians lost 20,000, it was said, compared to 1,500 Romans. The exact figures were academic, and were unknown anyway. The point was the difference.[14]

Hannibal had exhibited remarkable skill in how he had distributed his forces so as to counter the Romans' advantage. He had hoped the elephants would disrupt the Roman formation and cause confusion from the outset, planning that the opening assault by mercenary infantry would exhaust the Romans before the main confrontation with his best and most experienced troops, who would have saved their energy. Until then Hannibal had been undefeated. Polybius believed that a Roman victory only came this time because Scipio's conduct of the fight was better, yet his own description of the battle clearly described how luck had played a large part.[15] There can be no question that it was a brilliant victory, one for which Scipio deservedly took credit. But whether it was really the result of his generalship, or of happenstance in the chaos of battle, is a moot point.

Regardless, the Battle of Zama ended Carthage's role as a Mediterranean power and confirmed Rome's primacy in the region.

Not only did it earn Scipio immortality as one of the greatest Roman generals of all time but it also enhanced the reputation of the Roman army, as well as putting to bed the shame of Trasimene and Cannae. Scipio offered the Carthaginians remarkably moderate terms, based largely on the payment of reparations and the restriction on the numbers of their armed forces, though these had to be ratified by the Senate.[16]

Of the ordinary men who fought that day none is known to us by name, and nor are the anonymous feats of any individual. Even the celebrated Republican veteran Spurius Ligustinus did not enlist until two years after the battle. In 201 BC, after settling the peace, Scipio took his men home via Sicily for a triumph in which many must have participated, and carrying epic quantities of booty. How he acquired the name Africanus had been lost to history by Livy's time. Perhaps it was his men who gave it to him, or his friends, or even the mob – but he was the first Roman general to be named after a nation he had conquered, though none who came after, said Livy, were his equal. No wonder anecdotes about his skills, his views and his achievements were recounted for centuries.[17]

There was an amusing postscript to Zama. Some years later, in 192 BC, Scipio Africanus and Hannibal met in the city of Ephesus, on the Ionian coast of Asia (Turkey). Scipio was there as a member of a diplomatic delegation investigating the Seleucid king Antiochus III, Hannibal as the king's adviser. Allegedly they discussed generalship; Scipio asked Hannibal whom he regarded as the greatest general, privately hoping that Hannibal would name Scipio himself. Instead Hannibal gave first place to Alexander the Great and second to Pyrrhus. Scipio was sure Hannibal would name him third at least, but in fact Hannibal then named himself, citing his extraordinary march into Italy and the campaign that had followed. Scipio burst into laughter and asked Hannibal where he would have placed himself had he not been defeated at Zama. Hannibal said he would have been first, managing simultaneously to continue his self-flattery while implying that Scipio was greater than Alexander. The story is almost certainly fictional, but it added another to the range of tales and anecdotes about Scipio retold in later years.[18]

## MARIUS' OBSERVANT LIGURIAN

A single soldier's sharp eyes and quickness of wit could make all the difference at a crucial moment in a campaign. In the war against Jugurtha in North Africa (112–106 BC), the general Gaius Marius was engaged in the siege of a stronghold perched on a rocky outcrop that could only be approached from one direction down a narrow path. The track was far too narrow for siege engines to be moved up along it. On all the other sides there were steep precipices. The siege was starting to look impossible to maintain, not least because the stronghold was well stocked with food and even had a water supply from a spring. Marius began to believe he had made a serious mistake and considered giving up. But one of Marius' soldiers, an anonymous Ligurian, was out looking for water. He was also picking up snails for food, had climbed higher and higher towards the fortress up one of the precipitous slopes until he found himself near the stronghold. He climbed a large oak tree to get a better view and realized that by working his way through the tree and the rocks he had solved the problem of the Roman assault. He climbed back down, noting the exact path and every obstacle along the way, and went to Marius to tell him he had found a way up.

Instead of dismissing advice from an ordinary soldier Marius realized this might be the break he needed. He sent some of his men to confirm what the Ligurian had said. Based on their reports he was convinced and sent five of his nimblest troops, who were also trumpeters, led by four centurions up the incline again with the Ligurian. The men, who had left their helmets and boots behind so they could see where they were going and be as agile as possible, followed the Ligurian up the hillside through the rocks. To make the climb easier they strapped their swords and spears to their backs, and used straps and staffs to help them up. The Ligurian led the way, sometimes carrying the men's arms, and tying ropes to tree roots or rocks. When the trumpeters reached the rear of the fortress after their long and exhausting climb they found it undefended. No one inside had expected an attack from that direction.

In the meantime Marius was using long-range artillery to hit the fortress, but the defenders were not in the least concerned. They came out of the fortress accompanied by their women and children, who joined in as they taunted the Romans, convinced they were safe. At that moment the trumpeters at the rear of the fortress started up with their instruments. That was the signal to Marius to intensify his assault. The women and children fled at the sound of the trumpets, believing an attack from behind had taken place, and were soon followed by everyone else. The defence collapsed and Marius was able to press on and take the fortress, all thanks to the Ligurian.[19]


## Caesar in Britain

Sometimes soldiers were confronted with terrifying prospects simply for the purpose of gratifying the conceits and ambitions of their commanding officers, generals or emperors. When in 55 BC Julius Caesar began the first of his two invasions of Britain, he was the first Roman to attempt to do so. He had 80 ships built to carry two legions over the Channel from Gaul, and another 18 to bring the cavalry, but when his force arrived off the coast of Britain they were faced with cliffs that could not possibly be scaled. The ships had to be sailed 7 miles (11 km) further on so they could land on a beach.

Well aware of what was happening, the Britons positioned cavalry and charioteers along the coast to prevent the Romans getting ashore. It was already difficult enough for the invaders. Caesar's troop transports had to be beached in deep water, forcing the infantry to jump down into the water laden with their armour and weapons under a hail of missiles from the Britons. As a result the Romans became frightened and hesitant, not least because they had never experienced anything like it.

Caesar had to order his warships to move into position so his men could attack the Britons with artillery, arrows, and stones hurled from slings. 'This movement proved of great service to our troops,' he remembered. The Britons temporarily withdrew, but the Roman troops were still reluctant to risk all by jumping into the sea. Famously, at that moment 'the aquilifer of Legio X, after praying to heaven to bless the legion by his deed, shouted, "leap down, soldiers, unless you want to betray your eagle to the enemy. It shall be told for certain that I did my duty to my nation and my general".' Caesar's heroic aquilifer then jumped down from the beached transport into the foaming water and charged through the waves with his standard. The prospect of shame was too much for the others on the transport. They followed him, and one by one the men on the other transports followed suit.

Caesar went on to enjoy moderate success that year and the next, but the entire project had hung in the balance that day. His political career could have been destroyed by failure on that beach. The ignominy would have been too much to sustain, especially given the febrile politics of Rome at the time. One soldier had managed to turn the moment around in the nick of time.[20]

At least Caesar's standard-bearer had acted autonomously. Long before, in 386 BC, Marcus Furius Camillus, a military tribune, was also faced with his own troops holding back. He had physically to grab a signifer by the hand and lead him into the fray to get the others to follow, rather than be humiliated.[21]

## PHILIPPI: TRICKERY, BRAVERY AND SUICIDE

Playing tricks on the enemy was an excellent way of seizing the advantage, but it could have unforeseen consequences. In October

42 BC at Philippi, two years after the assassination of Julius Caesar, Antony and Octavian were determined to force Caesar's killers Brutus and Cassius into fighting a battle. In the end they fought two. Antony's idea was to have his men set up all their standards every day so it would look as if his entire army was in battle order, ready for the fight. It was a ruse. In reality an area of marshland lay between Antony and the enemy. Some of his troops were in the meantime working their way through the marshes to cut down reeds and build an earthen causeway embanked with stones, using piles to make their way across the deeper places. This all had to be done in complete silence, although Antony had the advantage that Brutus and Cassius' men were prevented by the reeds from spotting what was going on. After ten days Antony was able to send part of his army through to take up position and build redoubts (small fortifications) to reinforce their positions.

This of course gave the game away. Now Cassius knew what was happening. He was impressed and came up with the idea of secretly building a wall and palisade at right angles across Antony's causeway to cut off the advance force. It was an ingenious solution which enraged Antony when he discovered what had happened. He impulsively organized a charge, his men carrying tools and ladders, to bring the wall down and attack Cassius' camp. Meanwhile, Brutus' troops were watching, equally outraged 'at the insolence' of Antony's attack. They dived in without orders and killed as many of Antony's soldiers as possible, before turning on Octavian's army and causing it to run away, destroying Legio IIII in the process. Shortly afterwards they had Antony and Octavian's camp.

Antony kept up the attack in a reckless assault of exceptional bravery. He had Cassius' palisade torn down and its accompanying ditch filled up, killing the men on its gates and dashing forward under a hail of missiles. The men who had been working on the wall for Cassius were driven off and Antony headed for their camp. Cassius had failed to put more than a token guard on his camp, so Antony's men soon took it. The two sides both ended up in much the same position, but in the confusion neither was aware of what had happened. Antony had taken Cassius' camp, while Brutus had taken Octavian's. 'There was great slaughter on both sides', said Appian, Cassius losing 8,000 men and Octavian 16,000. In his shame Cassius ordered one of his freedmen to kill him.[22] The First Battle of Philippi was over.

On the same day, reinforcements were being brought from Italy to bolster Octavian and Antony's army against Brutus and Cassius. Domitius Calvinus was bringing two legions, including one known as Legio Martia, as well as 2,000 members of Octavian's personal praetorians, four cavalry wings and other unspecified troops. As they sailed across the Adriatic in troop transports they ran into an enemy naval force of 130 ships led by Lucius Domitius Ahenobarbus and Statius Murcus. The wind dropped and the transport ships were suddenly trapped. Ahenobarbus and Murcus sent their warships in and annihilated most of the transport ships. The beleaguered soldiers tried tying their ships together with ropes but Murcus ordered a hail of burning arrows to be fired, forcing the men to untie the ships again. Realizing there was no hope, the legionaries and other soldiers were furious at the thought of suffering pointless and humiliating deaths. Those of Legio Martia acted with particular bravery. Some took their own lives, others leapt across to the warships in suicidal attempts to fight back. In the event many men drowned or were washed up on the shore, while a large number capitulated and went over to the tyrannicides. But it was Legio Martia that was remembered that day.[23]

Meanwhile Octavian and Antony were in a bad way. The naval disaster was bad enough, but they were also running short of food and were desperate to force Brutus to fight. Brutus had no intention of obliging – he knew he had the upper hand – but his soldiers disagreed. They wanted a battle, believing that being kept back amounted to being 'idle and cowardly'. So did their officers, who thought the men were so whipped up that there was a good chance of victory. It was the officers who convinced Brutus he would have to fight, while Brutus was becoming worried that his men would go over to the enemy.

The battle, when it started, was a vicious close-combat affair with little in the way of missiles 'which are customary in war'. Instead the men killed each other with swords: 'The slaughter and the groans were terrible'. The dead were carried from the field to clear a space for reserves to march forward. When Octavian's men eventually pushed Brutus' army back into a steady, ordered and disciplined retreat which accelerated in speed, it was over. Brutus' army was routed but he escaped with four legions. The following day he ordered his friend Strato to kill him.

Later judgements varied. Valerius Maximus said Brutus had 'murdered his own virtues' before assassinating Caesar and that he had permanently associated his family name with 'abhorrence' (Lucius Junius Brutus, his ancestor, had been one of the most prominent men who threw out the last Roman king and established the Republic in 509 BC).[24] On the other hand, Appian called Brutus and Cassius 'two most noble and illustrious Romans' whose virtue was incomparable but for their one crime.[25]

The same could be said of many of their men, and those of Octavian and Antony, who fought two extraordinary battles exhibiting remarkable bravery, fortitude and discipline. Their victory set Octavian on the path that would lead in 27 BC to his becoming Augustus Caesar.

## CORBULO IN ARMENIA

Simply being on campaign could involve extraordinary levels of hardship, especially where remote territory and barbarians were part of the mix. In the reign of Nero two remarkable wars, only two years apart and on opposite sides of the Empire, found two of Rome's greatest generals facing their greatest challenges. Both overcame adversity in different contexts by relying on their leadership skills and the training and discipline of the men under their command.

In 59 the general Gnaeus Domitius Corbulo was leading the III Gallica, VI Ferrata and X Fretensis legions and their auxiliaries into Armenia to recover Rome's control over the region from the king Tiridates I. It was a bitter and piercing winter but the soldiers were camped out in tents made only of hide. To set the tents up the troops had had to cut through the ice that covered the ground and dig out pits into which they could erect them. The cold was so extreme that some died while standing out on watch. Tacitus alleged that 'one soldier was seen carrying a wood bundle whose hands were so deeply frozen that they stuck to his load and dropped off from the stumps of his arms'. Corbulo was forced to go about his men wearing only light clothing and without any head covering in order to bolster their spirits. He was not entirely successful. The climate was so severe and the duties so onerous that some deserted. As a result Corbulo instituted a zero-tolerance policy: any deserter, even a first-time offender, would automatically be executed. The desertions did not stop, but they were markedly reduced.[26]

Arrogance and a lack of discipline had nearly wrecked the campaign, but discipline and organization would later achieve great things. Corbulo decided that the legions would have to stay in camp until spring. He distributed his auxiliary troops in various strongholds and placed them under the command of a primus pilus called Paccius Orfitus. A man of Orfitus' experience, who had gained Corbulo's trust, ought to have been totally reliable. He turned out not to be. Keen to fight, he wrote to Corbulo to tell him that the enemy was being incautious and that the chances of a successful attack against the Armenians were therefore excellent. Corbulo to him to hold his fire and wait until reinforcements arrived. Orfitus took no notice. When auxiliary units turned up begging to be allowed to fight, he took them out to give battle. His defeat was total. Although he escaped, the remaining troops were completely demoralized and terrified. Corbulo was so angry he ordered Orfitus, the commanders of the auxiliary units and the men to camp outside the fortifications until the rest of the army pleaded for them to be let back in.[27]

As the campaign progressed, Corbulo became frustrated by what Tacitus called a 'roving enemy' that avoided either negotiating peace or facing the Romans on a battlefield. This was the sort of foe the Romans, who always preferred set-piece battles, hated. Corbulo had managed to fend off Armenian attacks on his supply routes by placing forts at key spots. He decided that he would have to bring the war to the Armenian strongholds. The climax was to be an assault on the city of Artaxata (modern Artashat). Corbulo's legions were unable to cross the river Araks by the nearest bridge because it was so close to the city they could be hit by Tiridates' defenders, so they had to take a circuitous route via a ford some distance away.

But Tiridates too had a problem. The city was difficult to defend, and any attempt to prevent the Romans blockading Artaxata would result in his cavalry foundering on impassable ground. He decided therefore that he would have to present Corbulo with the opportunity for a battle. That would either mean confronting the Romans or pretending to retreat and tricking them into pursuit. Corbulo had organized the legions so that they could either march or fight. Tiridates surrounded the Roman column, which was arranged with the best of Legio X's soldiers in the middle, the VI on the left and the III on the right, with 1,000 cavalry at the rear who had been given strict instructions not to be lured into chasing an Armenian retreat. Infantry archers and more cavalry were on the edges of the columns; these were presumably all auxiliaries.

Tiridates' tactics were to harry the Romans in an attempt to break up the formation. He achieved nothing. The Romans were, ironically, helped when a cavalry officer, advancing too far on his own, was 'transfixed by arrows' fired by the Armenians. That focused the attention of the other Roman soldiers on staying where they were. The day's fighting was ultimately inconclusive and Corbulo ordered his men to make camp where they were. He toyed with the idea of blockading Artaxata that night, believing that Tiridates was holed up there. However, his own scouts discovered that Tiridates had set out on a longer trip elsewhere. In the morning, when Corbulo surrounded Artaxata, the inhabitants realized their best chance was to surrender and save themselves. They opened the gates and let the Romans in. Knowing he could not commit enough troops to hold it, Corbulo ordered the city to be burned and razed to the ground.[28]

Corbulo's Armenian war was far from over. In 60 he headed for the capital Tigranocerta. Progress was a struggle because of the heat and a shortage of water, and Corbulo himself was nearly assassinated by an armed Armenian. Fortunately, when he reached Tigranocerta the city surrendered and he was able to place a Roman client (and puppet) king, Tigranes VI, on the Armenian throne.[29] Hostilities were to resume only two years later when the Parthians arrived to attack the city, but for the moment Corbulo's remarkable campaign had shown what the Roman army could achieve – and, in this instance, without ever fighting a pitched battle.

## SUETONIUS PAULINUS AND THE BOUDICAN WAR

At almost exactly the same point of Nero's reign another war broke out on the other side of the Roman Empire. Also involving several Roman legions, auxiliary forces, and a very experienced general, it came close to total disaster. The difference this time was that the Romans were caught unawares by a major rebellion when their garrison was widely dispersed. The setting was Britain, seventeen years after the conquest had begun in 43. Four legions – II Augusta, VIIII Hispana, XIIII Gemina and XX – were stationed in Britain at the time, as were at least the same number of auxiliaries, made up of infantry and cavalry.[30] The legions were distributed around the province, such as it was by that early date: Legio II Augusta was at Exeter, VIIII Hispana at Lincoln, XIIII at Wroxeter and XX at Usk in south Wales. They and the auxiliaries amounted to one of the greatest concentrations of Roman forces in the entire Empire at any time. The distances between locations were not great, but Britain's undeveloped state, its forests and numerous rivers, made it difficult to move fast, even though a road network radiating out from the new trading settlement of London was well under construction.

The governor, Suetonius Paulinus, who arrived in Britain or around the year 59, was determined to crush resistance once and for all. A highly experienced general, he was considered to be a rival of Corbulo's, both in terms of his military skill and his popular reputation. Paulinus identified the source of the problem as the native Druid priesthood, to whom all the tribal leaders deferred, and who were provoking rebellions and risings from their headquarters on the island of Anglesey just off the north-west coast of Wales. To make the short and shallow crossing from the mainland, the Roman infantry needed flat-bottomed boats and fords had to be found for the cavalry.[31] Doubtless Paulinus thought the campaign would be quick, brutal and easy.

Sailing across to the island was one thing. None of the Roman soldiers was, it seems, ready for what confronted them. They found armed warriors waiting, and mixed among them women dressed in black as furies, running around with torches, their hair streaming behind them. The Druids were also present, raising their hands to the sky and chanting incantations. This was so far outside the Romans' experience, either hitherto in Britain or on the Continent, that they were paralysed with fear. Paulinus had to urge them on by pointing out that they should not be scared of 'women and fanatics'. The Romans pulled themselves together and advanced. For all the Britons' noise and dishevelled hair they were hopelessly outclassed. The Romans easily cut down the warriors, the Druids and the women, and used the Britons' own torches to set them on fire. The island was garrisoned and 'the groves sacred to their savage rites cut down'. What seems to have provoked the Romans' disgust was the discovery that the Druids had indulged in human sacrifice of their prisoners in order to use the entrails to communicate with their gods. If the Romans thought they had exterminated the problem, they were wrong. Dispatches from the other side of Britain soon brought news of a major rising in the east of the province.[32]

The Iceni tribe of East Anglia had always been a serious problem for the Romans, but relations had settled down under the client king Prasutagus, who had come to an accord with the Roman government of Britain. Prasutagus knew that the Romans were more powerful and believed that if he made Nero, along with his daughters, his heir, his kingdom would be safe after his death. He was wrong. Centurions had been placed in charge of the region's civilian administration, performing tasks such as policing and the collection of tribute. Now these men spotted an opportunity to cash in when Prasutagus died, probably in 59. With the help of imperial slaves, they started ransacking the tribal lands, stealing estates and plundering property 'as if they were spoils of war'. For good measure they flogged Prasutagus' wife Boudica and raped her daughters, treating the family's relatives as if they were slaves. The Iceni, not surprisingly, rose up in rebellion, almost certainly with backing from the Druids.

The Iceni were joined by a neighbouring tribe called the Trinovantes, whose ancestral lands were in the vicinity of a new colony at Colchester, formerly occupied by Legio XX. The soldiers had been moved out around 47 and the colony established among the remains of the short-lived legionary fortress. With the typical and tactless arrogance of an invading army that looked down on the defeated as lesser beings, the veterans had enthusiastically helped themselves to Trinovantian land, urged on by serving soldiers who were hoping to do the same when they were discharged. The Trinovantes were already under pressure. Before the Roman invasion another tribe, the Catuvellauni, had taken control of their territory. After the invasion some of the Trinovantes had also been forced to spend their estates on funding compulsory priesthoods in the cult of the deified Claudius, founded at Colchester after his death in 54, while others may have done so voluntarily in the belief they would gain an advantage under Roman rule. Either way, it appears that some of them had borrowed money from Roman speculators to finance these positions. When the loans had been abruptly called in, some of the senior tribesmen faced ruin.

It was not entirely surprising that the veterans behaved as they did. Roman legionaries had been encouraged to believe they were entitled to such privileges since the time of Augustus. Neither had the complacent colonists bothered to construct any new defences, an extraordinary oversight for military veterans for which they would pay dearly.[33]

The notorious attack on Colchester soon followed. Knowing that Paulinus was too far away to help, and with time running out, the desperate colonists sent an emergency dispatch to the procurator of Britain, Catus Decianus, who was probably in London. His response turned out to be hopeless. He sent about 200 ill-equipped men to join the small unit of serving troops still based in Colchester. A few hundred Roman soldiers found themselves confronting a horde of tribal warriors that numbered thousands, without the slightest hope of relief. The soldiers, veterans, and colonists barricaded themselves inside the temple of Claudius, unable even to build a defensive ditch around the building and hoping that the structure would be enough to protect them. It was not. The Iceni and Trinovantes first burned the settlement to the ground and then turned their attention to the temple. The building and its terrified defenders held out for two days but the end was inevitable. The entire population of the colony was massacred.[34]

Enough time had passed for news of the emergency to reach Petilius Cerealis, then commanding Legio VIIII Hispana somewhere near Lincoln to the north. Cerealis headed south-east to find the rebels, but there were so many that thousands of his infantry were wiped out on the spot. Only he and his cavalry escaped. With one legion effectively incapacitated, the effective garrison of Britain had suddenly been cut by around a quarter, while one of the gravest crises in Roman provincial and military history was still taking shape. With an eye to his own survival, Catus Decianus abandoned Britain and headed for Gaul.[35] If the revolt really did involve around 120,000 rebels, as Dio claimed, then he could be forgiven for believing all was lost.[36]

Paulinus immediately abandoned the campaign in Anglesey when he heard the news which must have been brought to him by mounted military messengers. He had with him Legio XIIII Gemina and all or part of Legio XX as he headed down Watling St towards London in an effort to cut off the rebels. He sent orders to Legio II Augusta in Exeter to join him. London had no official status at that date but it was a major river crossing over the Thames as well as a road junction, and was rapidly developing into an important commercial centre. Paulinus seems to have hastened ahead to find out what was going on. He realized he had no choice but to abandon London and St Albans (Verulamium: the next city along Watling St as he retreated to the north-west) and their inhabitants to their fate, apart from those who were mobile enough to join him. The rebels were focusing their attention on killing as many people as possible and on amassing loot.* It was a fatal error. Their indulgences were beginning to slow them down, when it had been speed that had given them the initial advantage.[37]

Paulinus had, according to Tacitus, around 10,000 men. Together with Legio XIIII and part of Legio XX he also had some auxiliaries, all of which had been with him in Anglesey. These men had been following on behind him when he took his advance mounted force to London. Having abandoned London and St Albans, Paulinus met up with them somewhere in the Midlands and prepared for a final battle. In order to compensate for his lack of numbers, he chose to station his forces in a narrow gap with higher ground on both sides, while a wood to the rear would inhibit the Britons' chances of ambushing him. His men faced out across an open plain where the battle could be fought, with the cavalry on the edges. The Britons swaggered around, buoyed up with confidence and weighed down with loot.

---

* See Chapter 9 for the extraordinary tales of extreme violence that allegedly occurred.

They brought their women with them and placed them and their booty-packed wagons around the edge of the plain. Remarkably it is only at this point in Tacitus' account that Boudica appears as the leader of the Britons, rallying her army from her chariot with her daughters urging her fighters to vengeance which would have to be achieved if they were to avoid enslavement.[38] Dio, however, said she led and directed the whole war.[39]

Paulinus started the battle by ordering a launch of javelins, followed by a steady advance in wedge formation. If Tacitus can be believed, the Britons lost control almost immediately and started to beat a hasty retreat, only to run into their own wagons which prevented them escaping. The Romans had the day from that moment on, mowing down both warriors and women, and killing the baggage animals so the survivors could not dash away. The losses were colossal, though the figure of 80,000 Britons allegedly killed by the Romans is implausibly high; only 400 Romans were said to have died, with a few more wounded. This figure is a little more believable but neither total should be taken literally. They were supplied to provide the impression of a massive Roman victory, which indeed it was. Nevertheless, the battle was close-run, and even closer-run because Legio II Augusta had not joined in. The legion appears to have had no commanding officer at the time, and instead was in the charge of its praefectus castrorum, Poenius Postumus. Postumus had refused to march the legion from its base to join the war, probably out of fear. He committed suicide as the only reasonable course of action open to him, a humiliating end to what must have been a significant and successful career.[40]

In the aftermath, said Tacitus, 2,000 legionaries were sent over from Germany, along with eight auxiliary cohorts and 1,000 cavalry. These helped bring Legio VIIII back up to strength so it could participate in a punitive campaign to punish the other tribes and crush any further resistance. Evidently it had not been wiped out as Tacitus had claimed earlier.[41] Legio XIIII Gemina was awarded the title Martia Victrix and strutted into the future bearing that name for all time.

Its 'men had covered themselves with glory by crushing the rebellion in Britain', said Tacitus. Nero decided they were his best troops.[42] A few years later the legion would leave Britain and play a major part in the civil war of 69, returning briefly before being permanently reassigned in 70 to bases on the Continent. Legio XX may have been given the name Valeria Victrix on this occasion, though that is less certain. The sad fact is that virtually nothing is known for certain of the men who served in Legio XIIII at this time. The few tombstones that survive at the legion's Wroxeter base all seem to precede the Boudican War, commemorating men carried off by death before the legion's moment of glory (see Epilogue).

## IN THE FIELD

One of the great strengths of the Roman army was the range of available troops, but circumstances dictated which were best in any given situation. Sarmatian cavalry had a reputation for being among the most dangerous and terrifying troops; when they were on their horses they were virtually invincible, unless they had been victorious and were encumbered with booty. 'Scarcely a line of battle can stand up to them,' said Tacitus. But when they had to fight on foot they were 'utterly ineffective'. On one especially vile day in early 69 during the Civil War, 9,000 Sarmatian cavalry of the Rhoxolani tribe took advantage of the Roman civil war by invading Moesia. They were set upon by the experienced Legio III Gallica and its auxiliaries, then fighting for Otho. The frozen ground was becoming soft under a sudden thaw and it was raining. The enemy horses lost their footing. Forced to dismount, the Sarmatians were hopelessly weighed down by the leather body armour which made it impossible to wield their two-handed swords and lances. Trudging through the melting snow they were easy meat for Legio III, especially as the Sarmatians did not use shields. Most were cut down by javelins or by swords, while the rest died from their wounds or as a result of the severe winter weather.[43] In this instance the Roman infantry had turned out to have an enormous advantage over a very dangerous force that might in different conditions have cut them to pieces.

In his account of the Civil War of 68–9, Tacitus provides us with some of the most specific information about how a Roman army might be formed at any given moment. In one example, in the lead-up to the First Battle of Bedriacum (Cremona) in early 69, Otho's army was made up as follows as it marched towards his rival Vitellius' forces:

> A vexillation of Legio XIII, with four auxiliary cohorts and 500 cavalry, were placed on the left. Three praetorian cohorts in narrow formation held the high road. Legio I advanced with two auxiliary cohorts and 500 cavalry. In addition, 1,000 prae-torian cavalry and auxiliaries accompanied them to add force if they won and to act as a reserve if they were in difficulties.[44]

The opening skirmishes showed how unpredictable events could be. Some of the Vitellian forces were able to rush for cover in a vine-yard, where the trellises made it extremely difficult for Otho's men to attack them. They hid in a nearby wood, from which they were able to ambush Otho's Praetorian cavalry and kill most of them. But the Othonian counter-attack, when it came, turned out to be a success. The Vitellian soldiers had not all been brought onto the battlefield at once; a number had been left in the camp and they mutinied. After the mutiny was suppressed and the forces had regrouped, however, battle was joined: Otho's army was defeated and around 40,000 men were killed.[45]

## LOGISTICS AND PLANNING

There was no sense in running into a fight without first taking pre-cautions. Special tactics in the field, and opportunism, could work wonders. From early on the Romans had shown their capacity for brilliant pragmatic solutions in the field. During the Second Punic War in 213 BC the Roman general Fabius Maximus found the Italian city of Arpi occupied by Hannibal's Carthaginian forces. Taking advantage of the noise of pouring rain, he sent 600 soldiers with ladders to scale the city walls at points where they were most heavily fortified and therefore least well guarded. Hannibal's troops failed to hear the Romans, who were able to enter and break down the gates while Maximus attacked another part of the defences. The city soon fell.[46]

Later, in the Numantine War in Spain in 143–142 BC, Metellus Macedonicus ordered his men to divert a river, having spotted that there was an enemy fort on lower ground. When the river suddenly flooded the enemy camp, the enemy soldiers fled in panic straight into Metellus' soldiers, who had been lying in wait to ambush them.[47]

A marching army built whatever it needed as it went along. Quite apart from the camps (see Chapter 5), it also had to build roads and bridges to make sure the supply chain could be maintained. The most famous bridge in Roman military history was Caesar's bridge over the Rhine, which he ordered to be constructed in 55 BC. Claiming that he had been invited to help the Ubi tribe in their conflict against the Suebi, Caesar decided it was 'unworthy of his own and Roman dignity' to cross the river in boats. Nevertheless, the challenge of spanning such a deep, wide and fast-running river was obviously considerable, and daunting.

The bridge Caesar built was a remarkable example of Roman military engineering. It was laid on pairs of timber baulks which were dropped into two parallel rows across the river, separated by the width of the intended roadway. The sides slanted in towards the roadway and were joined across the gap with transoms, braced underneath. The idea, as Caesar proudly claimed, was that the force of the water would actually push the timber more firmly together. Each opposed pair of baulks with its transoms and braces made a single trestle; these were joined together across the river and a roadway laid down on top. Finally Caesar ordered piles to be driven into the river upstream, so that if the enemy tried to throw tree trunks into the river to wreck the bridge these would be prevented from being carried further. Caesar claimed that the entire project took only ten days to complete from the moment the collection of wood began. His description is so precise that it is hard to dispute his version of events.

As soon as the bridge was finished, Caesar's army marched over, leaving a garrison at either end.[48] The Suebi were sufficiently intimidated to withdraw from all their settlements and prepared to fight a pitched battle. But Caesar said he had done all he needed to. He had 'struck terror into the Germans' and saved the Ubi. He pulled back over the bridge after eighteen days, avoiding any further fighting or a major showdown, and ordered it to be destroyed.[49]

Caesar's achievement sounds remarkable, but perhaps it was not so unusual. The Roman army did such things all the time. In the year 90, Legio III Cyrenaica built a bridge at Koptos in Egypt in the name of Domitian, the work being carried out by Quintus Licinius Ancotius Proculus, the praefectus castrorum, and Lucius Antistius Asiaticus, the prefect of Berenice, under the care of the centurion Gaius Julius Magnus.[50] It was just one of the countless bridges built by the army for its own use and that of civilians. Over a century afterwards, Cassius Dio said that 'rivers are bridged by the Romans with the greatest ease, because the soldiers are always practising bridge-building, which is carried on like any other warlike exercise',[51] although Vegetius later recommended though that learning to swim was essential for soldiers because some rivers were unbridgeable: a pursuing or fleeing army might need to cross one in haste.[52]

Caesar's imaginative solutions to logistical problems reached a particularly revolting height at Munda in 45 BC during his Spanish campaign. Short of timber when he needed to build a rampart, he had plenty of enemy corpses:

Shields and javelins taken from among the weapons of the enemy were placed to serve as a palisade, dead bodies as a rampart. On top, decapitated human heads, impaled on swords, were set out in a row facing the town, the purpose being not only to surround the enemy with a palisade, but also to give him an awe-inspiring spectacle by displaying before him this evidence of valour.[53]

In 57 BC Caesar was assaulting the fortified stronghold of the Gaulish Aduatuci tribe. To begin with his soldiers were fought off. He therefore ordered his soldiers to start building siege machines. The Roman troops cut down trees, prepared the timber on the spot and assembled the machines before the enemy's eyes. This single occasion shows the Roman army's astonishing ability to create the equipment it needed to outclass an enemy from what was available in the immediate vicinity. Moreover, the men involved almost certainly knew what to do without having to resort to manuals. But such books certainly existed. Under Augustus, not many years later, Vitruvius wrote a treatise on architecture that included a section on how to build siege equipment such as mobile towers and the 'ram tortoise', following one devoted to the construction of artillery. He provided instructions and measurements of the components.[54]

Not having the slightest idea what they were, the Aduatuci made fun of the sight of Caesar's siege machines as they were manufactured. The smiles were wiped off their faces when they saw heavily armed Roman soldiers advancing towards the Aduatuci fortifications in the devices they had constructed. The result was panic. Attempts to appease the Romans followed, accompanied with offers of provisions. It was a trick. Waiting until they saw the siege machines standing idle and unmanned, the Aduatuci attacked the Roman forces at night. But Caesar was waiting. The Aduatuci were beaten and the whole population sold into slavery.[55]

## HEROES

Great awards awaited soldiers who pulled off remarkable feats. Spurius Ligustinus, a loyal old soldier of the Republic, proudly told his fellows in speech in 171 BC how he had been decorated thirty-four times and received the *corona civica* ('civic crown') six times for saving the life of fellow Roman citizens.[56] Such honours did, however, depend on the man's background, and on who was giving out the decorations. In around 46 BC, during the Civil War an opponent of Caesar's called Caecilius Metellus Scipio was handing out awards to soldiers who had excelled themselves. His associate, Titus Labienus, recommended that one particularly brave cavalryman deserved a gift of gold bracelets. Metellus Scipio declined on the grounds that the man had recently been a slave, and that the award would thus be degraded by giving it to someone of such lowly origins. When Labienus took gold from the booty captured in Gaul and gave it to the cavalryman concerned, Metellus was not to be outdone and said 'you will have the gift of a rich man'. Enticed by this, the trooper threw the gold back at Labienus. Metellus then said he would give the cavalryman silver bracelets; the soldier was delighted, preferring the glory of a decoration awarded by the commander to the intrinsic value of the gold.[57]

In the reign of Nero, one old soldier's brilliant career was set down in stone. Marcus Vettius Valens was a successful Praetorian guardsman who had the opportunity to travel all the way from Rome to take part in the invasion of Britain in 43, while serving as a beneficiarius on the personal staff of the praetorian prefect. His career was recorded in an inscription set up at Rimini in Italy in 66, during his later life, proudly proclaiming that in Britain he had been awarded necklets, armlets and medals for his achievements. He seems to have stayed on, despite reaching the end of his 16-year service term, winning a gold crown too. Later in his career he rose to the rank of centurion in Legio XIIII, not long after its success in the war against Boudica in 60–1. He must therefore have met some of the legionaries who fought in the final battle that destroyed the most serious provincial rebellion in Britain's history.[58]

Gaius Velius Rufus, primus pilus of Legio XII Fulminata, was awarded the *corona vallaris*, the 'rampart crown', along with collars, medals and armlets, by Vespasian and Titus for being the first man over the walls during the Jewish War of 66–70, though this did nothing to repair the legion's tarnished reputation. Nearly two decades later Velius Rufus was decorated again for his part in sieges during the war against Central European tribes including the Marcomanni. His remarkable career, which included other honours and military commands, was commemorated on an inscription set up at Baalbek in Syria.[59]

During the assault on the temple in Jerusalem towards the end of the Jewish War several Roman soldiers individually performed remarkable deeds. Pedanius was an auxiliary cavalryman in pursuit of retreating Jewish fighters who had attacked the Roman camp, He leaned down as he rode into them and managed to grab a fully armoured soldier and pull him up. With the prisoner grasped in his hand, Pedanius presented him to Titus, who ordered the captive's execution.[60]

The Jews were able to field their own heroes. A man called Jonathan once presented himself in front of the Romans and challenged them to single combat. One of the other auxiliary cavalrymen, named Poudes, stepped out and ran to fight Jonathan until he fell and was killed. The triumphant Jonathan mocked the Romans, swaggering over his kill, until an arrow fired by a centurion called Priscus killed him too.[61]

Another hero of the Jewish War was a Syrian auxiliary called Sabinus. He was so small and lean it was a surprise to everyone that he was a solder. His size was completely out of proportion to the heroism he showed. After the Jewish stronghold of Antonia was attacked, Titus exhorted the army to face the challenge ahead, pointing out that the first man to scale the wall – if he survived – would be 'envied by others' thanks to the rewards that he, Titus, would give him. Sabinus was the only one to speak up:

> 'I readily surrender up myself to you, Caesar. I will ascend the wall first. And I heartily wish that your fortune may follow my courage, and my resolution. And if some ill fortune grudge me the success of my undertaking, take notice, that my ill success will not be unexpected; but that I choose death voluntarily for your sake.'[62]

Eleven men were inspired to follow Sabinus as he led the way, his shield over his head, while missiles were fired at them and stones rolled down the slope. Three of the eleven were knocked over and killed but Sabinus kept going. He reached the top of the wall and maintained his attack, to the amazement of the Jews. Eventually he was isolated; having fallen over, he struggled back up, held his shield aloft and fended off his attackers, continuing to wound some of them until finally he was overwhelmed and killed. The other eight, all wounded, were rescued and carried back to the Roman camp.

## TRIUMPHS

The tradition of holding triumphs was well established in the Republic, and especially so during the period of heavy fighting following the Second Punic War. When a triumph was awarded, a major public celebration followed in which the general, now hailed imperator, decorated the best performing and bravest soldiers, handing out the booty to his men and sometimes also to the general public, any surplus paying for public works. Then the general would ride into Rome in his chariot with his family on horseback, followed by the rest of the army on foot. The climax was the execution of prisoners and religious ceremonies of thanksgiving at the temple of Jupiter Optimus Maximus on the Capitoline Hill.[63]

Thirty-nine such events were celebrated between 200 and 167 BC, an average of more than one a year, and a further 46 from then till 91 BC.[64] The consul Gnaeus Manlius Vulso was one recipient. He committed the outrageous act of going to war on his own accord on the Galatian Gauls in Asia so that he had the chance of a prestigious victory to brag about. And that was exactly what he got. Though he was censured for his actions, Vulso's opponents soon relented when he was voted a triumph by the Senate in 187 BC and was able to parade vast quantities of gold and silver through Rome while his soldiers sang songs in his praise.[65]

In 168 BC Aemilius Paullus held a three-day triumph to celebrate his defeat of Perseus of Macedon at Pydna that year. It was a particularly pointed gesture since he was the son of the consul of the same name killed at Cannae: he had restored his family's reputation. Enormous quantities of gold and silver seized from the defeated king were paraded before the Romans, after which Perseus' children and their attendants were forced to walk along and hold out their hands in supplication, followed by the bewildered and humiliated Perseus himself. The climax was the sight of Aemilius in his chariot, accompanied by his army, the soldiers singing traditional ballads interspersed with bouts of mockery, as well as songs celebrating their victory and Aemilius' leadership.[66] One can easily imagine the raucous and noisy exultation of Aemilius' troops as they pranced through Rome boasting of their achievements.

Some generals had an eye for a more permanent monument. In 119 BC the moneyer Marcus Furius issued an unusual silver denarius. Instead of the customary bust of Roma, the coin bore the head of Janus, and on the reverse a figure of Roma crowning a trophy. It commemorated a victory won over the Gaulish Allobroges and Arverni tribes in 121 BC by the consuls Gnaeus Domitius Ahenobarbus (consul in 122 BC) and Quintus Fabius Maximus (121 BC), the latter completing the campaign the following year and being awarded a triumph and the name Allobrogicus.[67] Out of the booty Fabius Maximus paid for a triumphal arch known as the Fornix Fabianus to span the Via Sacra in the Roman forum.[68] Fragments of the arch are still visible in the forum today, close to the remains of the later temple of the deified Julius Caesar. It was an interesting turnaround for Fabius Maximus. As a youth he had been considered the 'most disreputable' member of his generation, yet military success transformed his status. He ended up a 'most distinguished and respectable' old man.[69]

The prospect of a triumph was so attractive that every Roman general was desperate to have one to add to his own and his family's glory. In the aftermath of the Second Punic War so many were staged that eventually, at some point between 180 and 143 BC, a law was passed to place a minimum requirement that 5,000 enemy soldiers be killed before a triumph could be voted.[70] Triumphs were also expensive – hugely so. No wonder Suetonius said that Caesar paid for his with 'barefaced pillage'.[71] Some prisoners made a great show of their capitulation in advance of these great events. Vercingetorix, the Gaulish king who had led the war against Caesar's conquest of Gaul, surrendered at Alesia. In 46 BC, after 'putting on his most beautiful armour and decorating his horse [he] rode out through the gate. He made a circuit around Caesar, who remained seated, and leaped down from his horse, stripping off his suit of armour', well aware that because of his height 'he made an extremely imposing figure'. Vercingetorix then sat quite still beside Caesar until he was led off to await the triumph in Rome, where he was publicly executed.[72]

The Gaulish triumph turned out to be the first of five held by the vainglorious Caesar, four of them in little more than a month. The occasion was almost a disaster because he nearly fell from his chariot after the axle broke. Caesar's triumph celebrating his defeat of Pharnaces II of Pontus climaxed with 40 elephants carrying lamps to light his way, and his famously alliterative showcase inscription *veni, vidi, vici* ('I came, I saw, I conquered') which was displayed during the procession.[73] During the triumph to celebrate the war in Gaul one of the most memorable moments came when his soldiers burst into song:

> All the Gauls did Caesar vanquish, Nicomedes
>     vanquished him;
> Lo! now Caesar rides in triumph, victor over all the Gauls,
> Nicomedes does not triumph, who subdued the
>     conqueror.[74]

This was a salacious reference to rumours that as a young man Caesar had had a sexual relationship with Nicomedes IV of Bithynia. Such jibes seem to have been a common occurrence during triumphs (see below).

One of the greatest descriptions of a Roman triumph comes from Josephus, who saw the festivities put on by the emperor Vespasian and his son Titus in Rome in 71 to commemorate the Jewish War after Titus' return to the capital. Josephus depicts a scene like that from an epic motion picture, and the comparison is not inappropriate. The occasion was designed to be as magnificent and as theatrical a spectacle as possible. The two men spent the night in the temple of Isis, near the upper palace on the Palatine Hill. In the morning they donned robes of purple silk and laurel wreaths before going out to be greeted by the senators and the equestrians. The two men were seated on ivory chairs as the army hailed them, before Vespasian raised his hand as the sign for them to fall silent. He drew a robe over his head in the manner of a priest and said the appropriate prayers, followed by Titus. Vespasian made a speech – which must realistically have been almost inaudible to the assembled troops – and sent the men off to a celebratory breakfast. He and Titus then left to have breakfast themselves before changing into triumphal robes, and made sacrifices to the gods. Next they walked through the theatres of Marcellus, Balbus and Pompey in the Field of Mars so that the crowds gathered in the buildings could get a better look at them.[75]

The proceedings had of course barely started. Josephus said it was difficult to find words that could possibly describe the magnificence on display. He saw works of art and all manner of riches that provided a graphic illustration of the power and wealth of the Roman Empire. Given that, only a few years previously, Rome had been badly damaged by the fighting during the Civil War, the visual impact must have been even greater. Josephus compared the parade to a flowing river of floats laden with gold, silver and ivory as jewels, artefacts and statues of gods were carried past. As Vespasian and Titus had started the day wearing purple robes, so did purple feature prominently because of the enormous expense involved in antiquity in manufacturing the dye. There were tapestries of purple, and the men driving the animals in the parade wore purple uniforms. Even the prisoners had been dressed in expensive clothing, with the distasteful purpose of diverting attention from their injuries.[76]

Josephus was more intrigued by the way the floats had been constructed. Some had four levels, one above another, and curtains with gold and ivory fittings. The purpose was to provide a mobile depiction of the progress of the war. In this respect it too was almost cinematic, recreating in visual form the sort of scenes and events more normally experienced by ordinary Romans in the epic poetry of Virgil and Homer. These included battles, images of ravaged enemy territory and of people being captured and imprisoned, cities under siege, the Roman army bursting in through the walls and massacring the inhabitants, the destruction of houses and temples, and rivers flowing through a devastated land on fire. The crowd would have recognized instantly the echoes of the imagery of Rome's past wars, climaxing in Augustus' victory over Antony and Cleopatra and ending on civil war, on the mythical shield fashioned by Vulcan for Aeneas and described in book 8 of Virgil's Aeneid.

In the contemporary setting, the triumph was designed to celebrate and showcase the success of the new regime that had followed the Civil War of 68–9 and before that the chaotic and disastrous reign of Nero. It also bought into the Romans' vision of themselves as the divinely ordained conquerors of the world. The graphic depictions of brutality were entirely in harmony with the way the Romans venerated and enjoyed violence. Josephus, despite being a Jew himself, unhesitatingly blamed the Jews for bringing this all upon themselves.[77]

The centrepiece of Vespasian and Titus' triumphal procession was the treasure seized from the temple at Jerusalem, including a gold table and seven menorah candelabras. One of these at least was made of gold and was designed to hold lamps on each of its seven branches. The final component in the display of spoils was a book of Jewish laws. At the rear came figures of Victory made of ivory and gold, and finally Vespasian and Titus themselves in chariots, with Vespasian's younger son Domitian riding alongside. The procession's destination was the temple of Jupiter Capitolinus, which had been rebuilt since it was set on fire during the Civil War in 69 (it would be burned again in 80). The triumph had now reached its climax. Simon bar Giora, the principal leader of the Jewish forces, was beaten and dragged into the forum where he was executed. With this done, and sacrifices performed, Vespasian and his sons returned to the palace while the rest of the city enjoyed what remained of the day.[78]

Ironically, the war was not even yet over. Legio X Fretensis had been left behind to mop up any remaining Jewish strongholds, and it was not until the siege and fall of Masada in 74 that the conflict ended. Moreover, it would be a mistake to assume the soldiers were automatically cheering on Titus and his father. Soldiers were known habitually to poke fun at their generals during triumphs, as Caesar's men had done. In around 84, during the reign of Domitian, the poet Martial described how the new emperor had been subjected to jests during his own triumphs, and asserted that it was no bad thing for a general to be the butt of jokes.[79]

Vespasian followed up the triumph with the construction of a temple of Pax on a large site across the forum from the Palatine Hill and close to Augustus' temple of Mars Ultor ('the Avenger'). The temple became an art gallery but also displayed some of the Jewish treasures. Several years later, after Vespasian died and Titus succeeded him, the Arch of Titus was erected on the Via Sacra leading south from the forum, recalling Fabius Maximus' arch from two

centuries before. Today its reliefs depicting the triumph of 71, including the Jewish booty being carried in the procession, and Titus in his
chariot, are still in situ.

## LAUGHING STOCKS

Some campaigns risked humiliating the emperor and his army.
Caligula's notorious attempt to invade Britain in 40 – something
Augustus had also considered – was recalled in Roman military lore
as an occasion like no other. By that point in Caligula's reign a serious
illness and a complete psychological inability to cope with the extent
of his power were beginning to have a dramatic effect. Caligula
arrived at the English Channel on the north coast of Gaul and
ordered his soldiers to line up on the sand, equipped with ballistas
and other artillery. He boarded a trireme which sailed out to sea and
then returned. Once back on the shore he instructed his soldiers to
attack by gathering up shells from the beach. These were his 'booty',
which would be taken back to Rome. He was delighted, believing he
had 'enslaved the ocean', and in celebration set up a lighthouse to
guide ships. The soldiers must have been not only incredulous but
appalled at the lack of any opportunity to seize real booty, a blow
doubtless softened by a gift of 100 denarii each (about five months'
pay at the time). The episode was of course an absurdity – but at some
point serious preparations must have been put together, because after
Caligula's assassination in 41 it was possible for his successor Claudius
to carry out a real invasion.[80]

However, the Claudian invasion of Britain in 43 also came close
to falling apart before it had even started. Anxious to throw off his
reputation as the family idiot and the stooge of the praetorians who
had made him emperor two years earlier, Claudius had ordered
the expedition because he was keen to show the Roman army and
the Roman people that he was worthy of being emperor. Almost a
century earlier Julius Caesar had invaded Britain twice, but had not

succeeded in holding on to the island, if indeed he had ever intended to. Claudius believed a military triumph would be the best way of achieving the necessary prestige to consolidate his hold on power. Matching, and exceeding, his glorious forebear's achievements (Claudius was descended from Caesar's great-niece Octavia, sister of Augustus) would be an excellent means of doing that.

Claudius was handed a pretext on a plate when a tribal leader called Verica, ousted by his rivals, turned up in Rome asking him to intervene. Claudius placed the invading army under the command of Aulus Plautius, 'a senator of great renown', but despite the man's reputation he struggled to persuade the soldiers to take to the ships and cross over to Britain. Working back from later evidence dating to the time of the Boudican Revolt (see Chapter 7), it is possible to be reasonably sure that all or most of the II Augusta, VIIII Hispana, XIIII Gemina and XX legions were involved, together with vexillations of other legions, detachments of praetorians and about the same number of auxiliaries. This amounted to in total roughly 40–50,000 men, all of whom had to be gathered on the Gaulish coast to make the voyage. Roman soldiers were notoriously superstitious and the occasion could not, in their eyes, have been more inauspicious. They were, said Dio, 'indignant at the thought of carrying on a campaign beyond the limits of the known world' and refused to embark. Fortunately, Plautius' entourage included an imperial freedman called Narcissus, one of Claudius' closest advisers.

Claudius was still in Rome, waiting for news that the invasion had been successful, at which point he would travel to Britain to lead the army into the Britons' principal settlement at Colchester. In the meantime his freedman Narcissus loyally stood in for the emperor and decided to address the gathered force. This appalled the soldiers, who were disgusted at the thought of a former slave exhorting them to do their duty. They barracked and heckled Narcissus, who was unable to get a single word out. The climax came when they remembered that at the festival of the Saturnalia in December it was customary for slaves and masters to swap roles. They promptly saw the amusing side of the incident and began chanting 'Io Saturnalia', before deciding that they had better decide for themselves to get on with the job, rather than endure any more humiliation. They abandoned their protests and agreed to cross the Channel to Britain. The delay caused by their complaining had pushed the invasion back until late in the summer, but the soldiers were emboldened when a flash of lightning rose 'in the east and shot across to the west'. It seemed to be showing them the way to Britain, and since lightning bolts were thought to be sent down by Jupiter, it was all they needed to feel optimistic.[81] The invasion went ahead and Britain, or at least most of it, would remain a province of Rome until 410.

## A GRAND TRADITION

Although the later Empire is outside the scope of most of this book, there were incidents where Rome's soldiers managed remarkable feats in the face of adversity honouring the grand tradition of their forebears. From the mid-second century on Rome was increasingly on the back foot. Conquest and the opportunity for glory belonged mainly to the past, replaced by defensive and civil wars.

Ammianus Marcellinus is our best source for events in the third quarter of the fourth century. He recounted how in 359, during the siege of the city of Amida in Mesopotamia by Shapur II of Persia and where he was present, two Gallic legions were enraged by the way the Persians were carrying off Roman captives to their camp. They demanded to be allowed to attack the Persians, threatening to kill their own officers if refused. An agreement was reached and the Gallic soldiers burst out through a gate, and caused mayhem in the Persian camp, killing the sleeping enemy. Even when the remaining Persians woke up and fought back, the Gauls fought back bravely and with absolute determination. Despite their losses they only retreated slowly, and re-entered the city at dawn. The emperor Constantius II (337–61) ordered that statues of the Gallic officers in full armour were to be erected in their honour at Edessa.[82] The siege, however, ended in disaster when the Persians captured it, and killed everyone left.

A few years later Constantius' successor, Julian the Apostate (360–3), took the war into Persia, initially with great success. After crossing the Tigris with his army in 363, he soon after met the Persians in battle. Showing brilliant leadership Julian kept encouraging his men until the Persian line broke. The Persians broke into retreat and fled back to the city of Ctesiphon, the Roman soldiers having to be restrained from following them through the gates. Ammianus compared their heroics to those of Hector and Achilles, and recorded the Persian losses at 2,500 for the cost of only 70 Romans.[83]

Roman soldiers, under the right circumstances and leadership, were capable of extraordinary achievements and rightfully gained an unsurpassed reputation for resilience and resourcefulness. These were qualities that belonged to a tradition stretching back to Rome's earliest days, and explained how the city had risen from obscurity to unmatched power and success. Inspired by the feats of his predecessors, every Roman soldier knew he had a great deal to live up to, and many met the challenge. However, nothing could alter the fact that military glory and success were only won with violence, sometimes on a horrific scale. The Roman army's history was punctuated by innumerable tales of extreme brutality.

# NINE

# LIVING BY THE SWORD

## VIOLENCE AND ATROCITIES

*The soldiers ran their swords through every man they found,
and blocked the narrow streets with dead bodies, flooding the
city with so much gore that the blood extinguished many of
the fires.*

Josephus describing the fall of Jerusalem
to the army of Titus in 70[1]

R oman wars might theoretically have been conducted by
disciplined soldiers, but they were also often gratuitously
brutal, as will already be clear from some of the preceding pages in this book. Brutality was the stock-in-trade of both
the Romans and their enemies. 'Once the killing has started, it is
difficult to stop', said Tacitus in a blunt admission of an unpleasant
truth.[2] Fighting an ancient war was not for the faint-hearted. This
was face-to-face combat at close quarters with swords, knives and
spears, which could mean anything from massacring a rival Roman
army or wiping out a barbarian force to murdering women and
children in their homes and in the streets of their cities or villages.

The emperor Probus was said to have paid a gold *aureus* for the head of every decapitated barbarian brought to him.[3] Roman historians exulted in the stories, and Roman soldiers exulted when they were victorious. It was all part of Roman *virtus*, which meant honourable courage, manliness and heroism all dressed up in a semi-religious veneration of violence.

There is no question that extreme brutality was engrained in Roman society and warfare. It is hard to see how that made the Romans particularly unusual by the standards of their time, but it would be true to say that they were exceptionally diligent in their use of violence to get what they wanted, and in acting on the state's behalf to pursue its aims of conquest. It is also true that the people they fought were sometimes no less violent in their own way. The Romans were, however, usually better equipped and trained, and extremely persistent.

The world of the Romans was a place where life was cheap, though no one bothered to sit down and work out the price. Death could come quickly at the point of a spear or sword, slowly through starvation in a siege, or as a result of torture or devastating wounds. In the Republic, military power over Italy, Sicily, and then further afield had been won at enormous cost both to the Romans and their enemies. It set the pace for later Roman history. In 255 BC, in the First Punic War, the consul Marcus Atilius Regulus was defeated by the Carthaginians. He escaped with no more than about 500 of his troops. Of an original force of 15,000, just 2,000 survived the battle.[4] The disaster was merely one of Rome's expensive stepping stones to world domination. Some societies might have been so horrified by the losses they would have ended the war there and then. The Romans only became more determined.

The atrocities recorded during the days of the Roman army under the emperors had a long tradition. Polybius did not mince his words when he described Roman soldiers in action under the command of Scipio (later Africanus) in Spain during the Second Punic War in 209 BC. Scipio was only in his twenty-eighth year but was carving out a reputation for himself as a military leader of total ruthlessness. The target was Cartagena (Nova Carthago). Once the Romans were in, Scipio told his men 'to exterminate every form of life they came across, sparing nobody'. Polybius explained that the purpose was to provoke terror, and described how in a city taken by the Romans 'you can often see not only the bodies of human beings but also dogs sliced in two and the dismembered limbs of other animals'. As if that was not bad enough, Polybius mentioned that Cartagena's destruction was even more brutal than usual because of its large population.[5] Scipio was widely admired thanks to his later defeat of the Carthaginians at Zama, which brought the war to an end. No one considered him a war criminal, because the concept simply did not exist. He was a hero; someone to live up to.

Just a few years earlier, in 212 BC, the great Greek city of Syracuse in Sicily had fallen to a Roman force after a protracted siege. Syracuse had been defended with ingenious defensive mechanisms and machines invented by the brilliant engineer and mathematician Archimedes. The Roman commander Marcus Claudius Marcellus was so impressed that he wanted Archimedes' life spared. As the city fell, Archimedes was so completely absorbed in drawing diagrams in the dust on the floor of his house that when a soldier burst in looking for loot and asked who he was, Archimedes said, 'I beg, do not disturb this [work].' The soldier ran him through having no idea who he was, though it probably would have made no difference if he had. This example of gratuitous violence destroyed one of the ancient world's greatest geniuses, much to Marcellus' sadness. But then, as Valerius Maximus said, the whole scenario had only come about because of the fall of Syracuse, which had led to Marcellus' determination that Archimedes be saved; the same context resulted in his death.[6]

During the Second Macedonian War (200–196 BC), Philip V of Macedon and his forces witnessed the full horror the Roman gladius could cause. According to Livy, Greek and Illyrian bodies were 'dismembered ... with arms lopped off along with the shoulder, heads separated from the bodies with necks cut right through, intestines laid bare and other repulsive wounds'. No wonder there was 'widespread consternation'; Philip himself panicked at the thought of a battle against Roman soldiers armed with such a lethal weapon.[7] If that sounds like potential exaggeration, evidence of how brutal soldiers of this war could be turned up in archaeological excavations at Valencia, at a site dating to the city's capture from the rebel governor Sertorius by Pompey's troops in 75 BC. Some of the skeletons revealed that they had had limbs amputated, as had happened in Cartagena. The skeleton of one adult male showed that he had been impaled through his rectum on a Roman javelin, the weapon piercing his entire torso up to his neck.[8] His grisly remains demonstrate that the more graphic accounts written by ancient historians were not just gratuitously gory and rhetorical depictions of brutality.

In 82 BC Cornelius Cethegus, fighting for the dictator Sulla, succeeded in luring 5,000 men out of Palestrina in Italy and persuaded them to lay down their arms by promising to spare them. The men came out, gave up their arms and lay down. Sulla then gave his troops the order to kill all of them on the spot. In the event, 4,700 men were murdered and Sulla made sure the occasion was entered into the public record so no one would forget. This merely added to a catalogue of violence he ordered his soldiers to carry out, which included the killing of peaceable citizens for their wealth and the murder of women, 'as if killing men was not enough for him'.[9]

Sometimes Roman soldiers got as good as they gave. Not all civilians were terrified into submission. In 76 BC the Roman city of Lauro was sacked during the civil war in Spain between Sertorius and supporters of Sulla. When one of Sertorius' soldiers, who had taken the city, insulted a woman and tried to rape her, she gouged out her assailant's eyes; but instead of punishing her, Sertorius ordered an entire cohort of his men to be executed for allegedly committing similar acts of 'such brutality'.[10]

It was quite possible for a soldier, even one of senior rank, to be on the receiving end of violence from the civilian population during peacetime. Pollenzo in northern Italy was the scene of an ugly riot during the reign of Tiberius (14–37). The mob wanted a primus pilus centurion to stump up the funds to put on a gladiatorial show. He evidently failed to oblige, for the mob killed him and then refused to give up his body 'until their violence had extorted money from his heirs for a gladiatorial show', reported Suetonius. Tiberius had to send in two cohorts of soldiers to place Pollenzo under martial law, imprisoning many people including the town councillors.[11]

One only need look at Roman military sculpture to see the explicit glorification of brutality and violence. Publius Flavoleius Cordus, a soldier of Legio XIIII Gemina at Mainz, went to meet his maker aged forty-three armed with his gladius, his pilum and his pugio dagger.[12] The same was true of the auxiliaries, and perhaps even more so. Longinus was a Thracian cavalry trooper based in Britain shortly after the invasion of 43. At some point in the next few years he died at his base at Colchester, then a legionary fortress for Legio XX and its attached auxiliary forces. He was cremated and commemorated with a grand tombstone that depicted him on his horse, itself dressed in the decorative and ostentatious paraphernalia that the auxiliary cavalry forces loved. Beneath the horse a cowering barbarian shudders with fear, subdued and humiliated as Longinus and his horse stride over him with swaggering bravado.[13]

Longinus was far from alone. Gaius Romanius Capito, a cavalry trooper in a unit of Noricans based at Mainz in Germany in the mid-first century AD, died at the age of forty after 19 years' service. He was shown on a horse crushing a barbarian and aiming a spear at his victim while a servant stood at his side to pass him more weapons.[14] Longinus and Capito were provincial auxiliaries, not Romans by descent. They had joined the Roman world and brought their own love of fighting with them. Not many years earlier, their ancestors had been the 'barbarians' they now delighted in showing themselves annihilating.

Bragging about brutalizing the enemy came in a variety of forms. Longinus' posthumous image was perhaps generic, a metaphorical boast rather than a depiction of his personal exploits. Other military units used similar imagery. An inscribed and sculptured stone from the Antonine Wall in Britain shows Victory crowning the eagle standard of Legio XX, while on either side is a barbarian captive on his knees with hands bound.[15]

## CAESAR IN GAUL

Julius Caesar's conquest of Gaul in the 50s BC is one of the most celebrated military campaigns of all time. It was also one of the most vicious. His soldiers, ever loyal to their commander, would do anything to pursue his interests. In 52 BC they found themselves besieging the city of Bourges, a stronghold of the Bituriges tribe. Both the Romans and their enemies showed remarkable courage in the heat of battle. Caesar was hugely impressed by the ingenuity of the Gauls, who were determined to destroy the Roman wooden turrets and ramps with fire. One Gaul, throwing down lumps of grease and pitch into the conflagration started by his fellows, was hit by a dart from a Roman scorpion catapult and killed immediately. Without hesitation 'one of the party next to him stepped over his prostrate body and carried on with the same work', said Caesar. To his amazement a third man stepped in when the second was killed the same way, and then a fourth. This carried on till the fire on the ramp had been put out. It was a remarkable comment on what the Romans sometimes had to face, and Caesar made a point of saying that it was an incident he could not possibly leave out of his account.

Later on, aware that it was vital he break into the city because his supply train was being threatened by the Gaulish leader Vercingetorix, Caesar made a sudden attack during a heavy shower of rain which distracted the defenders. After the siege and a recent massacre at Cenabum, his soldiers' blood was up. Once in, 'no one was spared', recalled Caesar himself, 'the old, women or children'.

Of the 40,000 inhabitants, only around 800 managed to escape and reach Vercingetorix.[16]

In 50 BC Caesar's men were besieging the city of Uxellodunum, a fortified hilltop settlement near the modern village of Vayrac. The inhabitants exhibited exceptional bravery and fortitude in their resistance, but they had to give up when the Romans cut off the water supply by digging underground to divert the springs. To try and discourage any other settlement from holding out in the same way, Caesar ordered a terrible retribution, instructing his men to cut off the hands of any Gaul who had fought during the siege.[17]

## GERMANICUS IN GERMANY

Tiberius, on Augustus' orders, went to Germany to wreak revenge for the Varian disaster of AD 9. Tiberius 'penetrated into the heart of the country, opened fortified boundaries, laid fields waste, burned houses, routed those who came against him, and with great glory returned to the winter content without losing any of the troops with whom he had crossed [the Rhine].'[18] There was plenty more of that sort of violence to come. The golden boy of the Roman imperial family, Germanicus was hugely popular with the Roman mob, not least because he was married to Augustus' granddaughter Agrippina the Elder (15 BC–AD 33), who had already borne him several children. He was also good-looking, but this belied what he was capable of, as soon became clear on campaign in Germany in 14. The Roman army had mutinied on the death of Augustus (see Chapter 11), giving the German tribes the opportunity to believe they were going to be left in peace. But Germanicus' soldiers were desperate to make amends for their disloyal behaviour and also avenge the Varian disaster five years earlier. He sent 12,000 legionaries, 26 auxiliary units and eight cavalry wings over the Rhine. The villages of the Marsi, a Germanic tribe that lived north of the Rhine, were taken completely by surprise. According to Tacitus, Germanicus split his forces into four and ordered them to lay waste a strip of territory 50 miles wide; everything and everyone was to be wiped out, regardless of age or sex. One of the first targets was the tribal shrine of Tamfana. The killing roused some of the tribes into action but they were beaten off by the Romans, who forced them out into open country 'with great slaughter'.[19]

The following year Germanicus followed up with a war against the German Chatti tribe, taking them too by surprise. So sudden was his attack that 'those weak because of their age or their sex were immediately captured or butchered', said Tacitus. The men tried to stop the Romans building a bridge over the river Adrana (now the Eder) but were pushed back with a fusillade of artillery missiles and arrows. Their attempts to negotiate a peace were ignored. Germanicus pushed on to 'raze the tribal headquarters' before 'laying waste the countryside'.[20]

## LOYALTY IN THE HEAT OF BATTLE

One of the reasons Roman soldiers could heap so much violence and retribution on their enemies was their furious loyalty to one another, especially if they were in the same unit. Many years earlier, during Julius Caesar's Gallic campaign, two experienced centurions had earned themselves undying fame in the heat of another vicious battle. Caesar was so impressed that he made a special point of naming them and telling their story. Titus Pullo and Lucius Vorenus seem to have despised each other because they were senior centurions and bitter rivals for the best jobs. As a result they were always arguing. One day in 54 BC the legion was under attack from the Nervii tribe. As the fighting grew more intense, Pullo decided to goad Vorenus by accusing him of holding back and waiting for a better opportunity to prove his bravery. Pullo then dived into the attackers, leaving Vorenus no alternative but to follow in case he was thought a coward. Pullo threw his spear and struck one of the Nervii, who was knocked unconscious. Other Nervii tried to protect their companion with their shields, and all flung their spears at Pullo, who now had no chance of escaping. He had one spear stuck in his shield, another in his belt, and his scabbard had been pushed out of place. Struggling to unsheathe his sword, he was surrounded by men of the Nervii.

For a moment it seemed that all was lost. Then Vorenus dashed up to help, diverting the tribesmen's attention on to himself because they thought Pullo was dead; he killed one and chased off the others, only to fall into a small depression in the ground. Pullo had in fact been able to get away and bring up reinforcements. Amazingly neither man was hurt and both managed to escape back behind the Roman defences. As Caesar said, 'it was impossible to decide which should be considered the better man in valour'.[21]

Loyalty to one another could be at the expense of a commanding officer. In 41–40 BC, during the civil war fought by the triumvirs after Caesar's assassination, Octavian was besieging Antony's brother Lucius Antonius in the city of Perusia. Octavian laid on games to keep his troops amused. One of the ordinary soldiers decided to sit in the forward seating zone reserved for equestrians. Octavian ordered him to be removed, but the man then went missing, and a rumour erupted that Octavian had ordered him to be tortured and killed. Octavian was rushed by a mob of soldiers bent on killing him and was only saved when the missing man suddenly reappeared.[22]

Sometimes there was outright hatred between legionaries and auxiliaries, and it could turn nasty. In 69, during the Civil War, a civilian workman at Turin was being harangued by a Batavian auxiliary for cheating him. The workman had billeted on him a legionary from Legio XIIII Gemina, who stepped into his landlord's defence. Before long both soldiers were being backed up by their fellows. Blades were drawn and the argument was all set to explode into a full-scale battle between the rival supporters. Only when two Praetorian Guard cohorts stepped in on the legionaries' side was the fighting defused.[23]

## ABUSE OF PRIVILEGES

Roman popular culture often depicted soldiers as thugs who went around intimidating civilians. In Petronius' *Satyricon*, written around the time of Nero, the main character, Encolpius, is accosted one night by a legionary, or a man posing as one, who robs him.[24] The praetorians in Rome were among the worst offenders. The best paid, most privileged and prestigious troops in the Roman army, they were also sometimes the most arrogant. Juvenal wrote in one of his *Satires* how no civilian in Rome would 'dare to thrash' a soldier. Worse, he would have to keep his mouth shut if he was thrashed *by* a soldier, and not even consider going to a magistrate to show him 'the teeth that have been knocked out, or the black and blue bruises on his face, or the one eye left that a doctor holds out no hope of saving'. Pursuing legal redress was pointless, since the law forbade any soldier to be heard in a court outside the fort. As a result the case would be presided over by a 'hob-nailed centurion' with a jury of soldiers, who would then order the victim to be beaten up by the rest of the unit.[25] Juvenal was almost certainly talking about the Praetorian Guard; but as an officer himself, who once commanded an auxiliary unit, he had experience of the wider army too.

## ROMAN AGAINST ROMAN

Roman soldiers were perfectly capable of brutalizing each other. During the Civil War of 69 one of the imperial contenders, Vitellius, took the opportunity to visit the site of the First Battle of Bedriacum (Cremona), where his troops had defeated his rival Otho in April in a particularly vicious encounter involving hand to hand fighting (see Chapter 8).[26] 'There lay mangled corpses, severed limbs, the rotting bodies of men and horses.' Tacitus enjoyed telling how Vitellius exulted in the sight of the 'unburied bodies of his countrymen', unaware that fate was leading him to his own destruction.[27]

A few months later, in October, at the Second Battle of Bedriacum, Vitellius' soldiers found themselves fighting a Flavian force led by Marcus Antonius Primus on behalf of another contender for the throne, Vespasian. In the lead-up to the battle Antonius Primus sent his auxiliary cohorts out into the countryside to look for supplies, but the real purpose, said Tacitus, was to get them used to the idea of 'plundering Roman civilians'.[28] During the battle the fighting was characteristically vicious, both sides employing imaginative tactics. Antonius' legionaries had already helped themselves to axes and scythes from farmhouses in the area before organizing themselves into testudo (tortoise) formation with their shields above their heads. The Vitellian soldiers, in their fortified camp on higher ground, spotted their opportunity and rolled boulders down onto their opponents' heads, splitting the formation before reaching into the bedlam with their lances and pikes to break up the testudo formation and massacre the wounded men inside.[29] Despite that setback the battle was won by Antonius Primus. The city of Cremona, which had supported Vitellius, was his next target. Antonius' army fought its way in and sacked the town. The soldiers plundered everything they could find and set the city on fire, but it turned out that disaffected soldiers from Vitellius' army were already at work wrecking the place, using their local knowledge to target the richest houses.[30]

When the battle started, on 24 October 69, Vitellian artillery proved extremely dangerous to the Flavian forces. A Vitellian ballista belonging to Legio XV was proving particularly lethal. Two Flavian soldiers dashed forward, disguising themselves by picking up shields from dead legionaries of XV, and wrecked the machine by cutting its ropes. Having given themselves away the two men were killed immediately by the Vitellians. Even though their names remained unknown, their heroic effort seems to have gone down in Roman military lore, but since Tacitus refers to Antonius Primus bringing up his praetorians when describing this incident, they probably belonged to the Guard.[31]

A particularly tragic event which occurred around this time as a result of the terrible violence provides a rare instance of a personal story involving ordinary soldiers. Legio XXI Rapax supported Vitellius. One of its soldiers was a Spaniard called Julius Mansuetus, who had left a son behind at home. When the boy reached adulthood he joined Legio VII Gemina, formed in Spain by Galba in 68. By the time of the Second Battle of Bedriacum, VII Gemina was on Vespasian's side. During the battle, the young soldier fatally wounded an opponent; only when he was searching the man's barely conscious body did he realize that it was his own father. Profusely apologizing to Mansuetus before he died, he then picked up his father's body and buried it. Other soldiers noticed what was going on and they all ruminated on the pointless destruction the war had brought. Tacitus, however, reminded his readers that it made no difference. Nothing stopped them carrying on 'killing and robbing their relatives, kin, and brothers'. Calling it a crime, 'in the same breath they did it themselves'.[32]

The story of Julius Mansuetus recalls another unfortunate incident when father and son were on different sides in a Roman civil war. It took place a century before, shortly after the Battle of Actium in 31 BC, but had a much happier outcome. A man called Metellus had held a command in Antony's forces at Actium, while his son of the same name held a command under Octavian. After Octavian's victory, the elder Metellus was taken with all the other prisoners to Samos, where Octavian – together, as it happens, with the younger Metellus – was overseeing their processing. As Appian recounts:

> The old man was led forward covered with hair, misery, and dirt, his appearance completely changed by them. When his name was called by the herald in the array of prisoners the son sprang from his seat, and, with difficulty recognizing his father, embraced him with a cry of anguish. Then restraining his lamentation he said to Octavian, 'He was your enemy, I was your fellow-soldier. He had earned your punishment, I your reward. I ask you either to spare my father on my account, or

to kill me at the same time on his account.' There was much
emotion on all sides, and Octavian spared Metellus, although
he had been bitterly hostile to himself and had scorned many
offers made to him to desert Antony.[33]

## BRUTALIZING CIVILIANS

When a riot broke out in Rome around 40–39 BC due to a famine
caused by the civil war between Antony and Octavian and Pompey's
son Sextus Pompey, Octavian arrived in the forum to tell the mob
how unreasonable their complaints were and was promptly attacked
with a hail of stones. Antony burst in with a large force of troops to
crush the crowd and rescue Octavian. His soldiers did the job, but
followed it up by throwing the bodies of the civilians they had killed
into the Tiber; there they stripped the floating corpses, 'carrying off
the clothing of the better class as their own property'.[34]

The brutalizing of civilians continued well into the days of the
emperors. Pontius Pilate was the governor of Judaea under Tiberius
between *c.* 26 and 36. While in charge he decided to requisition the
Corban treasure, a fund raised by taxation on adult Jewish males
to pay for the Temple in Jerusalem and from which a surplus was
provided to benefit the city. Instead Pilate spent the money on a 50-
mile-long aqueduct, provoking popular outrage. But when a mob
surrounded him during a visit to Jerusalem they found that Pilate
was one step ahead. He had told his garrison soldiers to dress up as
civilians, wearing their armour underneath. They were equipped
with clubs which, on a signal from Pilate, they whipped out from
under their disguises and started beating the Jews. Some were killed,
provoking a stampede; others were trampled to death in the chaos.
'The fate of those who were killed stunned the crowd into silence',
said Josephus.[35]

It was not surprising then that in 66 the Jews rebelled against the
Romans. When the army of Titus burst into Jerusalem in 70 during

the war that followed, they first found whole families who had starved to death in their homes. In a rare moment of consideration, they gave up on the idea of looting. Josephus noted, however, that 'any compassion they felt for those who had died was not matched by what they felt for the living'. In a peculiarly graphic twist he described how the soldiers put to the sword every man they came across, killing so many that they 'flooded the city with so much gore that the blood extinguished many of the fires'. Titus tried to inject some restraint, ordering that only fighting men were to be killed while everyone else was to be taken alive. In the chaos, and with the red mist before their eyes, his men continued to kill the sick and old. Thousands of others were captured and sold as slaves, or sent to be massacred as entertainment in the amphitheatres of the Roman Empire.[36]

A few years later, in 72–3, the Roman army besieging the Jewish stronghold in the hilltop fortress at Masada was prevented from wiping out the inhabitants. Together with other Jewish families fleeing the Romans, an extremist Jewish sect called the Sicarii had seized Masada. The Roman forces built a massive ramp so they could run siege engines up to assault the fortress. They burst in on 16 April 73, only to discover that the occupants had all been killed. Josephus reported how, under instructions from their leader Elazar ben Yair, they had chosen

> ten men by lot out of them to slay all the rest, every one of whom laid himself down by his wife and children on the ground, and threw his arms about them, and they offered their necks to the stroke of those who by lot executed that melancholy office. When these ten had, without fear, killed them all, they made the same rule for casting lots for themselves, that he whose lot it was should first kill the other nine, and after all should kill himself.[37]

Although one woman had hidden herself with five children and survived, 960 bodies were found by the stunned Romans. With the exception of the last man, by being killed they had escaped the sin of suicide. Based on the experience of those the Romans had killed in Jerusalem, they had perhaps escaped an even more brutal end or a lifetime in slavery.

The Boudican War in Britain (see Chapter 8) was one of the most violent and destructive conflicts in Roman military history. This time it was the rebels who meted out violence on civilians, and the destruction of Colchester, London and St Albans provided a particularly graphic topic for Roman historians keen to depict barbarians as reckless maniacs. By Cassius Dio's time, what Tacitus had referred to as 'killing, gibbets, fire and crosses' had been suitably elaborated, though there may have been some truth in what he wrote. Dio recounted how the Britons 'hung up naked the noblest and most distinguished women and then cut off their breasts and sewed them to their mouths, in order to make the victims appear to be eating them'. If that was not bad enough, 'afterwards they impaled the women on sharp skewers run lengthwise through the entire body'. The violence was conducted against a backdrop of sacrifices, banquets and 'wanton behaviour'.[38] In the end the Romans got their revenge. At the final battle, when the rebellion was crushed, Tacitus said that 'the Roman soldiers did not hold back from the execution even of women'. The piles of bodies were made bigger by the Britons' horses, which were 'transfixed by weapons'. Such descriptions have to be treated with care. There was a long tradition in accounts by ancient historians of describing atrocities in the most graphic way. However, the impaled skeleton found at Valencia (see earlier in this chapter) shows that there may have been some truth in the stories.

Fabius Valens was commander of Legio V in Gaul, which sided with Vitellius in the Civil War of 68–9. With his legion and auxiliary troops he had around 40,000 men under his command. On the march to support Vitellius the men passed through friendly territory and were welcomed at Metz 'with every civility', said Tacitus. Evidently, though, the men were much twitchier than Valens realized. Tacitus described how a 'sudden panic gripped the troops'. They pulled out their weapons and ended up attacking the townsfolk. Before they could be stopped, 4,000 civilians had been killed. It was a total breakdown of discipline and order, belying the popular impression of the Roman army. Tacitus was fascinated by what had happened. He dismissed the idea that they had been intent on 'raping and pillaging'; instead they had been convulsed by some sort of hallucinogenic frenzy 'that defies analysis'.[39]

Other cities in Gaul, panic-stricken by the news, sent out deputations to plead with the soldiers as they advanced. Bizarrely, the citizens of Lyon begged them to sack its bitter rival, Vienne, instead. Fortunately, negotiations defused the situation and prevented the city being ravaged. Other regions were not so lucky. When the Helvetii tribe protested at money being stolen by Legio XXI, which had been sent to help them fund their defences, an army was sent against them. The Helvetii fled, but when they were found, thousands were killed and similar numbers enslaved.[40] In the days before the First Battle of Bedriacum in 69, Otho's troops savaged Italian farms and towns. 'One would never have guessed that the invasion was directed against Italy or the towns and homes of the mother country', said Tacitus. No one was expecting the savagery that ensued, and so no preparations or defences were in place. Farmers 'ran out with their wives and children' and were killed, the surprise victims of a war when everyone had thought it was a time of peace. Given that many legionaries were recruited in Italy, the violence was even more tragic.[41]

What had spooked the Roman troops in Gaul will forever remain unknown, but the phenomenon is attested in other conflicts in more modern times, such as the Vietnam War. It is possible the soldiers involved had been so brutalized that their behaviour was the result of a form of mental breakdown. Today we call it post-traumatic stress disorder and it can take on a variety of forms. Roman historians scarcely recognized the concept, and nor did Roman doctors. Only rarely do we find hints of what the violence could do to a soldier. In

218 BC, during the Second Punic War, the consul Publius Cornelius Scipio (father of Africanus) was wounded at the Battle of Ticinus and seems to have been traumatized. Sempronius Longus, the other consul, described him as 'sick in spirit' and said that just thinking about his wound 'made him dread a battle and its missiles'. Evidently the mental impact had been more serious than the physical injury.[42] Before the Battle of Pharsalus in 48 BC between Caesar and Pompey, the soldiers on both sides became demoralized at the thought of Italians fighting one another. 'Ambition ... gave way to fear', said Appian. The generals hesitated but the battle began, whereupon Pompey's cavalry, terrified by the violence, 'fled in disorder'. As his army collapsed, Caesar ordered his men only to attack Pompey's auxiliaries and spare the Italians. The defeated Pompey seems to have plunged into depression and retreated speechless to his tent before fleeing.[43]

## SEPTIMIUS SEVERUS' VIOLENT REIGN

In 193 another civil war broke out. One of the contenders, Septimius Severus, convinced another, Clodius Albinus, that he could be his heir if the two worked together against the third contender, Pescennius Niger. Albinus fell for the ruse and Severus went off to defeat Niger; then, in 197, he turned on Albinus. The two fought it out near Lyon in one of the largest and bloodiest battles of the whole Roman era. Cassius Dio seems to have been inspired by Tacitus, because he described in much the same way how both sides suffered gigantic losses with some bodies 'mutilated by many wounds, as if hacked in pieces, and others, though unwounded, were piled up in heaps, weapons lay scattered about, and blood flowed in streams, even pouring into rivers'. Albinus committed suicide and Severus carried the day, though it had been a close call. He had Albinus' head cut off and sent to Rome on a pole.[44]

Successful though Severus had been, as emperor he developed

a track record of sending his men into disastrous engagements in remote locations. In 198 he set out on campaign against the Parthians, whose empire was centred on what is now Iran. At Ctesiphon (22 miles/35 km south-east of modern Baghdad) he allowed his troops to sack the place and kill thousands of inhabitants. If that made him and his men feel pleased with themselves, the pleasure was short-lived. At Hatra (68 miles/109 km south-west of Mosul) he lost so many men, and so many were wounded, that a witty praetorian tribune called Julius Crispus pithily quoted Virgil's *Aeneid*. In the poem Aeneas arrives in Italy from Troy and secures his betrothal to Lavinia, daughter of the king of Latium, Latinus. Turnus, prince of the Rutulians and himself betrothed to Lavinia, takes exception, embarking on a costly war. As a result one of Turnus' men says, 'So that Turnus can marry Lavinia we are all perishing unheeded'. This was the line used by Crispus, who was promptly put to death by Severus when he heard, adding to the pile of dead Roman soldiers. But Crispus had been right. Things went from bad to worse. Not only were many soldiers to die attacking Hatra, but many more were lost when they went out on foraging trips, attacked by Arabian cavalry.[45]

Being stationed in remote frontier outposts was always a danger. Months or even years of inaction might pass for a garrison, only for lives to be turned upside down by a surprise attack. At the fort of Ambleside in northern Britain, a retired 55-year-old auxiliary centurion called Flavius Fuscinus and his 35-year-old son Flavius Romanus were buried under the same tombstone some time in the third century AD. The text on the stone includes a rare line, announcing that the two men 'were killed in the fort by enemies'.[46] Who those enemies were will remain a mystery. It is even possible that the father and son were murdered by other Roman soldiers supporting one of the many rival emperors of the time. At Simitthu in Africa, Lucius Flaminius of Legio III Augusta spent most of his time on garrison duty 'only to be killed by the enemy in battle in the Philomusian region' at the age of forty.[47]

## SOLDIERS AND THE IMPERIAL FAMILY

In the late Republic it became increasingly common for generals to order soldiers to kill their political enemies. In December 43 BC Antony ordered a squad of his men to assassinate Cicero, who had accused him of tyranny. The centurion who decapitated Cicero and cut off his hand was rewarded with a quarter of a million Attic drachmas in addition to the bounty already on offer. Both the head and hand were then displayed in the forum.[48]

Under the circumstances it is unsurprising that among soldier's more distasteful roles was their use by the imperial family to torture or murder those suspected of sedition or treachery – even if they were members of the same family. Augustus' youngest grandson, Agrippa Postumus (son of his daughter Julia), was killed in exile on the island of Pianosa by a hit squad consisting of a centurion and a tribune sent out on orders allegedly issued by Tiberius.[49] His sister Agrippina the Elder was Tiberius' daughter-in-law by adoption (he had adopted her husband Germanicus, his nephew). After Germanicus died in suspicious circumstances in Syria in 19 Agrippina did not hold back, complaining about how she was being treated by Tiberius and others. The emperor eventually exiled her to the island of Ventotene (Pandateria in antiquity). She continued complaining so he sent a centurion out to the island to beat her up which he did so badly 'that one of her eyes was destroyed'. Agrippina decided to starve herself to death, so Tiberius ordered her guards to force-feed her. She died anyway.[50] On another occasion Tiberius was being carried in a litter only for his entourage to be stopped by brambles. He ordered a centurion, probably from his Praetorian bodyguard, to clear them away, and then had the same man stretched out on the ground 'and half-flogged to death'.[51]

Since soldiers were highly trained in the mechanics of brutality, they were useful to emperors with a particularly perverted love of violence. Caligula notoriously enjoyed cruelty as a form of entertainment, for example eating lunch while watching torture. When he discovered that there was a soldier 'adept at decapitation', Caligula instructed him to cut off the heads of people being brought to him from prison for capital examination.[52]

The emperor Elagabalus (218–22) ordered a centurion to kill a senior senator called Sabinus. Or at least he thought he had. The jurist Ulpian, of whom Elagabalus disapproved, had dedicated some of his books to Sabinus, effectively signing the senator's death warrant. Elagabalus, not wishing to advertise his plans kept his voice low while he muttered his orders to the centurion. Unfortunately, he had overlooked the fact that the man was hard of hearing. The centurion misheard his instructions and thought he had only been told to throw Sabinus out of Rome, which he duly did. Sabinus was saved, and so was Ulpian, whom Elagabalus exiled.[53]

## THE ASSASSINATION OF CALIGULA

Because they were on hand, praetorians were always liable to be employed to murder people on behalf either of emperors or palace conspirators who set out to topple their rulers. Sometimes praetorian officers were themselves the prime movers behind the plots: Caligula's reckless and brutal reign came to an end in 41 when the praetorian tribunes Cassius Chaerea, Cornelius Sabinus, Papinius and Julius Lupus formed a plot, sick of the emperor's unpredictable and arbitrary behaviour.

The plotters believed that only the restoration of the Republic would solve the problems. They were joined by an imperial freedman called Callistus, the senator Annius Vinicianus and a praetorian prefect (which one is unknown). Chaerea was particularly aggrieved at the way Caligula had accused him of effeminacy, a powerful insult implying moral weakness and degeneracy. On 24 January 41, Caligula was watching the Palatine Games in a theatre in the imperial palace complex. He had been thoroughly enjoying himself, not least because an enthusiastic and large crowd had gathered to watch. Appropriately enough, the show included a mime in which a chieftain was caught and crucified; it involved, said Josephus, our most detailed source, the shedding of 'a great deal of artificial blood'. Eventually Caligula decided to head off to the baths, taking a short cut down a deserted alleyway to do so and inspecting on the way some boy singers and dancers who had come from Asia to perform for him.

It was the chance the plotters had been waiting for. Chaerea asked Caligula for the day's password. Caligula came up with something that humiliated Chaerea, who promptly produced a volley of verbal abuse before pulling out his sword and stabbing the emperor. Caligula staggered back in agony, the blow between his shoulder and neck not being fatal, and tried to escape. The other conspirators weighed in, and seem to have been joined by more people because the coup de grâce was delivered by a man called Aquila. The plotters made their escape but the violence had only just begun. Caligula's personal German bodyguards were the first to find their beloved emperor's body. To say they erupted into a frenzy of rage would be an understatement, largely because the assassination was about to bring to an abrupt end the gifts of money they kept being given by Caligula. Led by their tribune Sabinus, an ex-gladiator of enormous strength, they set out to exterminate Caligula's killers. The first they found was a man called Asprenas, whose blood-covered clothes gave him away. 'That was good reason to carve him limb from limb', they decided, before moving on to any others they could locate. Eventually the German soldiers arrived at the theatre, where it looked as if there was going to be a massacre of innocent spectators. Pleading for their lives, the audience managed to persuade the soldiers to back down, but in a gory gesture of triumph they stuck the heads of their victims on the theatre altar. The deed left the spectators terrified of what might happen next. The Germans finally backed off when they registered the futility of avenging the death of someone who was no longer in a position to reward them for their loyalty.

However, some of the conspirators were still alive and the violence was far from over. Chaerea in particular was exercised about the possibility that Caligula's wife Milonia Caesonia and daughter Julia Drusilla might survive. 'He was determined to do the job completely and to indulge to the full his hatred for Caligula.' Chaerea wanted to order a praetorian tribune called Julius Lupus to murder Caesonia and Julia, but the conspirators argued over the plan. Some thought it too cruel, others that Caesonia was largely to blame for her husband's behaviour. When eventually they agreed to go ahead with the killing, Lupus went to the imperial palace where he found Caesonia lying beside Caligula's bloodied corpse. She was covered in blood from having held his body, and her daughter was by her side. Only as Lupus approached did she realize why he had come; she bared her throat for his blade, screaming for him to get on with the job. He obliged and next killed Julia Drusilla, then aged around eighteen months. Meanwhile, the Praetorian Guard had kidnapped Caligula's uncle Claudius, the brother of the great general Germanicus, Caligula's father, and were in the process of making him emperor. The assassination did nothing, after all, to restore the Republic.[54]

All these episodes and events illustrate not so much the violence of the Roman army but the wider violence of society at the time. In battle Roman soldiers were usually up against enemies prepared to do exactly the same. Most spent the majority of their time in their barracks, or engaged in peaceable activities in the service of the state. But they were trained to fight and kill, and to do so in whatever way was necessary – though they seem sometimes to have been alarmingly willing to massacre civilians when told to.

For all the horrors they were capable of, there were rare occasions when Roman soldiers could be turned away from violence. In 86 BC, during the consulship of Marius and Cinna and the civil war with the supporters of Sulla, some of Marius' troops were sent to kill his enemy, the orator Marcus Antonius, grandfather of the future triumvir Mark Antony. Antonius was in hiding, but had been betrayed to Marius. What came next was remarkable. Antonius pleaded for his life with the soldiers, whose commander Annius was waiting downstairs. Unfortunately, the episode ended the way it so often did when the Roman army was involved:

> So indescribable, however, as it would seem, was the grace and charm of his words, that when Antonius began to speak and pray for his life, not a soldier had the hardihood to lay hands on him or even to look him in the face, but they all bent their heads down and wept. Perceiving that there was some delay, Annius went upstairs, and saw that Antonius was pleading and that the soldiers were abashed and enchanted by his words; so he cursed his men, and running up to Antonius, with his own hands cut off his head.[55]

# TEN

# QUINQUEREMES AND TRIREMES

## The Roman Army at Sea

*At Actium Antony's fleet held out for a long time against
Octavian. Only after it had been badly damaged by the high
sea which rose against it did it reluctantly, and at the tenth
hour, gave up the struggle. There were no more than 5,000
dead, but 300 ships were captured.*

Plutarch describes the last phase of the
naval Battle of Actium which brought
the Republic to an end in 31 BC[1]

U nlike the Roman army, whose origins lay at the beginnings of Rome's history, the navy came into existence in
piecemeal and haphazard fashion. But it was still on hand
to play a decisive part in Roman and even world history. Fleets featured in many campaigns, acting as transports for men, animals and
equipment, and sometimes even as the fighting platforms on which
fleet troops fought their opponents.

## FLEETS IN THE REPUBLIC

The Roman navy was created during the First and Second Punic Wars, though like the army at that time and throughout the Republic it was not a permanent institution. Since the Carthaginians were expert sailors and masters of the Mediterranean, war at sea was essential and unavoidable if they were to be challenged. The Romans had to learn and learn fast; yet until the First Punic War it had never occurred to them that they might need to become a naval power. 'Not only did they have no decked ships, but also they had no warships at all', explained Polybius. To begin with they borrowed ships to carry their troops over to Sicily, but when they captured a Carthaginian vessel they realized they had acquired a template whose specifications they could copy. Armed with a fleet of 100 quinqueremes and 20 triremes designed in imitation of their prize, the Romans were able to set about training crews.[2] They also developed the remarkable 'raven', which used a pole, ropes and a pulley to drop a gangplank with an iron spike from the Roman ship onto the deck of an enemy vessel. Roman troops could then dash across and fight the enemy crews and troops. It is an extraordinary fact that some rams from the front of Roman ships used in the First Punic War have been recovered from the sea off Sicily. [3]

Astonishingly, the Romans won their first naval battle against the incredulous Carthaginians, at Mylae off the north-east coast of Sicily in 260 BC. Further victories followed at Sulci (259 BC) and Cape Ecnomus (256 BC). Although problems were to come, it was the naval Battle of the Aegates Islands in 241 BC that finished off the Carthaginians and forced them to sue for peace. Rome was now not only a naval power but also pre-eminent in the Mediterranean. Like the army though, during the Republic fleets had to be formed on an as-need basis.

In the aftermath of Cannae in 216 BC during the Second Punic War naval forces were organized to protect Rome: 1,500 naval troops were sent from Ostia to the capital, and a naval legion was sent to Teano, a town in Campania.[4] When Scipio invaded North Africa in 205 BC his fleet included 50 men-of-war and 400 transports to carry not only the men and their equipment, but also over six weeks' supply of cattle, food and water (see Chapter 6).[5] This single instance gives an idea of how complex a Roman waterborne military operation could be. Their soldiers were brave and effective, but the relatively cumbersome nature of their ships continued to be a potential liability.[6]

The final destruction of Carthage in 146 BC in the Third Punic War removed the seaborne threat to the burgeoning Roman Empire until the emergence of Cilician pirates in the first century BC. The pirates' activities seriously compromised trade, and their strength grew with backing from Mithridates VI of Pontus between 76 and 63 BC because he knew the pirates were a useful means of damaging Roman interests in the Mediterranean. The pirates were also able to take advantage of Rome's civil wars, which enabled them not only to attack maritime trade but also to raid islands and coastal cities, plundering and taking away prisoners for ransom. Their numbers increased during the Mithridatic Wars because dispossessed people in Asia turned to piracy as the only option open to them, and support came from allies of Mithridates such as Crete. The profession was developing into a glamorous and ostentatious career with a network of support installations where the pirate crews could put in for supplies and to re-equip. 'It was', said Plutarch, 'a disgrace to Roman supremacy.' Eventually, crisis point was reached: trade on the Mediterranean had become crippled.[7]

In 67 BC a law was passed that gave Gnaeus Pompeius (Pompey Magnus, 'the Great', as he was later known) a three-year command to clear the seas. By requisitioning existing ships from Greek cities, Pompey was able to put together a fleet of 500 ships and a force of 120,000 men. Within an astonishing three months he had destroyed the pirate problem by dividing the Mediterranean into 13 zones and distributing the fleet among them.[8] Naval power was an important factor in the civil wars that followed, Pompey's son Sextus becoming a major threat to the triumvirs Antony and Octavian until he was defeated in 36 BC at the Strait of Sicily (Fretum Siculum) off Cape Naulochus by Agrippa with Legio X. The legion's achievements that day meant it was awarded the permanent title Fretensis in commemoration.

Fleets were often built on the spot to meet an immediate need. In 56 BC Caesar was fighting the Veneti in Gaul. The tribe lived in predominantly coastal locations and their strongholds were virtually impossible to attack by land. A naval assault was the only possibility, so Caesar built a fleet. But the tides made any attack by sea extremely challenging; the situation looked hopeless until Decimus Brutus arrived with a flotilla of vessels from the Mediterranean designed for speed, and much lighter and smaller than the ships of the Veneti. Even so, it was not till the wind died down and the heavy Gaulish ships were left unable to move that Brutus was able to attack them with great success.[9]

Fleet achievements were celebrated, though they were generally peripheral to the army's success. In the four triumphs he held in Rome after his war in Africa, Caesar celebrated the navy's contribution. As well as displaying the booty which he later distributed to the soldiers and the citizenry, the various military displays included a naval battle with 1,000 soldiers on each side, and 4,000 oarsmen propelling the vessels.[10] Over a century later Vespasian produced coins honouring naval victories, the only emperor ever to do so. Struck in his name and that of his son Titus, they bore the legend VICTORIA NAVALIS, and apparently commemorated a Roman victory in the Sea of Galilee during the Jewish War (see below).[11]

## ACTIUM

The final showdown between the forces of Antony and Cleopatra and those of Octavian at Actium off the north-west coast of Greece in 31 BC was primarily a naval battle. In fact it was the most decisive naval engagement in Roman history: the victory of Octavian's fleet, under the command of his general Agrippa, marked the end of the Republic. Plutarch likened the battle, though, to one fought on land. Neither side rammed the other's ships. Antony's ships were too heavy and could not build up the speed necessary to allow their bronze prows to pierce Octavian's ships. He had already deliberately destroyed the weaker ships in the Egyptian fleet and kept only the strongest, distributing 20,000 men and 2,000 archers among them. Antony's legions included Legio XVII Classica, whose name was derived from the Latin word for a fleet, *classis*; it must have been formed specifically to provide troops trained to travel and fight on ships.*

The conduct of the battle depended on the ability of the crews to row. Antony had chosen ships that had from three to ten banks of oars. Rowing was the only reliable way to control movement in battle; wind was far too unpredictable. The Roman soldiers, accustomed to fighting on land, were singularly unconvinced. A centurion was said to have protested to Antony, 'General, why do you distrust these wounds and this sword and (instead) put your trust in miserable logs of wood? Let the Egyptians and Phoenicians do their fighting at sea, but give us land on which we are used to standing and either conquer our enemies or die.' His ships' captains wanted to leave their sails behind, but Antony optimistically told them they must carry them so that no fugitive from Octavian's fleet could get away.

Octavian did not want to ram his ships' prows into those of Antony's vessels, while a sideways assault would have seen his rams broken off because of the huge heavy timbers used in Antony's vessels. Instead groups of three to four of Octavian's ships each attacked one of Antony's, the soldiers using spears, poles and missiles, while Antony's defenders fired catapults from towers. Then, when Agrippa on Octavian's left moved to encircle Antony's fleet, 60 of Cleopatra's ships took advantage of a favourable wind and made their escape.

---

* It is unknown if this legion was the same as the Legio XVII destroyed in AD 9 during the Varian disaster.

When Antony saw Cleopatra leaving, he followed her in a galley and abandoned the battle. Despite that, his fleet – largely unaware that their commander had left – held out for another ten hours, losing only 5,000 men before they gave up and handed over 300 ships to Octavian, followed soon after by Antony's land army.[12]

Just how dangerous it could be to rely on ships was highlighted that winter when Octavian faced a mutiny among troops he had sent to Brindisi in Italy while he wintered in Samos. Forced to sail back as a matter of urgency, on the voyage his fleet was struck by storms; some of the ships were sunk and his own vessel lost part of its rigging and had its rudder broken. Luckily he survived to suppress the rebellion; only then could he sail to Egypt to chase down Antony and Cleopatra.[13] Octavian had already survived a shipwreck as a young man, and lost two fleets to storms in the Sicilian war against Sextus Pompey before managing to defeat him in the end.[14] Had his ship been sunk this time, the entire course of western European history would have been altered.

Octavian later erected a monument at Actia Nicopolis ('the city of victory at Actium'), overlooking the maritime setting of Actium where he had camped before the battle. It was constructed as a sanctuary dedicated to Neptune, Mars and Apollo, and embellished with at least 23 prows (possibly originally as many as 35) of captured galleys fitted to sockets on the front of the wall of the monument. Since as Augustus he later claimed to have seized 300 ships, a number which probably omitted smaller vessels, each prow may represent ten of the seized vessels. Above the prows a monumental inscription with his titles for 29 BC declared that:

> Imperator ('the General') Caesar, son of the deified Julius, after the victory which he waged on behalf of the Republic in this region, when consul for the fifth time and declared imperator for the seventh, after peace had been secured at land and sea, consecrated to Neptune and Mars the camp from which he set forth to attack the enemy, now decorated with naval spoils.[15]

## FLEET COMMANDERS

Under Octavian – now Augustus – and the later emperors, the Roman navy consisted of a number of fleets berthed at key locations around the imperial coastline. A rough estimate of the numbers of men involved in the late second century AD is 30,000, the equivalent of about six legions. The two most important fleets were the Classis Misenensis and the Classis Ravennatis, based respectively at Misenum and Ravenna in Italy. Others dotted around the Roman world were probably both smaller and perhaps only made up to full strength when needed. They included those based in Britain (Classis Britannica) with a main fort at Dover, on the Rhine and the North Sea coast (Classis Germanica) and in Egypt (Classis Augusta Alexandrina). A number of others, such as the Classis Pontica in the Black Sea, are also known to have existed.

Each fleet was commanded by an equestrian prefect under the emperors. Marcus Mindius Marcellus is one of the earliest known and was a *praefectus classis* under Octavian some time between *c.* 36 and 27 BC, recorded on an inscription found at Velitrae, a few miles southeast of Rome.[16] Plinius Secundus (Pliny the Elder), in command of the Classis Misenensis in 79, took ships across the Bay of Naples to rescue people escaping the devastating effects of the eruption of Vesuvius. Famously, he lost his life to the toxic fumes on a beach while investigating the effects of the disaster.[17] Lucius Aufidius Pantera was prefect of the Classis Britannica in the late 130s. Appropriately enough he erected an altar to Neptune at another fort used by the fleet, Lympne in Kent.[18] An unnamed fleet prefect rescued Caracalla after he was shipwrecked en route from Thrace to Asia and had to climb into a skiff. The fleet concerned is unknown but the prefect was said to have been on board a trireme, presumably his flagship.[19] Tiberius Claudius Albinus was a *nauarchus*, 'captain of the squadron', in the fleet, and second in the chain of seniority after a prefect.[20] An individual ship (trireme or quinquereme) was commanded by a *trierarchos*, who presided over a crew that included a *proreta* in charge of the oarsmen, a *gubernator* (helmsman; our word 'governor' comes from the Latin) and a *medicus*, as well as centurions, the oarsmen themselves and marines.[21]

One of the most memorable fleet prefects was Quintus Marcius Hermogenes, commander of the Classis Augusta Alexandrina in the year 134 during the reign of Hadrian. He had time to head up the river Nile to explore the celebrated sights, as Hadrian himself had done four years earlier. Hermogenes crossed the Nile at Thebes (Luxor) and headed towards the famous Colossi of Memnon, which stood in front of what had once been the mud-brick mortuary temple of the Egyptian Eighteenth Dynasty pharaoh Amenhotep III (*c.* 1388–1349 BC). One of the statues had a natural fault in the stone. As the rising sun warmed the monolith each morning, it emitted a groaning noise. Visiting the statues and hearing the sound was considered to be a prime tourist attraction in antiquity. Hermogenes was lucky. 'At half past the first hour Quintus Marcius Hermogenes heard the Memnon', he proudly recorded in a third-person Latin graffito which he inscribed on the statue's lower leg, adding to a wealth of other Latin and Greek inscriptions still visible today.[22] He was in the nick of time. Within a few years an earthquake had toppled the statue. The enterprising Romans re-erected it, but the necessary repairs to the statue meant it never made the sound again.

The prefecture of a fleet could be one of the highest posts attainable in an equestrian's military career. Gaius Vibius Quartus started out as an ordinary soldier in Legio V Macedonica. From there he progressed to the rank of decurion in the Ala Scubulorum, prefect of the Cohors Cyrenaica, tribune of Legio II Augusta (in Britain) and prefect of Ala Gallorum, before becoming prefect of the Classis Augusta Alexandrina.[23] He appears to have had no experience whatsoever of naval affairs prior to his appointment to the fleet. Just as in the army, professional expertise does not seem to have been an essential component of a fleet officer's skill set. Tiberius Julius Xanthus, who lived to the remarkable age of ninety before dying in Rome, had two claims to fame. One was that he was the *subpraefectus*, possibly a deputy to the prefect, of the Classis Alexandrina at some time during his career. His tombstone, set up by his wife Atellia Prisca, also proudly recorded his role as a *tractatorus* of the emperors Tiberius and Claudius.[24] The principal meaning of *tractatorus* at the time was 'masseur', but the word also came to have other meanings such as 'inspector' or 'accountant of finances', because the root word *tractatio* meant the handling or management of almost anything. None of Xanthus' roles have an obvious nautical connection; the subprefecture seems to have been the only such position he ever held.

## Fleet bases

Fleet bases are not well known: either coastal erosion has destroyed them or their utility as harbours means they are now buried under modern ports. The Classis Britannica in Britain had a fort at Dover, fragments of which lie under the present-day port town. Built under Hadrian, Dover covered 2.5 acres (1 ha) and seems to have been broadly similar to an auxiliary cohort's fort, but with accommodation for as many as 600–700 men. The fort had easy access to the harbour and evidently operated in association with a pair of lighthouses (one of which still stands), guiding ships into dock. Excavations on the site suggested several periods of occupation, punctuated by years of disuse, presumably reflecting times when the fleet was stationed elsewhere.[25]

Across the Channel the fleet's main base was at Boulogne. Given the importance of London as a port, the presence of the governor and his garrison, it seems highly likely that the Classis Britannica had moorings there. It may even have been responsible for building some of the huge timber wharfs that have been found.[26] The Classis Germanica was based at Alteburg, 2 miles (3 km) south of Cologne on the Rhine, in a much larger fort (17 acres, 7 ha). In the third century the Classis Britannica probably used the new coastal forts of the Saxon Shore in Britain and Gaul, such as Reculver, Richborough and Portchester which were built to help in the campaign to fend off coastal raiders from northern Europe. The usurper Carausius (286–93), who used his command of the Classis Britannica to seize power in Britain and northern Gaul (see Chapter 11), may have played a role in commissioning additions to the series of forts.[27] Their remains are the most prominent relics of fleet bases anywhere in the Empire.

## FLEET MEN AND THEIR DUTIES

The fleets and their soldiers fought in wars, transported troops and goods, provided manpower for building military installations, and operated mines. The Roman Empire had spread so far by the mid-first century AD that seaborne military power was extremely important, especially for protecting the grain supply to Rome, the frontiers on the Rhine and Danube, and the coasts of Britain and Gaul. In 52 Claudius decided that the crews should be entitled to the same legal privileges as other veterans, and he issued a decree accordingly.[28] Fleet troops had however to serve longer than other soldiers at 26 years, rising to 28 under Septimius Severus, and were probably paid less too.[29] Whether the marines were treated the same as the crews, or were on equal terms with auxiliaries, remains a mystery.

The Classis Germanica is the only fleet for which any data about its size is known. During the Civil War in 69 it had 24 ships and all of them joined the revolt of the Batavian tribal leader Civilis. That makes it likely many of the crewmen had been recruited in the region, probably because of their knowledge of local waterways.[30] Few of the Classis Britannica's naval activities are known, unlike its land-based duties. In 70 it was poised to support Legio XIIII Gemina in the war against Civilis by raiding the Batavian homelands. In the event it was badly damaged by the Cannenfates, allies of Civilis, who destroyed most of it.[31] An inscription from Hadrian's Wall records the fleet's construction of a granary at the fort of Benwell between 122 and 126, during the period in which the first forts were being built along the new frontier. The fleet detachment was being used to perform a routine military construction task, reflecting the way fleet troops were so often used as any other detachment of the army might have been.[32]

An incomplete inscription concerning the Classis Britannica at the Beauport Park fleet bath-house in Sussex records a man called Bassianus who may have been an architectus; a small trace of the word possibly survives.[33] The iron-smelting activities of the Classis Britannica are almost entirely attested from the discovery of stamped tiles at installations found in southern England close to sources of iron ore.[34] The Beauport Park bath-house, which operated between c. 120 and 250, was buried by the collapse of a slag heap created by the smelting, which used vast quantities of charcoal obtained from the forests in the region. When the building was excavated, remains of almost the entire roof were found in and around the ruins. Although many tiles were smashed, a considerable number were not. This made it possible to calculate that the baths, covering 1,227 sq ft (114 sq m), had required a minimum of 5.7 tons (5,170 kg) of roof tiles, in addition to 7.1 tons (6,440 kg) of tiles of other sorts or indeterminate fragments.[35] The survival on a large floor tile of the impression of a tile comb, used to create a key for plaster on flue tiles, and bearing the CLBR mark of the fleet, shows that the tiles were made at a dedicated fleet works depot probably nearby.[36]

The fleet's involvement is easy to explain. Although today the Roman iron-working sites of south-east Britain are landlocked, in antiquity much of the area concerned could be reached by navigable inlets that stretched inland. The fleet ships were therefore able to transport out the iron pigs so that they could be carried to military sites in Britain and on the Continent where they were needed.

Other fleet vexillations turn up in a variety of places. A detachment of Saxons serving with the Classis Germanica worked in the quarries at Bröhl near Bonn, where the men set up a dedication to Jupiter Optimus Maximus and Hercules under Rufrus Calenus, the trierarchos.[37] This is particularly interesting because the context appears so incongruous for a sea captain, who might well have been surprised to find himself overseeing his men's efforts in a quarry rather than sailing on the Rhine.

Men from the Classis Misenensis were put to work on the awnings which covered the audience at the Colosseum. The reason was probably their expertise in handling ropes and sails.[38] A number of tombstones found in Rome of fleet soldiers probably belonged to those allocated to these and other official duties in the city. Titus Amydus Severus was one of them. He came from the Black Sea region, and served in the Misenum fleet on the roster of the trireme *Concordia*. He died at Rome aged twenty-five some time between the late first century and the end of the second.[39]

Detachments were also stationed at various locations on the Italian coast, such as Ostia and Puteoli. This information comes from a strangely amusing story in Suetonius' life of Vespasian. Suetonius calls the naval troops *classiarii* ('men of the fleet'), a word normally translated now as marines, though on their tombstones and other inscriptions they are usually called *milites*, 'soldiers'. The classiarii were annoyed at the cost of boots so the detachments at Ostia and Puteoli, the main ports serving Rome, went to Rome to see Vespasian and ask for a special allowance. They had not reckoned with Vespasian's legendary meanness and wit. He told them to march barefoot in future, and so they did. The anecdote also suggests that the marines spent most of their lives with their feet firmly on dry land.[40]

The ordinary soldiers of the fleet were allocated to centuries as legionaries and auxiliary infantry were but in their case centuria meant a ship's company. Like other auxiliaries, they shared the privilege that on retirement any existing children were enfranchised at the same time as their fathers.[41] The records of the ordinary classiarii of the fleet show that they could come from far and wide. Some of the tombstones found in Rome of Misenum fleet men show that they came from places as far apart as Cappadocia, Syria, Dalmatia, and Greece. Egypt was another major source of fleet recruits, like Apion (Antonius Maximus), the enthusiastic recruit of the Classis

Misenensis and Apollinarius (see Chapter 1). Fleet names were geographic descriptors of where they were stationed and not 'ethnic' labels in the manner of auxiliary units. Aemilius, son of Saenius, for example, was a Briton from the Exeter area, but he served as a soldier in the Classis Germanica under a captain called Euhodus (a Greek name) and died at Cologne.[42] Another Briton called Veluotigernus, son of Magiotigernus, was honourably discharged as a veteran from the Classis Germanica, then under the command of the prefect Marcus Ulpius Ulpianus, on 19 November 150, along with veterans from auxiliary cavalry and infantry units in Germania Inferior. His discharge diploma was found in Britain near the northern fort of Lanchester in County Durham, where he had perhaps retired and which might have been where he came from. He had enlisted in 124, around the time Britain's nearby northern frontier was being dramatically modified with the construction of Hadrian's Wall. He and his father's native British names both end '-tigernus', which means 'king' or 'master'. Magiotigernus meant something like 'great master', but the meaning of the Veluo- component of the veteran's name is unknown. On discharge Veluotigernus would have been Latinized into Titus Aelius Velvuotigernus (*sic*), taking the emperor Antoninus Pius' forenames as he became a Roman citizen.[43]

Surviving inscriptions record members of the fleets who, unlike other Roman troops, were sometimes inclined to mention both their Roman and their original names, with the formula *qui et* ('and who [were also named]'). Gaius Julius Victor was a soldier with the Classis Misenensis. He died at Misenum aged thirty, having served ten years but his tombstone adds that he was also known as 'Sola, son of Dinus'.[44] Lucius Antonius Leo, a Cilician who served with the same fleet, having signed up at nineteen and dying at twenty-seven, had been known as 'Neon, son of Zoilus'.

The division between the fleets and the regular army units is not clear, if indeed it really existed. Sometimes fleet troops were withdrawn from the navy and used to create a new legion. Soldiers from the Classis Misenensis were used by Nero to create Legio I Adiutrix ('the Rescuer'). Members of the Classis Ravennatis were used to form Legio II Adiutrix as part of the campaign to end the Revolt of Civilis. Within a year or two II Adiutrix had been moved to Britain, it remained there until 87, when it was sent permanently back to the Continent, ending up at Budapest. The tombstone of Valerius Pudens, a soldier of the legion who died only six years later in Britain, had a trident and a pair of dolphins carved into it to symbolize the new legion's origins.[45]

Where appropriate, a legion might include men with sailing skills, apparently independent of the fleets, only serving further to show how blurred the Roman military world was. Minucius Audens was a gubernator, a legionary helmsman, though he is the only such man known.[46] He made an offering to the mother goddesses of Italy, Africa and Gaul at York during his service with Legio VI Victrix, when he perhaps helped deal with the massive influx of men and materials during the Severan campaigns of 208–11. He may have served with a fleet before joining the legion, but on a religious dedication he would not have bothered to mention that. His role is a reminder that the army's colossal logistical requirements meant transportation of men and materials was an ongoing and essential part of its duties.

Fleet personnel were sometimes involved in major historical events. Anicetus, prefect of the Misenum fleet in 59, was one of Nero's freedmen and loyal stooges. He agreed to murder Nero's mother Agrippina the Younger by means of a specially designed collapsing boat in the Bay of Naples near the fleet base.[47] Indeed, the scheme was his idea. But it went disastrously wrong. Agrippina survived and had shortly afterwards to be murdered on land by Anicetus and a soldier from the fleet.[48]

It seems the navy was no less likely to be used for the emperor's personal purposes than the Praetorian Guard. In 69, during the Civil War, a dishonest centurion called Claudius Faventinus had a grievance. Having been cashiered by Galba, who had been toppled and murdered, Faventinus decided to do what he could to damage Vitellius, who was challenging Galba's successor Otho. He forged a letter, purportedly from Vespasian, offering the men of the Classis Misenensis a reward if they went over to him and abandoned Vitellius. The prefect of the fleet, Claudius Apollinaris, was not a man of reliable loyalty so he was easily bought. The upshot was that the cities of Puteoli and Capua decided to take sides too. Puteoli went over to Vespasian. Capua supported Vitellius and appointed Claudius Julianus, a former prefect of the Classis Misenensis, to lead some city troops and gladiators on their behalf. Julianus promptly changed sides and joined Vespasian. Faventinus' scheme had worked, and the fleet's actions helped hasten Vitellius' downfall.[49]

## Ships of the fleets

Ships of the fleets during imperial times are virtually unknown, either from sources or physical remains. There are however various references to quinqueremes, triremes, transports and even rafts. Some individual vessels are named. As we know, the Egyptian recruit Apion (Antonius Maximus) apparently joined the ship's company of the *Athenonica* at Misenum. Titus Memmius Montanus, a soldier of the Classis Ravennatis, served on the quinquereme *Augustus* in the year 150.[50] Tombstones of fleet soldiers in Rome give a variety of ship names drawn from the names of deities and also the personifications of virtues such as *Hercules*, *Apollinus*, *Minerva*, *Fortuna*, *Pollux* and *Fides*. One was called *Isis*, an appropriate echo of the Egyptian homeland of many of the recruits, served in by a Cilician called Gaius Mucius Valens.[51] A small bronze model found in London of a warship's prow with the inscription AMMILLA AVG FELIX probably names one of the Classis Britannica vessels, *Ammilla Augusta*.[52] A coin of the usurper Carausius, known from only one example, depicts a galley and the legend PACATRIX AV(G) ('Peacemaker? of the Emperor'), perhaps his flagship.[53]

It would be a mistake to imagine that fleets consisted exclusively of warships with multiple banks of oars and battering rams. The *liburna* was a light warship, designed for speed. Some of the Misenum fleet soldiers who died in Rome came from their crews. Marcus Ulpius Maximus was a Thracian who served on the liburna *Armata* ('The Armoured'). He died at forty-seven, after 28 years' service.[54]

The fleets also included transports. The Classis Misenensis might have possessed a raft (*ratis*) called the *Minerva*, though this relies on an uncertain reading of the tombstone, found in Sardinia, of the infant son of Valerius Frontus who was a soldier with the fleet; *Minerva* might be the name of a vessel, but it is an odd item to include on a child's memorial.[55] 'Heavily manned' rafts were used with great success by Vespasian in a small naval engagement on Lake Gennesaret during the Jewish War and indeed seem to have been the main vessels used to defeat the enemy boats and, apparently, kill the entire Jewish force.[56] In the winter of 214–15 Caracalla was preparing for war against Armenia and Parthia. He ordered the construction of 'two large engines' (siege or artillery machinery) to be used in the fighting. They were specifically designed to be taken to pieces so that they could be more easily carried to Syria by his naval transports.[57]

If a fleet was ever short of real ships there was always the possibility of pretending there were more. In the early second century BC Cato the Elder arrived with his fleet at Ambracia, a city at the time a member of the Aetolian League which was at war with Rome. Since his fleet had been blockaded by the Aetolians, Cato had arrived with only the ship he was sailing in. He resorted to making audible and visual signals as if the rest of the fleet was nearby and was now being summoned to follow him, and that his troops were on hand too. The Aetolians were fooled and called off the blockade out of fear that the Roman fleet was on the point of annihilating them.[58]

## FANTASTIC BEASTS IN THE STORM OF 16

In AD 15 Germanicus, during his campaign in Germany, ordered the II and XIIII legions to make a journey by land so that the fleet ships would be less heavily loaded and better able to negotiate the shallows. All went well until a storm flooded the land where the legionaries were, the water carrying off baggage animals and the soldiers' packs and causing havoc with the marching formations. Eventually the surviving men caught up with the fleet and were taken on board.[59]

No doubt this prompted Germanicus to avoid a repeat performance the following year. But instead of keeping his men safe, a spectacular storm hit Germanicus' huge fleet when he set out at the end of his campaign. With the intention of returning the majority of the legions to their winter quarters by ship, the fleet sailed out to the North Sea down the river Ems. But soon hail wiped out any visibility and was accompanied by a dangerous swell that prevented the steersmen from maintaining course. This terrified the legionaries on the ships; as ordinary soldiers, most of them had no idea what was going on or how to react. They obstructed the sailors by panicking or made inappropriate attempts to help. As if that was not bad enough, a severe gale blew the ships in all directions.

The crews were left desperately trying to avoid being washed up onto rocks, but all their efforts came to nothing when the tide turned and joined in with the wind. Desperate conditions meant desperate measures, so the men started throwing cavalry horses, pack animals, equipment and weapons overboard. Their efforts were futile. Some of the ships sank when the sea overcame them. Others were blown onto the shores of islands around the North Sea, where the soldiers starved unless they were lucky enough to find the rotting bodies of their horses washed up there too. Though the crew of Germanicus' trireme managed to bring the ship safely to shore on the German coast, his sense of devastation and disaster was so great that he contemplated suicide.

When the storm finally subsided the surviving ships managed to regroup, some of their crews having to use their clothing as sails. The ships were patched up and sent out again to find as many of the lost soldiers as possible. The coastal German Agrivarii tribe only gave up marooned soldiers when a ransom was paid, although tribal chieftains in Britain handed over any men who had been washed up there. Yet the terrible storm paid a form of dividend. The men who lived and were found came back with extraordinary tales to tell in the grand tradition of mariners of all ages. No doubt inspired by the adventures of Odysseus, Aeneas and the Argonauts that they had read or been told about as boys, they insisted they had seen whirlwinds, unknown varieties of birds, monsters from the deep, and fantastic beasts who might have been men or animals or both.[60]

It was not the first time Roman forces told tales of terrible creatures. In 256 BC Atilius Regulus had won the naval battle at Cape Ecnomus in the First Punic War before invading Africa. The following year his army was confronted at the mouth of the river Bagrada by a giant snake 120 ft (36 m) long. The animal seized soldiers with its mouth and crushed others with its tail. Spears proved useless. Only bombardment by catapults and with stones finally killed it. The snake's skin was removed and sent to Rome, while the reek from the decaying body was so repulsive the Roman army camp had to be moved.[61]

The role of naval ships at Actium made that occasion one of the most important and game-changing sea battles of all time. More often the Roman army turned the tide of history on dry land, and frequently in unedifying ways. With their shifting loyalties and willingness to be bought, Roman soldiers were often ready to join mutinies and rebellions that changed the course of history – especially once they discovered that an emperor was only as good as his word, while they were as good as their swords.

# ELEVEN

# MUTINEERS AND REBELS

## King-Makers for Sale

*As usual, most of the common soldiery were eager for any new commotion; and the darkness did away with the deference of the better men.*

Tacitus reporting how random
circumstances could lead to a mutiny[1]

Roman soldiers were supposed to reserve their unquestioning loyalty for the emperor. In reality they were often more faithful to their commanding officer, especially if he was a successful general who took care of them and promised success and glory. Likewise, soldiers who were ill-trained, badly led, badly treated, unpaid, defeated or humiliated were liable to have poor morale. Such men were easy prey for rabble-rousers, opportunists and chancers. Almost anything could provoke an outbreak of unrest, from outright boredom to the prospect of loot or the spontaneous gratification of bloodlust. The result was one-night mutinies, rebellions in frontier garrisons, and wars in which Roman soldiers fought other Roman soldiers, and even set up rival emperors.

## Mutinies in the Republic

Just how volatile Roman soldiers could be was shown in 88 BC when Marius, at the time a private citizen who held no office, was placed in command of fighting the war against Mithridates VI of Pontus. When Marius sent his legate Gratidius to Sulla, who had been in charge of the war, to take over the legions, the soldiers were so outraged at the thought of being led by a man with no official authority that – encouraged by Sulla – they killed Gratidius, defying the Roman state. They also stoned the tribunes to death. Sulla then marched on Rome with six legions, committing the outrage of bringing an army over the sacred boundary of the city.[2] Only the year before, Aulus Postumius Albinus was commanding the fleet during the Social War against Rome's rebellious allies when his men killed him by stoning. He had been charged with treason but he was apparently guilty of treating his men with cruelty. As a result the troops were not punished, Sulla saying they would fight more bravely in order to expiate their guilt.[3]

Boredom and greed were only two of the reasons Roman soldiers might end up mutinying. In 49 BC even some of Caesar's soldiers had had enough. At Piacenza, during Caesar's war against Pompey, they rebelled. They said they were worn out, but the real reason was that Caesar had ordered them not to plunder the land or do anything else they had decided they wanted to. The soldiers knew Caesar needed them and thought they were in a strong position, but they were getting ideas above their station. Caesar managed to suppress the mutiny, and rebuked them on the grounds that they were reliably paid in full, had plenty of food and yet were unsatisfied. He also scolded them for thinking they were better than others because they were soldiers, and that because they were armed they were entitled to do other people harm.[4]

## Percennius, the barrack-room troublemaker

Mutinies were sometimes provoked when the emperor and his commanding officers took soldiers for granted. One of the easiest ways of doing this was to ignore the terms of service instituted by Augustus himself. His death in AD 14 occasioned a mutiny amongst the Pannonian legions in the Ljubljana region (Slovenia) that summer, soon after the accession of Tiberius. Three legions, VIII Augusta, VIIII Hispana and XV Apollinaris, were based in one vast legionary fortress, a dangerous concentration of forces when grievances took root. At the time the Pannonian units were under the command of a former consul called Quintus Junius Blaesus, whose nephew happened to be an ambitious young man called Sejanus, the new praetorian prefect bent on worming his way into Tiberius' trust. Blaesus suspended normal military routines in observation of Augustus' death. It was a fatal mistake. Without the distraction of everyday duties the legionaries, who were already disgruntled, started listening assiduously to a barrack-room troublemaker. Named Percennius, he had a sharp eye for what were quite legitimate grievances over ludicrously extended periods of service, the substantial pay differential with praetorians, who had much more comfortable lives in Italy guarding the emperor, and the deductions for equipment. Blaesus managed to placate the malcontents and agreed that his son would lead a deputation to Tiberius on their behalf.[5] Unfortunately, this only achieved a brief respite before the mutiny erupted again, this time over Blaesus' personal troop of gladiators whom he employed to execute soldiers. Before long even Legiones VIII and XV were planning to fight each other.

When the news reached Tiberius, he sent his son Drusus with two praetorian cohorts and a group of 'first citizens' (these included senators who had reached proconsular rank) to find out what was going on and decide what to do. Drusus probably collected the praetorian cohorts from Aquileia, where some were stationed, on the route from Rome to Pannonia. He and the two cohorts were also accompanied by a unit of praetorian cavalry and the best troops in the German unit, described by Tacitus as 'the guards of the emperor'. Sejanus was sent too, so that he could watch over Drusus; this was a mark of his already conspicuous influence over Tiberius.

But dealing with the mutiny proved far from straightforward. A centurion called Clemens put forward the mutineers' demands: a denarius per day as their pay, a maximum of 16 years' service, and the release of veterans. Drusus prevaricated, insisting any decision was down to Tiberius and the Senate. The infuriated rebels openly threatened praetorian soldiers, while the praetorians themselves, along with the legionary centurions, killed any soldiers who strayed outside the camp. The agitation carried on until the superstitious soldiers were distracted by the way the moon kept being obscured by clouds before reappearing, uncertain whether the omen was in their favour or not. As dawn broke, Drusus seized the initiative and had Percennius and any other ringleaders killed. In the end the filthy weather of the winter of 14–15, seen by the men as a sign of the anger of the heavens, brought the mutiny grudgingly to an end.[6]

## MUTINIES IN THE CIVIL WAR

The Civil War that followed the death of Nero in 68 was effectively a series of rebellions against the state in which one emperor after another overthrew his predecessor. Galba was the first to take power. But in January 69 Galba had only been emperor for a few months when his reign came to a violent and bloody end at the hands of rival Roman soldiers who had decided to side with Otho, a senator and friend of Nero. Tacitus said 'Roman troops prepared to murder an old defenceless man who was their emperor' as if they were about to topple a foreign enemy. As it turned out, some of the soldiers involved had self-interest much at heart.

Soldiers might be motivated to rebel for some point of principle, but overthrowing Galba was down to money. While plotting to get rid of Galba, Otho put a freedman of his called Onomastus in charge of the arrangements. Onomastus knew that two of the praetorian guardsmen, a couple of ne'er-do-well non-commissioned officers by the names of Barbius Proculus and Veturius, were notoriously corrupt. They came to see Otho, who was delighted to discover that they were expert at bribery and securing favours. Otho promptly handed over enough cash to enable them to rustle up as much support as possible. 'In this way', said Tacitus, the two men 'agreed to dispose of the Empire which belonged to Rome'.[7]

In the confusion that followed the murder of Galba, no one was sure who had dealt the blow. It might have been a veteran called Terentius or someone called Lucanius, but the general verdict was that it was Camurius, a soldier of Legio XV, who had plunged his sword into Galba's throat. His fellow legionaries joined in the assault, slashing at Galba's unprotected arms and legs, continuing to hack into his body even once his limbs had been cut off. Another legionary, Julius Carus, thrust his sword right through Titus Vinius, one of Galba's chief henchmen. Sempronius Densus, a praetorian solidly loyal to Galba, took exception to what had happened to his master. 'With his dagger drawn, he denounced the mutiny and advanced towards his armed enemies.' That 'real act of heroism' allowed another of Galba's cronies, Piso, to make his escape, only to be killed on the steps of the temple of Vesta in the forum by Sulpicius Florus, an auxiliary cavalryman based in Britain but currently in Rome to receive an award of citizenship from Galba, and the praetorian Statius Murcus.[8]

It was a 'long day of villainy', said Tacitus. There was a bloody conclusion to this phase of the rebellion which showed how fragile the role of emperor could be, and what could happen to an emperor's associates during violent regime change. Tacitus described how 'the victims' heads were impaled and carried in a procession', the standards of the praetorian cohorts and a legionary eagle coming behind them. The mutineers for their part then competed by showing off who had the most blood dripping from their hands, while arguing who had actually done the killing and who had only witnessed it, or whether they were telling the truth. Revolting though the story might seem now, it has to be remembered that the Romans were essentially an Iron Age people. At the time the head trophy was seen as a powerful symbol of an enemy's destruction.

Why did they do it? The answer came in the form of 120 applications made to Otho, now emperor, for a reward for their services. The mob dashed to the Castra Praetoria in a desperate bid to side with the troops to save their own skins. No wonder Tacitus said 'after that, the troops got their way with everything'. The benefits turned out to include the praetorians' ability to choose their own prefects and no longer had to bribe centurions to be exempted from duty. Since that meant a rich soldier could get away with doing far less than his fellows, and ending up hopelessly out of condition and unskilled, it seemed like a genuine reform and showed how Roman military rebellions were easily provoked by corruption in the army. Unfortunately, Otho could not afford to alienate the centurions, so he ended up compensating them for the loss of their illicit income. But the reform survived, thereby genuinely removing an abuse.[9]

Out in the provinces the legions, one by one, began to declare either for the rival emperors Otho or Vitellius as the year 68 came to a close. Some soldiers were reluctant to give up on Galba. Four centurions of Legio XXII – Nonius Receptus, Donatius Valens, Romilius Marcellus and Calpurnius Repentinus – tried to stop the soldiers ripping up portraits of the former emperor on the standards, but the legionaries hustled them off and locked them up. As Tacitus said, the real point was that the soldiers had completely overturned their oaths of loyalty to Galba – a dangerous development.[10]

A simple misunderstanding and too much alcohol could also lead to a mutiny spontaneously breaking out. After becoming emperor, a position he held on to for only three months in 69, Otho instructed Cohors XVII Urbana to move from Ostia to Rome. The praetorian tribune Varius Crispinus was told to open the armoury at the Castra Praetoria and hand out weapons to the urban cohorts. Unfortunately,

Crispinus decided to do this at night when the camp was quiet. He had overlooked the fact that there were drunken praetorians loafing around who became excited when they saw weapons being loaded onto wagons for the urban soldiers.

The praetorians decided this was an outstanding opportunity to go off looting in Rome, so they beat up Crispinus and any centurions who tried to stop them. The armed drunks made their way to the palace, Otho's dinner guests fleeing at their arrival. With absolutely no idea what they were doing, the inebriated praetorians fulminated generally about the Senate, the centurions and their officers. The mutiny was turning into a farce, albeit a dangerous one. Otho pleaded with them to go back to their barracks, and an offer of 5,000 sestertii per man magically defused the crisis. When Otho turned up to address them the rebels were stripped of their uniforms by their centurions and tribunes; crushed with shame, they asked to be discharged from the Guard. No doubt sobered up and with sore heads, they said no more except to denounce anyone who had been a ringleader, and went back to their normal routines.[11]

## THE REVOLT OF JULIUS CIVILIS, 69–70

Batavian troops played an important role in Rome's wars in the middle of the first century. Exempted from paying tribute to Rome, they instead made a substantial contribution of infantry, cavalry and armaments, being 'reserved for war [only]'.[12] Having fought for Tiberius, Drusus and Germanicus in Germany from 12 BC to AD 16, and over a generation later in the conquest of Britain, they were greatly admired for their fighting prowess, especially their ability to cross rivers while fully armed, but there was always the danger such troops would change sides and fight for their own people.[13] Gaius Julius Civilis, a member of the Batavian royal family and a Roman citizen, remains a slightly mysterious figure who rose to prominence under Nero and during the Civil War of 68–9. He had 'a cunning intelligence beyond what was normal for a barbarian', said Tacitus.[14] When Vitellius ordered that Batavian troops be forcibly levied to support his army, the soldiers in charge of the levy disgraced themselves by coercing bribes for exemption out of the sick and old, and debauching the best-looking young men.[15] Civilis' extremely dangerous rebellion, in which he was joined by other tribes, started in late 69 and lasted into 70. Legio V Alaudae was one of the units that suffered badly. Before long the confederation disintegrated and Civilis was defeated by a Roman army. Although it was a sideshow in the turbulence of the time, the Revolt of Civilis came close to destroying Rome's hold on north-east Gaul and the Rhineland, and showed how dangerous a trained ally could be if he turned.

## RECKLESS MUTINEERS

Some recruits quickly changed their minds when army life turned out to be tougher than they had anticipated. In around 80–2 an auxiliary cohort was raised from the German Usipi tribe and sent to Britain to take part in Agricola's campaign. This comprehensively backfired when the men swiftly took exception to the brutal training regime. They mutinied, killing at least one centurion and other soldiers who had been detailed to instruct and lead them. Three galleys with a pilot apiece were stolen and the reckless mutineers set off, probably from the Clyde estuary in Scotland, to sail home, though they had not the slightest idea what they were doing. Distrustful of the pilots, they killed two, after which the voyage degenerated into disaster as they sailed north around Scotland and down the eastern side. Lost, dispersed and disoriented, they were beaten off by the Britons when they landed having hoped to requisition supplies, and ended up resorting to cannibalism. The survivors were eventually captured by other tribes and sold into slavery, their short careers as Roman soldiers well and truly over.[16]

They had, however, incidentally demonstrated that Britain was an island. Agricola was inspired to order the voyage to be repeated so as to confirm the discovery.[17]

## The Rebellion of Lucius Antonius Saturninus

There was always the danger that a concentration of Roman forces might give an aggrieved provincial governor or legionary commanding officer ideas about mounting a rebellion. In 89, during the reign of the increasingly paranoid and cruel emperor Domitian, Lucius Antonius Saturninus, governor of Germania Superior, had used the fact that he had Legio XIIII Gemina and Legio XXI Rapax together at Mainz to lead a revolt. The rising only collapsed because the frozen Rhine thawed out unexpectedly, preventing Saturninus' German allies from crossing the river to support him. Saturninus was defeated by another Roman general called Lucius Appius Maximus Norbanus. Domitian then moved to prevent other governors or generals from taking similar action. One of his precautions was to ban soldiers from having more than 250 denarii per head in the unit's savings, so that a would-be rebel could not take advantage of the money. Domitian also followed the rebellion up with a purge, preventing any record being made so that no one would know how many victims had been executed. For his part, Maximus had destroyed all of Saturninus' papers to prevent Domitian blackmailing anyone, earning the admiration of the historian Cassius Dio, who hated Domitian. Dio said, 'I do not see how I can praise Maximus enough'.[18]

## The Maternus revolt

In around 187–8 a soldier called Maternus was another to turn against the army in which he had started his career. Disaffected, he deserted from one of the Rhine garrisons and was able to inspire other soldiers to leave the army and join him, forming not so much a rebellion as a criminal gang of malcontents who proceeded to raid villages and farms. The robberies provided him with so much money that other criminals were attracted to join him on the promise of a share of even more loot. Maternus' men attacked the larger cities, broke into prisons and released prisoners who joined them out of gratitude at gaining their freedom. Eventually, settlements across Gaul and Spain were affected, or so Herodian reported some years later.

Maternus' crime wave must have happened quickly to reach so far, because as soon as the emperor, Commodus, heard about it he furiously castigated the governors of the Gaulish provinces Aquitania (Pescennius Niger), Belgica (Clodius Albinus), and Lugdunensis (Septimius Severus) for letting things get out of control and ordered a force to be put together immediately to crush Maternus. Hardly believing how much he had been able to achieve, Maternus was however far too clever to risk a major confrontation. He also had designs on trying to seize the whole Empire. He ordered the plunder to stop, split his men into much smaller groups and told them to make their way surreptitiously into Italy.

On 25 March the Romans, including the emperor, would be celebrating the feast of Hilaria (Joy). This involved a parade in which anyone could dress up in whatever costume they liked. Maternus planned that he and his men would disguise themselves as praetorians, work their way into the crowd undetected and at an opportune moment attack Commodus and kill him. The plan might have worked had not some of Maternus' men, bitterly resenting the idea that their robber chieftain was about to become emperor, gone to Rome and given the game away. Maternus was arrested on 24 March and beheaded. Other conspirators were caught too and Commodus was able to organize a public thanksgiving.[19]

The most astonishing aspect of the whole story is how close Maternus was able to come to assassinating an emperor after an extraordinarily public and successful career as a crime gang leader. The incident went down in Roman history as the 'War of the Deserters' (*bellum desertorum*). Ironically, each of the three governors ordered to suppress Maternus tried to become emperor themselves after the murder of Commodus on the last day of 192, although only Severus succeeded. The three were not alone. The story of how the Praetorian Guard triggered the events of 193 is one of the most remarkable in Roman military history.

## Auctioning the Empire

Military forts and camps were potentially dangerous places where rumour and imagination ran riot, and where soldiers could find themselves believing almost anything. 'The greatest potential for imagination and credulity', said Tacitus, 'was in the camps themselves, with the soldiers' hatred and fears, and their self-confidence when they considered their strength.'[20] A concentration of soldiers, especially disaffected ones, could spell trouble. In early 193 the praetorians were to change the course of history in one of the most degrading spectacles ever to involve the Roman army. The entire scheme was cooked up in their camp, the Castra Praetoria, out of sheer greed and a complete abrogation of responsibility.

After Commodus was murdered on 31 December 192, the experienced senator and general Pertinax became emperor. He took care to win the Praetorian Guard's loyalty with an offer of 12,000 sestertii per man, which backfired because it seems he may only have ever paid half of it.[21] In any case, the people of Rome had been so worried the Guard would refuse to be ruled by Pertinax – who had earned their respect thanks to his military reputation and his service as prefect of Rome – that a huge crowd had gone to the camp to coerce the soldiers to accept him.[22]

The praetorians were horrified when Pertinax told them he would be setting right all sorts of distressing circumstances. These were part of his plans for a more widespread programme of reform. Used to behaving more or less as they wanted to under Commodus, and outraged at the thought of their privileges disappearing under

Pertinax's clampdown,[23] the praetorians and their prefect Aemilius Laetus became embroiled in yet another plot, this time to get rid of the new emperor. The first scheme they dreamed up was to seize power while Pertinax was at the coast inspecting the grain supply, although the conspiracy was uncovered in the nick of time and Pertinax was saved.[24]

Laetus turned on the praetorians so that they would be blamed rather than him. He executed some of them, claiming that he was following Pertinax's orders. As a result 200 infuriated praetorians went straight to the imperial palace and forced their way in. Pertinax tried to negotiate with the angry praetorians. One praetorian stopped him from leaving. Meanwhile palace staff let the soldiers in through the other entrances. When they confronted Pertinax one praetorian leaped forward and stabbed him, announcing that 'this sword is sent you by the soldiers'. Pertinax's body was beheaded and the trophy displayed on a spear.[25]

Terrified of the consequences of what they had done the praetorians dashed back to the Castra Praetoria and locked the gates.[26] Strangely, nothing happened and the praetorians realized not only that no one had come after them but also they were the ones who were really in control. Emboldened by this startling discovery, they decided to post a notice outside the camp offering the Roman Empire for sale.[27] Most of the senators were suitably disgusted,[28] but the greedy and ambitious Marcus Didius Julianus had spotted an opportunity. So had his equally greedy and ambitious wife and daughter, Manlia Scantilla and Didia Clara, and two praetorian tribunes called Publius Florianus and Vectius Aper. They raced round to the Castra Praetoria. They were not the only ones. Titus Flavius Sulpicianus, who was the prefect of Rome as well as being Pertinax's father-in-law, arrived too. When Didius Julianus reached the camp he found himself locked out while Sulpicianus busily secured his position with the Guard.[29] Only by means of placards advertising his promises, and thanks to a Sulpician supporter called Maurentius who changed sides, was Julianus able to attract the praetorians' attention.

The praetorians now knew no one could hope to rule without their backing. Didius Julianus and Sulpicianus, both desperate to become emperor, started making rival cash offers to the Guard. The soldiers enthusiastically threw themselves into the auction, running across the camp from one candidate to the other to tell each how much he would have to raise his bid by. Sulpicianus was about to win with an offer of 20,000 sestertii per praetorian when Julianus seized the day with a reckless counter-bid of 25,000. Julianus added for good measure the warning that Sulpicianus might seek revenge for the death of Pertinax and also that he, Julianus, would restore all the freedoms the praetorians had enjoyed under Commodus. So delighted were the praetorians by the new offer they promptly declared Julianus to be the new emperor.

The Praetorian Guard had brazenly created an emperor purely on the promise of a huge cash handout, nakedly abusing their position and power. Didius Julianus could not possibly afford the money, but for the moment the greedy praetorians did not realize that. They gathered up the new emperor and carried him to the forum. This blatant public display of their power had the desired effect. Julianus indulged the conceit that he had come 'alone' to address the Senate, though in reality a large number of armed troops secured the building outside and some came with him into the Senate.[30] Under the circumstances it was hardly surprising that the Senate should confirm by decree that Julianus was emperor. He also tried to ingratiate himself with the Guard by accepting their personal nominees for the praetorian prefecture. Unfortunately for Julianus, not only did he lack the funds to pay the handout he had promised, but the imperial treasuries were empty after Commodus' reckless rule. The praetorians started publicly humiliating Julianus.[31]

Didius Julianus' reign fell apart almost as soon as it started. In the east Pescennius Niger, governor of Syria, had been declared emperor by his troops after they heard about the murder of Pertinax. In Rome a frustrated mob began demonstrating in Niger's favour, angered at how Pertinax had been prevented from reforming Rome. Soon the crowd started fighting with 'soldiers', which must mean the urban cohorts and the praetorians. Niger was not alone in having designs on becoming emperor. The garrisons were spoiled for choice. Lucius Septimius Severus, governor of Pannonia Superior, had supported Pertinax but after his death had been declared emperor on 9 April 193 by his own troops.[32] In the west, Clodius Albinus, the governor of Britain, had also made it clear he wanted to be emperor. Another civil war broke out, driven by the allegiances of the army.[33]

Julianus told the Senate to declare Severus a public enemy. Next he had a defensible stronghold constructed and readied the whole of Rome for war, fortifying his own palace for a last stand. The preparations turned into a farce. Years of easy living cultivated by Commodus had left the praetorians without much idea of what they were supposed to do. The Misenum fleet troops stationed in Rome to work on awning machinery in the Colosseum had even forgotten how to drill. Used to being unarmed, and completely out of condition, the praetorians were told to equip themselves, get back into training and dig trenches.[34]

The praetorians were overwhelmed by the work they had to do. They were also bothered by the prospect of confronting Severus' Syrian army, which was approaching through northern Italy. Severus sent letters ahead promising the praetorians they would be unharmed if they handed over Pertinax's killers. They obliged, which meant that Julianus was finished. Severus also sent an advance force to infiltrate the city, disguised as citizens. The Senate, realizing that the praetorians had abandoned Julianus, voted for his execution. The sentence was carried out on 2 June 193, after which they declared Severus to be emperor. Didius Julianus had reigned for a little over two months.[35] In the space of five months in 193, working out of the Castra Praetoria, the Praetorian Guard had toppled and murdered an emperor and installed another, only to abandon him in short order.

The praetorians were to pay a heavy price for their behaviour. Their tribunes, who had gone over to Severus, ordered them to leave their barracks unarmed, dressed only in the *subarmilis* (under-armour garment) and head for Severus' camp. The new emperor stood up to address them, but it was a trap. The praetorians were immediately surrounded by his men, who had orders not to attack them but to contain them. Severus then ordered the execution of those responsible for murdering Pertinax.[36] He slated the praetorians for being disloyal to Didius Julianus, in spite of his shortcomings; in his eyes what really damned them was that they had completely ignored their own oath of loyalty to an emperor.

Severus was being clever. He avoided creating any impression he was buying the Empire as Didius Julianus had, even though previous emperors, including those as respectable as Marcus Aurelius, had paid a donative. He said he would spare the praetorians' lives, but only if they were stripped of rank and equipment and cashiered on the spot. Sacking them en masse saved him the expensive prospect of an accession handout or a retirement gratuity.[37] The praetorians' uniforms, belts and military insignia were forcibly removed by Severus' legionaries, and they were made to part with their ceremonial daggers. To prevent the humiliated praetorians rushing back to the Castra Praetoria in their underwear and arming themselves, Severus had sent a squad ahead to secure the camp. The praetorians had to abandon their horses and then ignominiously dispersed into Rome. One was in such despair he killed his animal before committing suicide.[38]

Severus replaced the cashiered Guard with his own experienced legionaries. Lucius Domitius Valerianus from Jerusalem joined the new praetorians soon after 193. He had served originally in Legio VI Ferrata before being transferred to Cohors X Praetoria, where he would stay until his honourable discharge on 7 January 208. Valerianus cannot have been in the Guard before Severus reformed it in 193, which means that his 18 years of military service, specified on the altar he dedicated, must have begun in the legions in 190 under Commodus. Valerius Martinus, a Pannonian, lived only until he was twenty-five; by then he had already served for three years in Cohors X Praetoria, following service in Legio XIIII Gemina. The legion had declared early on for Severus, so Martinus probably benefited from being transferred to the new Guard in 193 or soon afterwards.[39]

The new praetorians were not necessarily any more reliable, though it depended on who was emperor. Severus' son Caracalla lasted six years after a turbulent and murderous reign that more or less guaranteed a coup. When the revolt came it was led by Macrinus, the praetorian prefect, who had made sure Caracalla's guard escorts would support the challenge. On a journey between Carrhae and Edessa Caracalla made the fateful decision to dismount in order to relieve himself. His bodyguards were all around him but did nothing when his groom stabbed him to death with his dagger.[40]

## THE CARAUSIAN REVOLT

One of the most remarkable rebellions took place on the fringes of the Empire. In 284 Mausaeus Carausius, a soldier from Menapia (a region roughly equivalent to Belgium), excelled himself in a civil war in Gaul. He fought for the joint emperors Diocletian and Maximianus, leading soldiers against the Bagaudae, landless outlaws whose livelihoods had been amongst the casualties of the disorder of the age. By 286 Carausius had crushed the Bagaudae. Maximianus was so impressed that he appointed Carausius to lead the Classis Britannica against pirates in the North Sea. Carausius did well. Too well, in fact. Stories were put about, perhaps by Maximianus himself, that Carausius had let the raiders through his blockade so that they could land in Britain or northern Gaul, where they would attack villas and towns. He had then apprehended them on their way home, their ships laden with loot, and either helped himself to the treasure or charged a commission. Either is possible, but it is no less likely that by successfully clearing the sea of pirates Carausius was gaining a degree of popularity that infuriated Diocletian and Maximianus. No emperor could afford a military commander with that sort of fame, especially among the troops.

In 286 Carausius declared himself emperor in Britain. He may have been planning to do so all along, or he may have made the decision on the spur of the moment to save himself. Though far from being the first Roman soldier to become an emperor or rebel emperor in the third century, Carausius had an eye for the moment. One of his most successful predecessors was Postumus, who had established the so-called Gallic Empire in Britain, Gaul and Germany in 259, lasting till 268 when he was murdered. Growing up during those days, Carausius must have been influenced by the experience.

Meanwhile Diocletian and Maximianus and their supporters had subscribed to a reactionary form of propaganda. They were habitually praised by imperial panegyricists in terms and phrases borrowed from the literature of the Augustan era and even earlier. It was a way of attempting to lay claim to being the restorers of the good old days.[41] They ruled the Empire jointly as a more efficient way of controlling the borders: Diocletian in the East, Maximianus in the West. Carausius imitated and emulated the legitimist regime from the outset, including its pretensions to being the noble defender of a past paradise when Rome had been great. Although he only held Britain and part of Gaul, Carausius posed as the all-time original Roman ruler, the man who in his own words would be 'restorer of the Romans'. Carausius was not in any sense attempting to create a revived pre-Roman independent Britain. Instead, starting in Britain, he was promising to revive Rome and all it stood for.

Almost all that we know about Carausius' regime has had to be pieced together from a few scattered references in imperial histories, from imperial panegyrics, and from his coinage. One of the best pieces of evidence that the rebellion had not been long in the planning was Carausius' initial reliance on over-striking existing coins with his own image and slogans. The original designs are often still visible on these coins. Soon completely new coins were being made in his mint at London, in its first era as a mint town, and at one other uncertain location. Carausius borrowed numerous coin types from recent emperors such as Probus and the rebel Postumus. Around 80 per cent featured the banal reverse type PAX AVG, which recalled the Augustan Peace (Pax Augusta) and promised liberation from years of war. However, it is his special issues that demonstrate the regime's flamboyant and imaginative command of Roman iconography and imagery. Utilizing phrases, words and images from the works of the Augustan poet Virgil, Carausius plundered the classical Roman tradition to portray himself as a messianic figure who would revive Rome's 'golden age'.

Some of Carausius' silver coins bear what looks like a mint mark, the letters RSR. These are now recognized to stand for a line from Virgil's Fourth Eclogue, *Redeunt Saturnia Regna*, an idiomatic Roman expression that means 'the Golden Age is back'. A pair of unique Carausian medallions confirms this. One bears the same three letters, the other the letters INPCDA. These represent the next line of the same poem, *Iam Nova Progenies Caelo Demittitur Alto*, 'Now a new generation is let down from Heaven above'. It was a messianic promise, suggesting that Carausius was a saviour. Given the tone and nature of the legitimate regime's propaganda, there is a real possibility that the lines were taken by Carausius from panegyrics composed in his favour but which no longer exist. If so, then one can imagine Carausius' troops chanting them when he arrived at a fort to speak to them.[42] There is nothing in the least incongruous about this. There are several instances in the sources where Roman soldiers are attested using literary references as well as their general familiarity with them (see Chapter 13).

Diocletian and Maximianus made no such use of coinage, despite their propaganda efforts. Indeed, no other official Roman coinage ever carried such explicit literary references. Carausius even styled himself Marcus Aurelius Mausaeus Carausius in order to manufacture a spurious lineage from the second-century Antonine emperors. As a means of displaying his slogans, he issued the best-quality silver coins produced for 220 years, although these were swamped by his abundant bronze coinage that repeated many of the same themes and added a host more. Some of these bronze coins included types commemorating not only the legions stationed in Britain but also several others on the Continent, and were clearly minted to appeal for the support of their men. Carausius is recorded on a single inscription, on a milestone found near Carlisle at the western end of Hadrian's Wall.[43] He had spread his power from one end of Britain to the other.

Some Carausian coinage was targeted specifically at the army. Carausius evidently knew his history, or at least his advisers did. The promise of good, reliable bullion coinage virtually amounted to legitimacy in the eyes of the Roman army, especially in the late third century when one regime after another had paid out donatives to buy military support. The result had been chronically debased coinage. The new Carausian reformed silver would have gone a long way to establishing Carausius' control over Britain's garrison, though where the silver came from remains a mystery. Perhaps he had indeed been helping himself to the pirates' loot.

By 292 Carausius had decided to adopt a more conciliatory approach. With breathtaking cheek he issued coins in the names of Diocletian and Maximianus as well as his own, adding for good measure a type that showed the three of them together, with a legend that read 'Carausius and his brothers'.[44] He had, in effect, appointed himself to be their co-emperor.

Carausius was a source of monumental embarrassment to Diocletian and Maximianus. Bad weather wrecked their plans to invade Britain in 289. In 293 they formed the Tetrarchy ('the Four Rulers'). They would continue to rule jointly, but each appointed a junior partner who would work with and eventually succeed him. Galerius was appointed by Diocletian, and Constantius Chlorus by Maximianus. In the same year, however, Constantius recaptured Boulogne, causing terminal damage to Carausius' prestige. Before the year was out, Carausius had been murdered by his finance officer Allectus, who became a new rebel emperor. But Allectus had none of Carausius' flair. His coinage was produced with greater mechanical competence, but no more silver coins were minted; his troops and fleet would not have been impressed.

Construction of the fort at Portchester in southern Britain seems to have begun around 293, which suggests it was commissioned by either Carausius or Allectus to bolster their defences. It augmented the Saxon Shore chain of fortified sites around eastern and southern Britain and on the coast of Gaul. In 296 a new imperial fleet set out in two waves. One flotilla, commanded by Constantius Chlorus, embarked from Boulogne. The other, under the praetorian prefect Asclepiodotus, left from the Seine estuary. Asclepiodotus approached the south coast of Britain via the Isle of Wight. Shielded by fog, he landed somewhere in the Solent area and headed north. Asclepiodotus must have had spies in Britain. When he and his forces landed they ambushed Allectus, who had to flee inland before he had had time to organize his forces. Having caught up with whatever army Allectus was able to pull together, Asclepiodotus' men defeated it and killed him. Constantius, meanwhile, sailed up the Thames and seized London, to the obsequious gratitude of the population. Or so at least it was claimed. Britain had been 'restored to the eternal light', as a celebratory gold medallion issued later by Constantius bragged.[45] This was far from the last military rebellion in Britain. A succession of revolts followed in the fourth century in support of various would-be emperors, including most notably Magnus Maximus (383–8) and Constantine III (407–11), the withdrawal of troops for use in their bids eventually running down the garrison to such an extent that by the time Britain was given up by Rome in the early fifth century it was little more than a skeleton force.

Rebellions and mutinies were a constant reminder to any emperor just how febrile the army's mood could be. However well-disciplined they were, soldiers were potentially volatile and dangerous. Any problems with pay and conditions, or poor leadership, were compounded by a gradual realization of how much power they wielded. One of the best solutions to avoiding discontent in the ranks was to keep the men busy, especially in peacetime. The Roman world relied almost entirely on military manpower for countless tasks that meant soldiers, whether individually or in squads, were to be seen in every province working on the state's behalf.

# TWELVE

# PEACETIME DUTIES

## Jacks-of-all-Trades

*Owing to the situation of the free city of Byzantium . . . in*
*conformity with established precedent, I have decided to send*
*them a legionary centurion to protect their privileges.*

Pliny the Younger, as governor
of Bithynia and Pontus[1]

Roman soldiers, whether serving or retired, were some of the most productive and essential agents both of the state and of wider society in general. They became involved in all sorts of building projects, collecting the emperor's taxes, sorting out property disputes, policing country roads and going on journeys of exploration. No wonder then that Roman soldiers and veterans turn up in Pompeii, in Egypt seeking the source of the Nile, circumnavigating northern Britain, smelting iron, supervising quarries and serving as lead-workers, among numerous other tasks. The fourth-century biography of the emperor Probus recounted how many 'works' still to be seen in the cities of Egypt which Probus had ordered soldiers to build, including bridges, temples, porticoes and basilicas. 'The labour of the soldiers' had also drained marshes so that they could be used as farmland.[2] 'You were a bonny labouring boy more than you were a fighter', remembered one veteran of the Great War of 1914–18. He might as well have been talking about life in the Roman army.[3]

## POLICEMEN AND TAX COLLECTORS

Soldiers were the closest thing Rome had to a police force, which meant they could often find themselves keeping the peace in city streets and at public events. In Rome praetorian guardsmen, the urban cohorts and the Vigiles all found themselves supervising public entertainments from time to time, and even unwillingly participating. In the provinces, and especially in unpredictable frontier areas, carrying out such tasks was a risky business. In AD 9, one of the reasons Quinctilius Varus lost three legions in one of the greatest catastrophes in Roman history was because, believing the province of Germania to be at peace, he had agreed to send out soldiers at the request of various communities to cover policing duties (see Chapter 7).[4]

Under Tiberius events in Rome's theatres were becoming notoriously dangerous because of unruly audiences – Tacitus later called them 'the squalid lower classes haunting the circus and theatres'. In the year 15 a praetorian tribune, a centurion and several soldiers were killed on one such occasion.[5] A few years later, when the Theatre of Pompey caught fire, praetorians were ordered there by the praetorian prefect, Sejanus, to act as firefighters (successfully, as it happens). [6] In 38 Caligula carried out one of his few 'good and praiseworthy acts' when he helped soldiers, who must have been praetorians or Vigiles, put out a fire in Rome.[7]

Soldiers were being paid anyway, so it made sense to order them to carry out any duties the emperor wanted. In 28 soldiers were responsible for collecting a tax paid in ox hides from the Frisians, in this instance because it had been decided the hides would be useful to the army.[8] Caligula made it his business to impose a regimen of new taxes, but he bitterly resented having to do so through the professional tax farmers known as *publicani*. Instead he decided on a much more logical solution: he instructed the tribunes and centurions of the Praetorian Guard to collect the new taxes.[9] The soldiers were available, while in theory the collection ought not only to have been more secure but also to have saved the commission paid to the tax farmers. Responsibility for enforcement was therefore handed over to the praetorian tribune Cassius Chaerea (who would later lead the assassination of Caligula). But this backfired. Chaerea was deliberately inefficient because he felt the taxes were such unfair impositions.[10]

There was nothing subtle about how the Praetorian Guard and the urban cohorts kept control in Rome. Elagabalus, the short-lived religious fanatic emperor, held a festival for his sun god Heliogabalus in about 220. The sacred stone (probably a meteorite) which represented the deity was carried by chariot from a temple in the city to a new summer shrine on the outskirts of Rome. Elagabalus threw gold and silver, clothing and food to the mob which led to a mass scramble. The soldiers weighed in with their spears, killing a number of the crowd, adding to those who had already been trampled to death.[11]

Roman soldiers also found themselves serving as police officers out in the wider civilian community. Rome had the urban cohorts, some cities had militias, and Augustus had created the stationarii, armed police installed in specific locations. Praetorians were dispatched to the Bay of Naples area in 58 after the town council of the city of Puteoli had sent representatives to the senate to Rome to complain about the behaviour of the city's mob, while the townsfolk sent their own deputation to complain about the greed of the magistrates. Puteoli was at boiling point. With rocks being thrown and 'threats of arson', there was a real risk that worse violence would break out. In spite of that the population refused to accept the strident measures the senator Gaius Cassius, appointed to deal with the dispute, wanted to impose. The commission was transferred to two other senators, the brothers Scribonius Rufus and Scribonius Proculus, who with the assistance of a praetorian cohort set out to terrorize the people of Puteoli by executing a few people as an example. When confronted with the praetorians, the mob calmed down. The Guard was clearly being used in this instance as a militarized police force for the ruthless maintenance of public order. What is not clear is whether the cohort was sent from Rome or was already in the area, perhaps at Nocera. The presence of a number of tombs of praetorians at Pompeii suggests that they were a well-established presence in the region by 79, if not long before;[12] at an unknown date in Pompeii before 79 a soldier called [Jul?]ius Crescens recorded in the town's basilica how he had been posted to guard duty as a *stationarius*.

Like Puteoli, Pompeii already had a history of civilian violence. In 59 a riot had broken out in the amphitheatre between rival gladiator fans from Pompeii and visitors from nearby Nocera resulting in a number of deaths and serious injuries. Gladiator bouts in the city were banned by the Senate for ten years. Such outbreaks were, said Tacitus, 'typical of the brazenness of country towns', but the occasion showed what could happen when soldiers were not on hand to keep control.[13]

Soldiers were constantly finding themselves being seconded to policing duties. This could put undue pressure on garrisons in provinces. Pliny the Younger was concerned about the prisons in his province of Bithynia and Pontus. The guards in the prisons were state-owned slaves, but Pliny thought they were unreliable and was considering using troops instead. He even started allocating soldiers to share the slaves' duties but became worried this could end up using too many soldiers. Before going any further he wrote to Trajan to ask his advice. Trajan told Pliny to leave the prisons in the care of the slaves, keen that as few soldiers as possible 'be called away from public service'. Both he and Pliny were worried that if anything went wrong in the prisons the soldiers would blame the slaves and vice versa.[14] This is interesting because it suggests that when soldiers were used in everyday capacities to keep a province going it was as much a result of the lack of any alternative as of deliberate policy. This could cause conflict with protocol. Almost a century earlier, in 23, a procurator in Asia called Lucilius Capito was supposed only to be taking care of Tiberius' imperial slaves and money in the province, and not impinging on the authority of the proconsular governor and his staff, whose jurisdiction covered everything else. When Capito overstepped his remit and used soldiers to enforce his decisions, he was punished by being brought before the Senate.[15]

A stationarius could file a report that ultimately went all the way to the emperor. In 111 Appuleius, one of the stationarii based at Izmit (ancient Nicomedia, the capital of Bithynia), sent a report to Pliny about Callidromus, a former slave of Laberius Maximus, Trajan's legate in Moesia. Captured in Moesia, Callidromus had been sent by Decebalus, king of Dacia (whom Trajan had fought and defeated in two wars), to be a slave of Pacorus, king of Parthia, from whose service he had escaped (this part of the story is the least plausible). He had ended up working in Izmit for a firm of bakers who had then arrested him, perhaps because they had discovered his former status. Having escaped from the bakers too and 'taken refuge' by a statue of Trajan, Callidromus had been apprehended there and taken to the magistrates. Pliny decided to send him on to Trajan to deal with.[16]

Over half a century later, between 169 and 172, praetorian prefects served as commissioners of police. There is evidence for this in a letter from Cosmus, an imperial freedman, to the praetorian prefects Bassaeus Rufus and Macrinius Vindex.[17] The purpose of Cosmus' letter was to appeal to the prefects for their help in stopping the magistrates and stationarii of the cities of Saepinum (Attilia) and Bovianum (Boiano) from troubling lessees of sheep flocks on an imperial estate. It is not clear in this instance if praetorians were being used as stationarii, though some inscriptions show that praetorians could be used in this role and sometimes far beyond Italy. Titus Valerius Secundus of Cohors VII Praetoria, for example, was a stationarius at Ephesus, where he died in service at the age of twenty-six.[18]

Cosmus alleged that the magistrates and stationarii had accused the lessees of being runaway slaves and had appropriated the sheep accordingly. This meant they were stealing imperial property. Rufus and Vindex obliged by writing to the magistrates, attaching a copy of Cosmus' letter; it is the text of this which has survived. It was a warning to the magistrates and other suspects to desist, on pain of further investigation and punishment. The document is one of the most specific records of the praetorian prefects operating in a way more akin to the chiefs of a civilian police force.[19]

In Britain the centurion Gaius Severius Emeritus worked as a police officer in charge of the region around the spa shrine settlement at Bath, probably in the late second or early third century. Best described as a district centurion, he represented the reliance of the Roman Empire on centurions for dealing with civilian administration, policing and justice. Severius Emeritus was disgusted to find that one of the sacred places had been wrecked 'by insolent hands', as he called them, clearly expressing his frustration at the gratuitous vandalism and the oafs responsible. He had the place restored and set up an altar to commemorate the fact, dedicating it as a loyalist to the 'virtue and divinity of the emperor'.[20]

Other such centurions are attested in Egypt, one of the few provinces to produce the original texts of letters, including many involving soldiers. Gemellus Horion, who lived in Egypt in 198 during the reign of Septimius Severus, was engaged in complaining to the local administrator, the epistrategos Calpurnius Concessus, about a tax collector called Kastor. Kastor had rather overstepped the mark, so Horion claimed, by beating his way into his house and attacking his mother. He wrote to Concessus, 'I request, if it seem good to your Fortune, that you write to the centurion stationed in the Arsinoite nome [the local administrative district] to send the defendant for your examination and that you hear my complaint against him, in order that I may obtain justice.'[21] Horion's case illustrated the everyday involvement of Roman military personnel in ordinary civilian life. He expected that the man sent to step in and resolve his grievance would be a centurion.

In another instance, the appeal was made directly to the centurion in person, in this case on 18 April 193, the same year as Horion's complaint:

> To Ammonios Paternos, centurion, from Melas, son of Horion, of the village of Soknopaiou Nesos, a priest of the god who is in the village. There belongs to me and also to my cousins Phanesis and Harpagathes held in common and equally in the same village as an inheritance from our maternal grandfather a vacant plot surrounded with a wall where we stack our annual supply of hay. Now the one (cousin) Harpagathes died recently and although his share was inherited equally by both of us, yesterday, which was the 23rd, while I was stacking my hay in the place, Phanesis violently and shamelessly assaulted me and appropriated my hay, not allowing me to stack it in our share (but) attempting to exclude me there-from and to claim for himself alone what belongs to me. Not only this, but he also offered me the most brutal ill-treatment. Wherefore I beseech you to command him to be summoned so that I may be able to obtain the just judgement which comes from you. Farewell.[22]

Not surprisingly, it seems to have been a good idea to keep centurions on side. Early in the second century, a centurion of Legio XXII Deiotariana wrote to someone called Sokration to thank him for a gift:

> Julius Clemens, centurion, to his most esteemed Sokration, greeting. I thank you for your kindness with the olive oil, as Ptolemaios wrote to me that he had received it. And do you write to me about what you may need, knowing that I gladly do everything for you. [In another hand] I pray for your good health, my most esteemed friend.[23]

The potential for centurions or other soldiers placed in charge of civilian administration to abuse their positions is obvious, especially in a remote province where the authority of a centurion on the ground was a great deal more tangible than the abstract notion of imperial authority.[24] Since it was entirely possible for praetorians to saunter around Rome and intimidate members of the public within the vicinity of the imperial palace, how much easier would it have been in a village in Egypt?

In 48, during the reign of Claudius, Vergilius Capito was prefect of Egypt. On 7 December he issued an edict which was sent out to the province's strategoi, ordering that 'soldiers, troopers, orderlies, centurions, military tribunes and all others' passing through the nomes of Egypt were prohibited from taking anything or requisitioning transport unless they were carrying one of his permits. Even then the amounts would be limited to those specified by a previous prefect of Egypt under Augustus. Capito was clearly trying to put a stop to the practice whereby soldiers of all types helped themselves to whatever they wanted while on the move, whether food, accommodation or transport. Even worse, the soldiers had evidently been accustomed to inventing additional expenses for the purposes of making claims. Capito ordered that anyone caught doing so would have to pay back ten times the amount to the public purse. The edict was so important that it was inscribed in stone on the gateway of a temple near Thebes.[25] But Capito's orders seem to have had little long-lasting impact. In the latter days of the reign of Hadrian, ninety years later, Marcus Petronius Mamertinus, then prefect of Egypt, was told that soldiers making journeys through the province were habitually helping themselves, sometimes by force, to transport such as boats and animals.[26]

The two documents paint a picture of soldiers acting as bullies and thugs who habitually exploited their privileged status to intimidate civilians and extort money and goods. Such abuses were probably ubiquitous. It had been clear in Britain in 60 that soldiers who cheated and stole from civilians could provoke dangerous rebellions;

the Boudican Revolt was in large part triggered by such behaviour. The problem was clearly endemic and lasted as long as the Roman army existed. Nonetheless, the Roman world could not function without soldiers being on hand to police and maintain control, and enforce the state's legal decisions.

The most famous centurion attested in a policing context is referred to in the Bible, participating in the execution of Christ during the latter years of the reign of Tiberius. Unnamed, the centurion is described in the Gospels as examining Christ's body and announcing that 'truly this man was the son of God' (Mark), or 'certainly this was a righteous man' (Luke).[27] The centurion was evidently overseeing the proceedings as a matter of routine on behalf of the procurator of Judaea, Pontius Pilatus, to whom he confirmed Christ to be dead. In Petronius' *Satyricon*, written during Nero's reign about thirty years later, a soldier is described as standing guard over crucified robbers to prevent their bodies being removed at night.[28] Centurions and soldiers appear in the Acts of the Apostles, again in a policing capacity to suppress civil unrest. During this encounter the future St Paul challenged a centurion about his right to scourge him, on the grounds that he (Paul) was a Roman citizen.[29] A centurion called Julius 'of the Augustan band' was subsequently involved in Paul's transportation to Rome as one of 276 prisoners, not only identifying a suitable ship but also going on the voyage with an escort of soldiers. He consistently singled Paul out for special consideration as a Roman citizen, an indication that ordinary provincials could normally expect harsher treatment by centurions acting as state police. The ship was wrecked off Malta, but after several months, and passage on other ships, the party reached Rome where the centurion 'delivered up the prisoners to the captain of the guard'.[30] The references are incidental – the story is primarily about Paul – but they surely represent largely routine activities for centurions of provincial garrisons who must constantly have had to take on similar tasks.

Sometimes there was no centurion on hand and a request had to be put in to have one seconded to a specific location. Calpurnius

Macer, probably while he was governor of Moesia Inferior, was told by Trajan to send a legionary centurion to Byzantium (later Constantinople, and now Istanbul) to deal with the vast amount of through traffic the city experienced. Having heard about this, Pliny the Younger decided to ask Trajan to send a centurion to the city of Juliopolis in Galatia (now central Turkey). Pliny pointed out that the city was a 'frontier town of Bithynia with a considerable amount of traffic passing through' but being such a small place it found any problems that arose difficult to cope with, making for a 'heavy burden'.[31] Although Pliny says nothing about what these problems were, he almost certainly means issues arising from the collection of tariffs and tolls on goods, as well as the ability to control the movement of people and goods.

Trajan was not sufficiently impressed to accede to the request, pointing out that Byzantium was 'exceptional'. He was worried, he said, that he would then have to deal with all sorts of cities the same way and that Pliny should take charge of making sure justice was seen to be done. His most interesting comment was to tell Pliny that where soldiers had committed offences too serious to be dealt with on the spot, 'you must tell their officers what they have been accused of'. It was another instance of how soldiers were able to abuse their positions at the expense of civilians, although it was they upon whom law and order generally depended.[32]

The general public could also be sufficiently concerned about their own security to request that a soldier be sent to act as a policeman. Two communities in Asia, Anossa and Antimachaeia, were involved in a long-running dispute about the amount of money they were supposed to pay towards the keep and transport of persons passing through on official business. The controversy, which had erupted in c. 200, was still rumbling on thirteen years later. The Anossans sent a letter to the procurator Philocurius at Prymnessus asking that a soldier be sent to sort things out. Philocurius agreed.[33]

Soldiers were liable to find themselves detached from their main garrison forts and sent out to man little fortified compounds, or fortlets, or even the watchtowers that controlled certain road routes and frontiers. The so-called milecastles on Hadrian's Wall are some of the best known; since they served to police movement across the frontier rather than prevent it, they are better described as fortified gateways. In such places, troops numbering only a few dozen at most found themselves with orders to sit it out. In Egypt, there seems to have been a small military base along the road from Koptos (a city north of Thebes, modern-day Luxor) to the oasis of Phoenikon. Among the men based there was Gaius Papirius Aequus, a centurion detached from Legio III Cyrenaica, a legion originally based in Egypt which had permanently moved to Arabia by *c.* 140.[34]

Legio III Cyrenaica shared such duties in Egypt with Legio XXII Deiotariana (until the latter's disappearance from the record) and with various auxiliaries. Dida, who served as a trooper with a cavalry wing of the Gaulish Voconti, set up a dedication to an unnamed emperor in which he noted that he had been at the little fort of El Moueh for five months.[35] The remote fortlets in which these men found themselves eking out their days were often as little as 0.6 acre (0.25 ha) in area, arranged in the standard playing-card layout with rooms around a small courtyard and few other facilities to speak of.

Other men found themselves allocated as pairs to man watch-towers. These were accessed by ladder and were close enough to one another to transmit visual signals and warnings, if danger or threats – most likely resulting from banditry in the region – occurred.[36] Such men supervised the passing traffic, which in Egypt included soldiers and men going to and fro from the marble quarries at Mons Claudianus, north-east of the modern city of Qena, where a more substantial fort was sited.[37] The traffic also included luxury goods from India passing up from ports on the Red Sea on their way to the Mediterranean coast where they could be shipped on to Rome and other places where people could afford the prices.

Sometimes the 'men' were boys. One watchtower was 'manned' by the unnamed son of a man called Balaneos and who was under-age, the word used being νεανίσκος (*neaniskos*, 'youth'). This angered a decurion called Herennius Antoninus who wrote to Amatius, presumably a superior, to make sure that a suitable replacement was sent. The message was scrawled on an undated pottery sherd (ostrakon).[38]

In Numidia in North Africa, at a place called Skikda, Aelius Dubitatus made a dedication to Jupiter during the reign of Claudius II (268–70). He was then serving in Cohors XI Praetoria, but far from having a ringside seat to history by participating in the endemic warfare of the third century, Dubitatus belonged to a detachment of praetorians sent out to guard the grain route. He was based at a *statio* (staging post) on the road at Veneria Rusicade and had been there for nine years when he set up his altar.[39] Such records are extremely rare, but they show that references to praetorians on campaign with the various emperors of the era can be misleading. Evidently members of the Guard, like legionaries, were dispersed on a number of different and innocuous duties, as they had been for years. This inevitably meant that praetorians became ever more distanced from their original purpose of guarding the emperor's person.

## Mining and engineering

The supervision of slaves or convicts, either on engineering projects or in the mines that proliferated throughout the Empire wherever suitable minerals were found, was yet another common task for soldiers that on the face of it had nothing to do with soldiering. From 104 to 101 BC, during his war against the German Cimbri and Teutones tribes who were threatening Gaul and Italy, Marius gained an advantage from the enemy's slow and chaotic advance. Marius was training up and organizing his new troops as part of his radical military reforms. The extra time gave him the opportunity to order his men to dig a canal that diverted the Rhône into a deep bay, bypassing the silted-up estuary. This brilliant tactic and use of his men provided Marius with a harbour where transport ships could be moored, allowing supplies to be taken off and carried upriver by barges.[40]

Keeping soldiers busy even when there were no wars to fight was vital. Their labour was essential to engineering projects and the extraction of resources. During the reign of Claudius, Curtius Rufus, governor of Germania Superior, put his legionaries to work on silver mining in a region occupied by the Mattiaci tribe. It was a costly exercise that produced little profit and did not last long – Tacitus commented that 'the legions lost heavily in the work of excavating water channels and also creating underground mines that would have been hard enough on the surface' – yet Claudius gave Rufus a triumph. The work was so shattering the exhausted soldiers decided to write to the emperor, begging him next time to award a triumph from the outset to any general he put in charge of an army.[41] It would save them having to go to so much effort simply in order that their general might be awarded a triumph.

Curtius Rufus' project might have amounted to little, but other silver mines were vital to paying for the army, and indeed to the whole Roman state. Extracting silver moreover often involved silver-bearing lead ores, which produced vast amounts of the lead that was another vital component of Roman water systems, roofing, and numerous other applications that needed a malleable metal. Rome's exploitation of these natural resources really began to get under way in Spain after the Second Punic War, when the silver denarius coin was introduced. It became the staple unit of currency used to pay soldiers.

Polybius visited mines near Cartagena in Spain sometime around the middle of the second century BC when Roman mining was in full sway. He was the only contemporary to describe what he saw. His original account is lost, but luckily he was quoted by Strabo. Polybius described the mines as being about twenty *stades* (about 2–2.6 miles or 3.2–4.2 km) from Cartagena and extending 'in a circle for 400 *stades*' (about 40–50 miles or 63–83 km), producing enough silver daily to make 25,000 drachmas which went to the Roman state. It was an onerous task. Extracting silver from lead ore involved five stages of crushing and sieving in water, at a ratio of ore to silver of about four thousand to one.[42] Daily output was equivalent to about 22,500 denarii (enough to pay 100 legionaries in the first century AD), or 8.2 million per annum (enough to pay 36,000).

The silver extracted from the mines at Cartagena alone could have funded the army's costs in their entirety at the time with plenty to spare, though the costs of operating the mine must have been considerable. These included the fees paid to the publicani, the state contractors or tax farmers, who operated at least some of the mines on leases; they certainly were doing so in Macedonia by 167 BC, though state ownership at that period was probably the main method of running mines, especially in Spain, with soldiers in charge of overseeing the work.[43] The physical infrastructure required was enormous, including large-scale installations for smelting the lead ore with tall chimneys to carry the poisonous fumes away.[44] We can assume that the so-called 'Silver Mountain', named for all its silver mines, and referred to by Strabo as being near Castalo, was operated on a similar scale.[45]

Some strong evidence for a military association with mining comes from Britain. It is clear that metal extraction was one of the motives for the invasion. The garrison seems to have become involved with the mining of silver-bearing lead ores soon after the invasion in 43, around the same time that Curtius Rufus was putting his men to work. One of the earliest lead pigs known from the province was found a few miles from the site of a Roman lead mine at Charterhouse-on-Mendip, and bears the titles of Claudius for the year 49.[46] No military involvement is mentioned, but the date is so soon after the invasion there is no question that the army was responsible, an assumption reinforced by the existence of an earthwork at Charterhouse which is probably a small fort. Some of the other lead pigs discovered were produced by private commercial concerns holding leases from the state. However, an undated lead pig from Blagdon in Somerset bears a stamp reading 'British (lead).

Legio II Augusta'.[47] Legio VII Gemina was raised by Galba in Spain in 68. After the Civil War it was based permanently in Spain at León (the name is derived from Legio), close to the mines of north-western Spain, for example the gold mines at Las Medulas and Montefurado in Gallaecia. Iron was also mined in the area.[48]

During its time in Egypt the men of Legio III Cyrenaica were found all over the province, allocated to numerous different jobs: in the year 82/3 another of its centurions, Titus Egnatius Tiberianus, erected a dedication to Zeus Megistos noting that he was 'in charge of the quarry for the paving stones of Alexandria'.[49]

Massive hydraulic engineering works were built by soldiers in Germany in the mid-50s. Paulinus Pompeius and Lucius Vetus, commanding the army in the region, were concerned that the soldiers would become 'sluggish' now that the region was at peace. Some were told to complete an embankment begun in the late first century BC that would help contain the Rhine, while Vetus had plans for a canal between the Moselle and the Arar so that there would be a continuous maritime route between the North Sea and the Mediterranean. The plans seem to have been thwarted by the jealous Aelius Gracilis, governor of Gallia Belgica, who was annoyed that Vetus might be planning to bring the German legions into his province and blocked the works from proceeding further.[50]

When the general Corbulo was told by Claudius in 47 to abandon a campaign against the German Chauci tribe and pull his troops back across the Rhine, he was disappointed but did as he was ordered. Instead he instructed his soldiers to dig a 23 mile (37 km) long canal between the Meuse and the Rhine to give them something useful to do. The canal would enable water transports to avoid having to pass through the dangerous North Sea. Claudius was so impressed he gave Corbulo a triumph as compensation for refusing him a war.[51]

Soldiers also toiled on essential works for civilian communities. During the reign of Vespasian in 75 a massive project was begun by Marcus Ulpius Trajanus, father of the later emperor, who was governor of Syria. He ordered Legiones III Gallica, IIII Scythica, VI

Ferrata and XVI Flavia, as well as 20 auxiliary units and the militia of Antioch, to build a canal 3 miles (4.8 km) long not far to the north of the city. It seems to have been intended to provide a link between the rivers Orontes and Karasou, which join nearby, and was to include a number of bridges. The fact that so much of the garrison was sent out on the work indicates both that the project was urgent and that the soldiers were not otherwise engaged.[52]

This kind of work might also involve repairing or maintaining existing facilities. In Judaea during the reign of Hadrian a detachment of Legio X Fretensis was detailed to repair an aqueduct supplying the city of Caesarea Maritima, work recorded by the soldiers on at least five inscriptions. At other times detachments of at least two other legions, VI Ferrata and II Traiana Fortis, were also seconded to work on the structure. In Africa Proconsularis in 239 a team of soldiers from Cohors I Syrorum Sagittariorum (a specialist unit of archers) was sent far south from the Mediterranean coast to repair an aqueduct which until it collapsed from age had supplied the colony at Magna.[53]

There was a good reason for using soldiers in a range of engineering capacities. They were highly trained and included among their number men with a variety of professional skills on which the wider community depended. Such men included Blesius Taurinus, a praetorian land surveyor (*mensor agrarius*) serving with the praetorian cohort during the reign of Antoninus Pius (138–61). Taurinus was sent by the emperor to Ardea, 22 miles (35 km) south of Rome, where he was to work out the boundaries of the settlement.[54] With that done, Tuscenius Felix, serving for the second time as primus pilus of an unspecified unit, delivered a decision (what that was is unknown). As so often in the Roman world, military personnel exhibited the greatest concentration of professional expertise available. In this case it seems that Antoninus Pius had used Taurinus and Felix to resolve a matter, perhaps a property dispute that had been brought to his attention.

The praetorians, being always on hand at the emperor's side, were particularly liable to find themselves ordered to carry out random or eccentric tasks on his whim. It rather depended on who the emperor was, but Claudius and Nero were particularly imaginative. By the time of Claudius' reign in the mid-first century, conditions at the port of Rome at Ostia were close to crisis point. The estuary of the Tiber was silting up and it was becoming impossible for grain ships from Egypt and Sicily to dock, especially in winter. The ships were forced to moor off the coast in the hope that lighters could unload the grain and bring it into the docks for processing and sending on to Rome in barges. The huge mob in the city had become so dependent on the state grain handouts (known as the Annona) that any disruption to its supply could be extremely dangerous to the emperor, who would be held responsible; indeed Claudius was attacked by a furious mob in the forum during a spell of drought, an episode which only further emphasized the practical need for a bodyguard.

The construction of the harbour involved creating a mole to protect the entrance, using a ship that had brought an obelisk (now standing in the Vatican piazza in Rome) for Caligula. The germ of the idea may well have come from the occasion when, after a killer whale became stuck in the sand while foraging in a wreck, Claudius ordered nets to be thrown across the mouth of the harbour. He then sailed out with members of the Praetorian Guard so that the soldiers could throw lances at the unfortunate animal. The whole event seems to have been presented as a public entertainment, with the Guard centre stage.[55]

A few years later, in 52, Claudius came up with the idea of using praetorian soldiers in a way Augustus is unlikely ever to have thought possible: as part of an imperial performance. Claudius had commissioned drainage works for the Fucine Lake designed to control its level. To celebrate their completion he ordered that a naval battle (*naumachia*) be held for the benefit of a gigantic crowd that gathered around the lake to watch. Battleships with 19,000 convicts aboard to fight floated on the water, hemmed in by a ring of rafts manned by praetorian infantry and cavalry to contain the action and prevent any ship from sailing away and escaping.[56] The event may well have inspired Claudius' adoptive son Nero, who acceded in 54 after Claudius was murdered by Nero's mother Agrippina, Claudius' wife at the time (see Chapter 10).

## INDULGING THE EMPEROR

Nero was committed to only one thing: self-indulgence. Soldiers, willing or unwilling, were brought into enforce or act upon his whims. This included regular performances as a singer, dancer and chariot racer, activities that disgusted the senatorial class because they were considered demeaning for a member of the aristocracy. The mob, on the other hand, seemed to have enjoyed the notion of having a ruler who shared some of their passions. Not long after Nero had his mother Agrippina the Younger murdered in 59, he decided to put on a show with himself as the star attraction. In case the audience failed to appreciate his performance, he arranged for 5,000 soldiers to be present in the audience to lead the applause. Most of them must have been praetorians. The Guard may by this date have been made up of 12 cohorts, each quite possibly 1,000 strong. If so, plenty of them would have been available, as well as the urban cohorts and the Vigiles.[57] At a later event, praetorians disguised as civilians were placed in the audience so that they could spot any disaffected theatregoers complaining about Nero and report them.[58]

Nero's tutor Seneca and his praetorian prefect Burrus decided to allow him to ride privately in a chariot across the Tiber in the Vatican valley.[59] The plan went wrong because a crowd turned up and, far from being appalled at the indignity of their emperor taking part in a profession regarded by the senators as little better than semi-criminal, they were greatly impressed that he shared their tastes.

Nero was delighted by the mob's response and proceeded to embark on even more eccentric thespian and musical activities. This plunged the praetorians into one of the most humiliating episodes in their history. Nero arrived on stage with his lyre and voice-trainers, while a cohort of praetorians complete with centurions and tribunes packed out the audience. Tacitus said that even a 'sorrowful Burrus', the praetorian prefect, had been compelled to come along and add his praises. Dio suggests that Burrus and Seneca put on a rather more convincing performance of guiding and supporting Nero, orchestrating the crowd to join in.

In 66 Nero set out for Greece to take part in games and to perform. During the tour he came up with the idea of having a canal cut through the Isthmus of Corinth, which would shorten the journey to Athens from Italy and also avoid the dangerous voyage round the Peloponnese. It was a project Julius Caesar had considered, and would undoubtedly have been a useful facility had it been a realistic prospect in an age when dynamite did not exist. Nero's praetorian escort was ordered to start digging out the canal, though in practice this meant merely that they were to supervise the necessary slave labour. Vespasian, then in command of the war in Judaea, sent '6,000 of the strongest young men [prisoners]' to do the work, after executing 1,200 who were too old or sick.[60] In an attempt to rally the praetorians and inspire them to get on with excavating the canal, Nero had a trumpet sounded. This was his signal to break ground himself with a tool, gather up the earth and carry it off in a basket.[61] Needless to say, the project was never finished, though evidence of the work executed under the praetorians' control remained visible until the nineteenth-century canal was begun. The existing canal, 26 ft (8 m) deep, with walls 295 ft (90 m) above the water level, took from 1881 to 1893 to complete and entailed the use of explosives.

Over two centuries later Probus supposedly had soldiers tear living trees out of the ground, including the roots, and transport them to the Colosseum in Rome. There they were laid out in the arena on a platform made of wooden beams and covered with earth to create an artificial forest for an especially ostentatious wild beast hunt in which thousands of animals were killed.[62] Probus was to discover that the over-use of soldiers on imperial indulgences could have the most dramatic consequences. Disliking the thought of idle soldiers, he allegedly decided in 282 to order thousands of troops to drain a marsh and build a canal to enrich his homeland, Sirmium in Pannonia. The men refused, rebelled and chased Probus to a lookout tower, where they killed him. Despite that they still summoned up the energy to build him a large tumulus, with a tomb on top and embellished with an honorific inscription.[63]

## Explorers

One curious episode involving the Praetorian Guard seems to belong to the latter part of Nero's reign, around the same time as the Corinth Canal plans. Nero apparently decided that a campaign against Ethiopia might be a good idea. To prepare for this he sent out to Egypt a reconnaissance party of praetorians under the command of a tribune, an expedition we know about because of a comment by Pliny the Elder. The voyageurs were sent up the Nile and reached as far as Meroë in the Kingdom of Kush in Sudan, but reported back that there was nothing but desert. The expedition was an example of the Guard being deployed not as a bodyguard, but as a state-owned force carrying out the emperor's wishes.

The Neronian expedition up the Nile was not the first time an emperor had sent Roman soldiers to that region of Africa. 'Roman arms reached that area in the days of the divine Augustus under the command of Publius Petronius, of the equestrian order, and prefect of Egypt' – where, added Pliny, a number of cities were conquered after a journey from the city of Aswan that covered 870 miles (1,400 km).[64] Cornelius Balbus, a wealthy Spaniard from Cadiz, had been awarded Roman citizenship in the mid-first century BC in return for his contribution to the Roman war against Sertorius in Spain, and later supported both Caesar and Augustus. In 19 BC he took an army deep into the desert of Libya, reaching the city of Germa, capital of the Garamantes, an ancient Berber tribe. Germa and other Garamantian cities were then 'overwhelmed by the Roman arms of Cornelius Balbus, who was granted a triumph in the chariot of a foreigner, the only of all to be so honoured', and given full rights of a Roman citizen.[65]

## DUTIES OF THE COHORTES VIGILUM

The Cohortes Vigilum (night watch) operated in Rome and the port at Ostia, working effectively as a civic fire brigade under the command of an equestrian prefect. They rarely participated in what might be regarded as normal soldierly duties, though under Tiberius in 31 they were involved in providing muscle for the toppling of the ambitious praetorian prefect Sejanus.[66] In 7 BC Rome had been divided by Augustus into 14 districts for the purposes of fighting fires and it was still organized the same way two centuries later.[67] After a major fire in Rome in AD 6, the Cohortes Vigilum was created using freedmen to replace an earlier 6,000-strong body of watchmen. The exact numbers of the Cohortes Vigilum are unclear. The epigraphic evidence shows that by 205 there were seven cohorts that theoretically had around 1,120 men, divided into seven double centuries of 160 each. In practice the numbers varied considerably, from 85 to 178.[68] These may have been the arrangements from the start, but if the numbers had been increased by Septimius Severus there was good reason for doing so. In 191 a massive fire broke out in Rome after an earthquake was followed by a lightning storm. The temples of Pax and Vesta were destroyed, along with a huge number of other public and private buildings.[69] Though in reality the Vigiles had little chance against major conflagrations once they were out of control, it appears that after coming to power two years later, Septimius Severus provided them with additional watch-houses (*excubitoria*) in Rome, one of which has survived. It was a small building with a shrine, its walls covered with graffiti referring to the firemen's equipment which was stored there; that way the tools were available to deal with small local fires before they got out of hand.[70]

As freedmen the Vigiles were looked down on by the Praetorian Guard as a lesser form of soldier though there was some movement between them, especially involving centurions and tribunes. Gaius Gavius Silvanus was primus pilus of Legio VIII Augusta, apparently during the Claudian invasion of Britain in 43, before going on to command Cohors II Vigilum; he then took command first of Cohors XIII Urbana and then of Cohors XII Praetoria.[71] Lucius Laelius Fuscus had a stellar career in the centurionate. He rose from serving in the Praetorian Guard through other posts to become centurion of Cohors I Vigilum; having finally become centurion trecenarius of Legio VII Claudia, he died at the age of sixty-five after 42 years in service.[72]

The barracks of the Vigiles in Ostia reflects the port town's unique blend of local officialdom and the state's interest in keeping the place safe and secure. It was a practical precaution. Grain in hot and dry storage buildings could easily catch fire in a trivial accident. Claudius stationed one of Rome's urban cohorts at Ostia to fight fires.[73]

The surviving Vigiles barracks in Ostia was built *c.* 117–38, replacing an earlier structure, and was restored and enlarged in 207, in response to the fire of 191. In the middle was an open courtyard, probably used for firefighting exercises and ceremonies, surrounded by a portico and rows of rooms which backed onto the surrounding streets. The firefighters were ultimately under the emperor's personal control and, appropriately enough, the ruins at Ostia contain a number of surviving loyalist dedications made to the emperor and members of the imperial family. These were arranged by the commanding officers of the barracks in whose names they were made. The most conspicuous feature of the building today is nevertheless a shrine dedicated to the imperial cult, dominating the western side of the building and facing the eastern entrance. Entering the shrine meant crossing a mosaic depicting the sacrifice of a bull before going up a step into the shrine itself. Here survive several inscribed blocks dedicated to Antoninus Pius, Marcus Aurelius and his short-lived co-emperor Lucius Verus, and Septimius Severus.

The existence of the shrine and the inscriptions shows the importance of observing a cycle of religious statements of loyalty to the regime, reflecting a pattern followed by all the Roman armed forces. After a long series, inscriptions from the building cease by 244, suggesting that the firefighters had been transferred to the increasingly important Portus where new harbour facilities, built as the river Tiber silted up, had made the old Ostia obsolete.

Firefighters were sent to Ostia from the Rome cohorts on four-month tours of duty, changing over on 15 December, 15 April and 15 August. They were commanded by one or two tribunes, but by 211 all had come under centralized control: the Rome sub-prefect in charge spent at least part of his term at Ostia. Firefighters received a grant of state corn, and those of Latin status earned full Roman citizenship after three years in the job. The work included all-night patrols armed with axes and buckets, and the operation of a basic fire engine called a *siphon*. It was potentially dangerous work and they sometimes needed help. In the early first century AD a praetorian of Cohors VI Praetoria, whose name is unknown, was killed at Ostia while putting out a fire. The grateful city provided a place for his burial and granted him a public funeral, which meant it was paid for at civic expense.[74]

The Vigiles did not always operate in the manner expected of a dedicated public service. During the terrible fire that broke out in Rome in 64 during Nero's reign, they were supposed to be putting out the flames. They failed to do so because they were more interested in plunder. In order to speed things up they set more houses on fire themselves with the help of 'the soldiers' – which must mean the Praetorian Guard as well as the urban cohorts. As the flames took hold across the city Nero came up with the eccentric idea of going on to the roof of the palace to perform the 'Capture of Troy' with his lyre.[75] In the end so much of the city was destroyed that Nero was able to use a huge area on which to build his vast new 'Golden House'.

## The price of idleness

Though Josephus portrayed the Roman army as always alert and at the height of its powers, the reality was sometimes rather different. It was well known that inactivity could result in some soldiers degenerating into layabouts and discipline breaking down. When in 58 Corbulo went out to the East to recover control of Armenia, originally taken by Rome in the Third Mithridatic War (74–63 BC), he was shocked to discover that the Syrian-based legions, Legio III Gallica and Legio VI Ferrata, had become 'sluggish from a lengthy peace'. Some of the soldiers were veterans who performed no duties, including standing watch, and thought building ramparts and ditches was some sort of novelty. This was not surprising since apparently they did not arm themselves, instead spending their whole time in towns trying to make money. Corbulo threw the lot out and had to embark on a quick recruitment campaign to make up the numbers ready for the campaign.[76]

In spring 69 during the Civil War, the First Battle of Bedriacum (Cremona) was imminent between the emperor Otho and his challenger Vitellius, who was approaching with his forces over the Alps. Otho's army already consisted of five legions and five cohorts of praetorians, amounting to around 30,000 men, to which Otho brought the rest of the Praetorian Guard, praetorian veterans, and a contingent of naval troops. One of Otho's generals, Vestricius Spurinna, held the city of Piacenza with three praetorian cohorts, a detachment of 1,000 legionaries and some cavalry. Worried that his men lacked experience, Spurinna decided to keep them safe behind the city's walls rather than lead them out to fight. Unfortunately, the inexperienced and ill-disciplined garrison panicked when they heard that some of Vitellius' army was on its way towards them. They had never seen active service and evidently had no intention of doing so, certainly not right then. They 'ran amuck', said Tacitus, grabbing their standards and ignoring their centurions and tribunes before marching off 'recklessly', abandoning Piacenza and leaving Spurinna with no choice but to follow.

Things went from bad to worse. The men reached the river Po as daylight faded and realized they would have to build an overnight camp. To read Polybius or Josephus is to believe that such a task was a walkover for Roman soldiers. Not on this occasion. The praetorians were far too accustomed to an easy life in Rome, where physical effort played little part in their everyday activities. The back-breaking labour of digging the trenches that night totally demoralized them. The praetorians became convinced defeat was inevitable, which under the circumstances was a reasonable conclusion. With the help of his centurions and tribunes Spurinna managed to rally the men, who conceded that safety behind Piacenza's walls had been a much better idea. Meekly, they marched back to the city, reinforced the defences and waited for the Vitellians to lay siege – which followed shortly afterwards.[77]

It is easy to believe that life on the remote frontiers of the Roman Empire was a bitter and arduous affair. Sometimes, no doubt, it was. However, the reality was that whole decades might float past with little or nothing in the way of warfare apart from the occasional skirmish or exercise. In 121 Hadrian arrived in Germany during his tour of the Empire. A stickler for military discipline, he made a point of living among the soldiers as if he was one himself (as indeed he had been), and insisted on training being maintained to a traditional standard 'as if war was imminent'. Believing that military discipline had slackened since the days of Augustus, he ordered that the forts on the Rhine frontier be rid of 'banqueting rooms, porticoes, grottos and bowers'. He 'banished luxuries', as well as making sure that centurions and tribunes were men of appropriate experience and abilities. Having left Germany for Britain, 'he corrected many abuses' in the province, a cryptic phrase that almost certainly means he dealt with the garrison there in the same way.[78]

The decision to build Hadrian's Wall, construction of which followed shortly afterwards, was almost certainly inspired not only by strategic considerations but also by Hadrian's belief that the task would force the legionaries out of their fortresses and into a regime of hard work. Detachments of the II Augusta, VI Victrix and XX Valeria Victrix legions, as well as auxiliaries, found themselves on the northern fringe of the province laying out and building what we now know was called the Aelian Frontier, with its milecastles and turrets, rear and forward ditches and later a series of forts. Their work was all recorded in inscriptions, some of which recorded lengths of walls built and the name of the centurion in charge to engender competition and also act as quality control. At the fort of Carvoran 'the century of Silvanus built 112 [Roman] feet of rampart under the command of the prefect Flavius Secundus', a length equivalent to 33.2 m.[79] Meanwhile the legionary fortresses, far to the rear in the more settled regions of the province and still unfinished, had to wait for completion. Major structural work at Chester, the fortress of Legio XX Valeria Victrix, was mothballed for years while the legion's men built the Wall.

Hadrian's reforms, though, could only ever be temporary and localized. Keeping soldiers disciplined meant maintaining an eagle eye on their activities. This did not always happen and the fallout could be disastrous, as Onasander had warned in the mid-first century when he pointed out that soldiers who had been allowed to spend periods in idleness were not only unwilling to fight an enemy but easily became disheartened and were likely to retreat.[80] Marcus Cornelius Fronto, an orator and senator from North Africa, was appalled at how the garrison in Syria had become 'demoralized with luxury, immorality and prolonged idleness', as he told the emperor Lucius Verus (161–9) in letters written in 163 and 165. According to Fronto the soldiers at Antioch spent their time gambling, watching theatrical performances and loafing about in their gardens. They slept all night long, and if they did stay up to keep watch it was only because they were drinking wine. Even worse, these popinjays indulged in manicures, and saddled their horses with goose-down cushions while leaving the animals neglected. The indolent fops were moreover so unfit that they could barely mount their horses, while they had no idea how to throw their lances and regarded battle trumpets not as the signal to start fighting but to run away. The governor of Syria, Pontius Laelianus, 'a disciplinarian of the old school', was able to rip apart some of the armour with his bare hands, so bad was its condition. Lucius Verus had to step in, Fronto claims, to sort out the indiscipline, inspired by some of the great figures from the old Republic including the general Scipio Africanus and Cato the Elder.[81]

Every community in every province of the Roman Empire, then, was accustomed to seeing soldiers at work in all walks of life. The Roman state would have been unable to function in any meaningful way without the army. Government and keeping order would have been impossible. War, although it dominates the historical record, was in fact a relatively minor part of being a Roman soldier. Over the course of his career a legionary or an auxiliary might find himself representing imperial authority in countless different ways in communities of all sizes, from one end of the Empire to the other. The modern stereotype of the Roman as a soldier is therefore not inaccurate, but perhaps for reasons that would surprise most people today.

# THIRTEEN

# LEISURE AND LEAVE
## HUNTING WILD BOAR AND OTHER DIVERSIONS

*Gaius Tetius Veturius Micianus, prefect commanding the
cavalry wing of Sebosians, willingly set this up to the Divinities
of the Emperors and Unconquerable Silvanus [in return] for
taking a wild boar of remarkable fineness which many of his
predecessors had been unable to turn into booty.*

An altar found on Bollihope
Common, Durham, Britain[1]

Roman soldiers spent most of their time doing almost anything except fighting wars. Peacetime duties kept them busy, but they also enjoyed other distractions. Sometimes any spare time was spent purely on pleasurable diversions such as hunting, though the main beneficiaries of such opportunities were generally the officers. Other activities involved taking part in showcase parades, fighting in the arena, or even malingering on the sick list and taking advantage of the army's medical support system. Some sought to supplement their income with second jobs.

## POETS AND WRITERS

Aristocratic officers, who were educated and often only serving as officers in the army as part of their senatorial career, might have rather refined tastes when it came to relaxation, even in the most astonishing circumstances. In the summer of 54 BC Cicero received a letter from his brother Quintus, who was on campaign with Caesar in Gaul and at one point commanded Legio XIIII. The letters of Quintus do not survive but Cicero's excited responses, written at the end of August, do.[2] Cicero was fascinated by the terrifying prospect of the vast unending sea the Romans called the Ocean, which it is clear Quintus had graphically described to him. Evidently part of the appeal of being involved in Caesar's war was the prospect of writing home with tales of high adventure.

Quintus crossed from Gaul to Britain in the late summer of 54 BC on the second of Caesar's two invasions of the island (the first had been the previous year). Cicero was excited at the thought of hearing about the landscape, the local customs and tribes, about the battles and about Caesar himself. On 13 September he heard again from his brother, whose letter had left Britain five weeks before on 10 August.[3] Quintus was more interested in a literary composition of his own called *Erigona*. Erigone was the daughter of Icarius, who had been killed by his shepherds after Dionysus gave them wine in return for Icarius' kindness. Erigone committed suicide when she found her father's body, after which Dionysus forced all Athenian girls to commit suicide too. Such writings were the pastime of an educated man, as interested in esoteric pursuits as in fighting a war. Nonetheless, it seems a remarkable diversion when facing a dangerous mobile enemy that used chariots to infiltrate the Roman columns.[4]

The poet Horace was the son of a freedman engaged in the mundane businesses of collecting money at auctions and dealing in salted foods. He managed the unusual feat for a man of such modest background of serving as a military tribune in the army of Brutus at Philippi in 42 BC.[5] He recalled the occasion in his poetry and letters, focusing on the experience of defeat, 'leaving my shield ignominiously behind' as he fled, but was proud the battle had not destroyed him.[6] He was pardoned by Octavian and Antony and released. He had been, he said, 'ignorant of war', and returned home in poverty to write poetry, eventually rising to move in Augustus' circle.[7] (Horace's brief military career is, incidentally, an example of how an officer, albeit a junior one, might easily find himself in the thick of a major battle without any proper experience or training.) Over a century later in *c.* 80, Martial wrote one of his epigrams in praise of a soldier called Marcellinus, perhaps a friend of his, admiring how Marcellinus had been in the far north and seen the stars rotating around the sky.[8]

A poet once found his way to the northern frontier of Britain. The officer Junius Juvenalis is known to have commanded Cohors I Delmatarum, an auxiliary infantry unit which was certainly in Britain by the year 122 and based at the fort of Maryport by 138.[9] This man was likely the poet Decimus Junius Juvenalis, known to us as Juvenal. He can be identified not only by the name, but because the inscription recording his command was found at Aquino in Italy, Juvenal's birthplace. In his second *Satire*, moreover, Juvenal referred to how Roman arms had conquered the 'Orcades [Orkneys] and short-nighted Britons', which reads like a personal anecdote. The short nights would only have been apparent in the summer, the principal campaigning season.[10] Various references in Juvenal's poetry suggest that he wrote during the reigns of Trajan and Hadrian, by which time he was already of advancing years. If he was therefore born sometime around 60–70, his position as the commanding officer of an auxiliary unit must belong to the 90s. Clodius Albinus, Severus' sometime co-emperor – and later rival in the civil war of 193–7 – was a career soldier in his earlier days. He was said to have composed poetry, as well as erotic stories, the latter being dismissed by his biographer as 'mediocre'.[11]

There was nothing particularly unusual about soldiers quoting poetry or other lines from well-known writers, like Septimius Severus's witty praetorian tribune Julius Crispus (see Chapter 9). Ordinary soldiers might also be sufficiently educated to be able to quote from literature. In the closing days of Nero's reign in 68, news arrived that armies around the Empire were abandoning him. Attempting to flee, Nero wanted the tribunes and centurions of the Praetorian Guard to accompany him. Some refused, others were non-committal; but one said, 'is it so dreadful a thing then to die?' quoting a passage from the last book of Virgil's *Aeneid*.[12]

The *Historia Augusta*'s account of the emperor Carus and his sons Carinus and Numerian described how Diocletian had quoted the *Aeneid* when ordering the execution of Numerian's father-in-law Aper, the praetorian prefect who had allegedly murdered Numerian. 'By the hand of Aeneas you die', he is reported to have said.[13]

The biographer went on to observe, 'I know certainly that many soldiers make use of sayings in Greek or Latin from the comic writers or other such poets.' He added that it was normal for comic writers to introduce military characters in their works by having them quote well-known phrases such as 'You are a hare, yet you are looking for game?' In other words, the quoting of such passages was a means of identifying the character as a soldier and was recognized as such by the audience. The biographer attributed the line to Livius Andronicus, whose works are only now known as fragments, but it was later used by Terence in *The Eunuch*.[14] Unreliable as the *Historia Augusta* may be when it comes to later emperors – the biographies of Carus, Carinus and Numerian are mainly a mixture of copying and outright invention – that shortcoming mainly applies to the 'historical' information; this particular passage reads more like a personal and general observation about soldiers. Even if it was copied from someone else it would be no less relevant since there is no basis on which to argue it is invention. The evidence from Vindolanda suggests that education in classical literature was available, but possibly only for the children of officers (see Chapter 14).

Soldiers were not, however, reliably literate, though they seem to have been more so than the general population on the evidence of their inscriptions, graffiti and surviving documents. When Gaius Julius Apollinarius wrote to his father in Egypt, he specifically refers to having written 'very often' in the past 'through Saturninus the *signifer* and Julianus the son of Longinus'. This appears to suggest they wrote the letters for him. Alternatively, he might mean they acted as couriers on his behalf if they were travelling to Egypt, but the 'very often' reference makes that unlikely.[15]

## HUNTERS

Officers with less refined tastes took themselves off hunting. Flavius Cerealis and Aelius Brocchus, the officers commanding auxiliary units in northern Britain around the year 100, were friends and went hunting together. A letter from Cerealis has survived in which he wrote to Brocchus to ask for some hunting nets as a favour, though he added a request that they be strongly repaired first.[16] A century or more later, little had changed for officers commanding these units in the far north as they idled away their spare time hunting. Gaius Tetius Veturius Micianus was the commanding officer of the Gaulish Ala Sebosiana in northern Britain. He triumphantly hunted down a boar that apparently had fought off any attempts by his predecessors to capture it and commemorated his kill on an altar that he set up on Bollihope Common.[17]

## PARADES AND GAMES

Parades and games were an important aspect of the Roman army's annual calendar, providing a focus that helped keep the men busy and reinforce a sense of competitive pride. There were two locations where these were most likely to take place. The first was the parade ground, a suitably flat or levelled area close to the fort where the soldiers could form up not only to be addressed by their commanding officer but also to put on displays of their fighting prowess and, if relevant, their horsemanship. One of the best known is the remote auxiliary fort at Hardknott in Britain's Lake District, a well-preserved site where much of the fort, along with its parade ground, is still visible. A fragmentary inscription found in 1964 shows that during Hadrian's reign a Cohors (IIII?) Delmatarum was based here.[18] Even in this remote upland location time and effort had been expended on clearing and levelling an area of 459 ft by 262 ft (140 m by 80 m). A large sloping mound rising to 30 ft (9 m) in height may well have been the site of the temple of the Matres Campestres ('Mother Goddesses of the Parade Ground'), though it might alternatively have been the place from which the commanding officer spoke to the cohort and made religious dedications to the state cults on behalf of the unit.

For auxiliary units in particular, parades seem to have been an especially impressive sight and occasion. In around 136–7, a book written by Arrian about Roman cavalry and military tactics includes a description of an auxiliary cavalry regiment's parade. Each trooper wore a special parade helmet with elaborate decoration and a full face covering, often in a style that emulated mythological warriors, and rode a horse that also wore decorative bronze plates and pieces,* while displaying noise-emitting standards:

At the start of the parade ... the best cavalrymen ride past wearing gilded iron or bronze helmets. This makes sure they get the watching crowd's attention. These special helmets completely cover the head and face, with only small holes to look through. Their first charge onto the parade ground is well rehearsed to make a spectacular show. Each group has its own standard including ones that look like serpents. They are made by sewing pieces of dyed cloth together ... and when the horses are moving, they are plumped out by the wind and look exactly like living things and even hiss as the air whistles through them.[19]

---

* See Chapter 3 for an example from Xanten, inscribed with the commanding officer's name.

Some of the helmets have survived, including those found in Britain at Ribchester and at Crosby Garrett in Cumbria, not far from Hardknott. The Crosby Garrett helmet is in the style of a Phrygian cap with a griffin on top; its most striking feature, shared with other examples like that from Ribchester, is the glassy and expressionless stare of the face covering. This must have made the troopers look almost robotic as they made a display charge. The Ribchester helmet decoration features six cavalrymen in combat with eleven infantrymen, while the face framed by locks of hair developing into serpents was clearly intended to suggest a Gorgon, a popular decorative theme on military equipment.[20]

The other location for parades, and for performances re-enacting either the unit's greatest successes or battles drawn from history or myth, was the amphitheatre. These were located close to forts and fortresses and resembled their civilian counterparts in cities. Unlike the parade ground, an amphitheatre provided seating for an audience, which might include visiting dignitaries or in exceptional instances the emperor himself. They could also be used for gladiatorial or beast fights, sometimes conducted by the soldiers. There was an amphitheatre in London, located outside the fort occupied by soldiers seconded from around the province of Britain to serve on the governor's bodyguard. It was built in timber as early as *c.* 70, and its stone replacement, constructed forty years later, had an estimated seating capacity of 8,000, way in excess of the garrison's needs. London's amphitheatre must therefore have also provided entertainment to some of the townsfolk.

One of the best-known military amphitheatres is that built at Caerleon in Britain for Legio II Augusta during the reign of Domitian in the late first century outside the south-western sector of the legionary fortress's defences. Immediately to the north was the legion's parade ground. The amphitheatre was created by soldiers from the legion who, under the supervision of their centurions, dug out earth to create a sunken elliptical arena, using the spoil for the embankments that supported the timber seats. The working parties installed inscribed stones commemorating the sections of wall for which each was responsible. The facilities included a small baths, a seating area for the higher-status members of the audience, mainly officers, and a shrine dedicated to Nemesis (Fate). Appropriately enough a lead curse was found on the site. Neither the author nor the intended victim, who had apparently stolen a cloak and boots, are named:

> Lady Nemesis, I give you a cloak and pair of boots. Let him who wore them not redeem them except with his life and blood.[21]

Legio XX's fortress at Chester had a similar, though larger facility which also featured a shrine to Nemesis. This one contained a small altar, inscribed:

> Sextus Marcianus, centurion, (dedicated this) to the Goddess Nemesis (as the result) of a vision.

It would seem that Sextus Marcianus had tried his hand, or was about to, in the arena. If he did, a small piece of evidence has survived at Mainz that might refer to this man later in his career, showing that he survived. A fragmentary dedication to a number of gods, including Jupiter Optimus Maximus, was made in 192 by a centurion whose name can easily be restored as Sextus Marcianus of Legio XXII Primigenia.[22]

Neither of Britain's known legionary amphitheatres was anything like as elaborate as some found elsewhere in the Empire. Legio III Augusta's base at Lambaesis had a more sophisticated masonry amphitheatre, complete with subterranean features that allowed gladiators and animals to be prepared before being elevated to the arena floor. At Dura-Europos vexillations from two legions, III Cyrenaica and IIII Scythica, were involved in 'building the amphitheatre from the ground up' in 216 under the charge of a centurion called Aurelius Mam[mius?] Justinianus, as recorded on a dedication to Caracalla and his mother Julia Domna. It was a small building and could only accommodate around 1,000 spectators.[23]

## SECOND JOBS

Quite apart from carrying out an almost infinite variety of peacetime tasks on the emperor's behalf, or that of local communities, Roman soldiers also frequently had second jobs, private businesses, and property interests that included owning slaves, which they carried on – sometimes illicitly – at the same time as soldiering. Obviously, these were easiest to organize if the soldiers concerned were on hand in a fairly settled garrison. None, though, were as conveniently located as the Praetorian Guard in their camp on the outskirts of Rome. Praetorians either stole the money they needed for the bribes or took second jobs, with the unfortunate effect that the better off they were, the more duties were allocated to them in order to coerce them into paying up.[24]

In Justinian's law code there are some clues to the kinds of second jobs soldiers might take on. Evidently such activities were tolerated to some degree. The Codex records a variety of judgements made by previous emperors and which were still enshrined in law. Severus Alexander determined that it was legal for soldiers 'to attend to their own affairs without committing a breach of discipline'.[25] Under Diocletian and Maximianus, a man called Martial was offered compensation by the emperors if he had engaged a soldier to act as an attorney on his behalf and paid a sum of money to the soldier. The principle here appears to be that the emperors were accepting liability for a soldier, a state employee, who stepped outside his normal sphere; the implication is that this was not an especially unusual state of affairs.[26] Severus Alexander found himself hearing a case concerning a soldier called Avitus and his share in a vineyard:

*The Emperor [Severus] Alexander to the Soldier Avitus*

If it should be proved before the Governor of the province that your brother gave in pledge certain vineyards owned by you in common, as he was unable to encumber to his creditor your share in said vineyards, the Governor shall order it to be restored to you, together with any crops which the creditor may have gathered from the same. The Governor must also provide for the division of the vineyards between you and your brother's creditor, and order him to deliver to you the portion which he received from your brother, after having been paid the price which he decides that your brother's share is worth; or he must order it to be transferred to your brother's creditor after your share has been appraised, and he has paid to you the amount of its valuation.[27]

In another case heard by Severus Alexander, a soldier called Florentius, classified as a minor under twenty-five years of age, had clearly been involved in selling land.[28] Around the same time another soldier, Crescens, was in dispute with his parents about the shared ownership of a slave:

*The Emperor [Severus] Alexander to the soldier Crescens*

If the ownership of the female slave, with reference to whom you have brought an action, belongs to your mother, she could not lawfully have been sold by your father; and if you claim her for yourself. The Governor of the province shall order her to be produced in order that the truth of the matter may be judicially ascertained.[29]

## LEAVE

'You have ten days' leave, Ammonas. You have two extra days in which to return.' So says a short letter found in Egypt advising a soldier called Ammonas about a few precious days off.[30] A career in the Roman army was generally a full-time occupation with the chances of being able to get away limited and only on an ad hoc basis. The following short letter, thought to be from a praefectus castrorum, concerns a grant of leave to a cavalry officer called Teres in an unknown unit. It was probably written in 103 during the reign of Trajan:

> [Name unknown] to the decurion Teres, greetings. [Chry?]ster-mus, strategos of the Coptite nome has asked me to grant you thirty days' leave so that you may visit your possessions which you have in the Arsinoite nome. I have written not only to him but also to the centurion Petronius Fidus that I grant this .[310]

The exact locations are not mentioned, but the journey between the nomes would have involved a lengthy journey down the Nile and out to the Fayum. The leave in this case was clearly a special grant so that Teres could deal with his personal affairs, which may well have included other sources of income. At the same time the document seems to be routine and untoward, even though it clearly involved administration and the requirement to inform several relevant parties. Since Teres was an officer, perhaps it is not surprising that he was allowed so much time off, but there is no reason to assume this was unusual.

Among the finds at Vindolanda are a couple of scrappy fragments of writing tablets on which are written soldiers' personal requests for leave. One says 'I ask that you consider me a proper person to whom you might grant leave at Ulucium'. The location of Ulucium is unknown today, but it was clear the soldier making the application intended to travel to a different part of the province, probably somewhere on the northern frontier. Another used the same formula: 'I, Messicus, ask lord that you consider [me a proper person] to whom you might grant leave at Coria'. Coria was Corbridge, a major military town about 12 miles (20 km) east of Vindolanda along the Stanegate, the Roman road that served as the frontier at the end of the first century AD, a generation before Hadrian's Wall was built. It probably offered far more opportunities for 'relaxation' than the remote and windswept Vindolanda. Sometimes a group of soldiers might ask another to intercede on their behalf with the commanding officer as did several Raetian tribesmen serving in Cohors I Tungrorum at Vindolanda in the late first century. Masclus, a soldier of unspecified rank, wrote the letter to request leave for them.[32]

The tone of the requests implies that leave could not be taken for granted. While he was governor of Germania Superior, Galba felt the need to toughen up the garrison. As well as introducing a rigorous training regime to get the troops into condition, he ordered an end to requests for leave.[33] It was a mark of the strictness he would bring to his spell as emperor in 68, and which would contribute to his murder in early 69. Not all governors were so firm. Galba's contemporary Caesennius Paetus, governor of Cappadocia in 62, was so lax with granting leave applications from the garrison of the province that he seriously compromised his ability to fight the Parthians when they invaded Armenia. The army was nearly wiped out at Rhandeia later that year.[34]

In one unusual instance *c.* 90–100 a pair of soldiers at Vindolanda decided to compose a letter using the formulae normally found in leave applications on behalf of one of their friends for a different purpose. Andangius and Vel ... (the rest of the name is lost) wrote to their commanding officer Julius Verecundus, prefect of Cohors I Tungrorum, asking if their fellow countryman and *mensor* Crispus could be given lighter military duties. His job description is not qualified which means we do not exactly what he did. What occasioned this request is unknown. Perhaps his friends felt Crispus was overworked and entitled to join the ranks of the *immunes*.[35]

Most soldiers spent the majority of their time tied to their duties. Roman soldiers usually had to make the best of the facilities at their bases, or in the civilian settlements that grew up around most forts or the towns where they were stationed. Those who had been ordered to man remote fortlets or watchtowers would have had little or nothing to do in the way of relaxation. The troopers of Ala Veterana Gallica in Egypt were spread out all across the Nile Delta on various duties in the first three months of 179.[36] Conversely, soldiers were sometimes able to visit spas, which were often associated with religious healing shrines, and it seems the Roman army may have been instrumental in establishing some of these (see below).

Bathing was a ubiquitous feature of Roman life. Forts and other military establishments were no different, even in the remotest parts of the Empire. Chesters on Hadrian's Wall had small baths with a full range of facilities, including a changing room and a latrine as well as hot, warm and cold rooms, all covered by several vaulted roofs. On the lonely northern frontier its warmth must have made it a particularly appealing refuge. Similar baths are often found close to auxiliary forts. One of the most remarkable finds from the legionary baths at Caerleon was a copper strigil used for scraping cleansing oil from the skin of a bather. Inlaid with silver, gold and brass, it bears an inscription in Greek that told the user, 'it washed you nicely'. The handle was decorated with images of six of the Labours of Hercules. Hercules was a popular figure among soldiers and also worshipped as a deity. His appearance in this context is therefore not surprising, though no other such strigil is known. Six of the Labours are featured (numbers three to eight): the Kerynitian Stag, the Stymphalian Birds, the Queen of the Amazons, the Augean Stables, the Cretan Bull, and Diomedes and his Mares. It would therefore make sense if originally there had been a set of two or possibly three strigils, the first and third bearing images of the remaining Labours and perhaps slogans or additional motifs.[37] Since no others have ever been found it is impossible to say if such decorative implements were common in a Roman military base, but the Caerleon strigil evokes a picture of legionaries visiting the baths and being oiled before being scraped down by a slave.

Physical exercise often took place in the baths' exercise area, the *palaestra*. Vindolanda's astonishing anaerobic deposits have preserved two leather boxing gloves, found in a cavalry barracks dating to around 120. Like the decorated strigil from Caerleon, nothing like them has ever been found before. Better described as padded knuckle protectors, they were wrapped round the hands and must originally have been used by members of the garrison, perhaps in competitive displays of combat practice.[38]


## SPAS AND SHRINES

Military involvement with spa locations is well attested at the shrine of Sulis-Minerva at Bath in Britain, and at Baden-Baden and Wiesbaden in Germany. Some of these sites have produced a wealth of inscriptions and archaeological evidence for visits made by soldiers, sometimes to recuperate, but also to make religious dedications. Bath is a particularly interesting example because the army may well have played a substantial part in developing the site in the first place. Early in the history of Roman Britain, soldiers were engaged in building the trans-province highway now known as the Fosse Way, which connected the new (and short-lived) legionary bases at Lincoln and Exeter and passed through the Bath area. The military road builders must have discovered the bubbling hot waters in a swamp by the river Avon. The area was clearly already sacred to the Britons, but they had left the natural setting as it was. By the late first century, the cult of the native deity Sulis had been conflated by the Romans with Minerva and the hot spring had been contained within retaining walls and equipped with sluices and drains. Nearby, a newly built classical temple of Sulis-Minerva presided over the comings and goings of pilgrims who threw offerings and their messages to the god into the bubbling waters. On the other side of the spring, a vast complex of baths was already operational.

Many of the visitors were soldiers. If the army had been responsible for the initial development of the site, it must have been because Bath offered the chance to rest and recuperate – such places offered leisure as well as medical treatment – or to go in the hope of a final cure from injury and disease (see below). Soldiers were also the only people capable of providing the necessary architectural, engineering and hydraulic expertise to create the facilities in such a remote setting. The Roman army had developed several other healing spas along the Rhine in the early imperial period, from the reign of Tiberius.[39]

## HEALTH

Spas were closely linked to the more general subject of soldiers' health. The army was well supplied with doctors (*medici*), often of Greek origin and who frequently turn up on the dedications they made to Asklepios, Greek god of medicine and healing. Sextus Titius Alexander, probably from Greece himself, was a medicus with Cohors V Praetoria in 82 when he made a dedication to Asklepios. Marcus Naevius Harmodius was the medicus attached to Cohors X Praetoria when he died at the age of fifty-five, probably in the late first or early second century. His name is more obviously Greek, though it may have suited a doctor to adopt a Greek name for the sake of image.[40] Tiberius Martius was *medicus castrensis* ('doctor of the fort') at Legio III Augusta's base of Lambaesis, where he made a dedication to Asklepios in 146–7.[41]

Doctors presumably worked for the most part in the fort hospital (*valetudinarium*). The word was derived from *valetudo*, 'health and the condition of the body', so a more accurate translation would be 'place of health'. Not all forts, especially temporary ones, had such a dedicated facility, but legionary fortresses and permanent auxiliary forts usually did. Identifying them is usually based on a process of elimination, since inscriptions referring to them are rare. The typical plan was of four wings arranged around a small courtyard. They have been found at the legionary fortresses of Neuss and Inchtuthil, and on Hadrian's Wall at Housesteads and Wallsend (see the Plan on p. xiii). In 108–9 under Trajan a hospital was built for Cohors IIII

Lucensium at Aleppo in Syria,[42] while an inscription set up in 179 at Stojnik in Moesia Superior recorded the building of the hospital of Cohors II Aurelia Nova Milliaria Equitata under the prefecture of Titus Bebenius Justus.[43] Such inscriptions are rare, but these fairly random examples suggest that for the most part any soldier who fell sick or was injured, and had a fairly permanent base, could expect to be treated in such a facility.

The doctors were assisted and supported by medical orderlies such as the *capsarii* and the *optiones valetudinarii*, although exactly what these men did is unknown. Aurelius Munatius was a capsarius with Cohors Hemesenorum at Dunaujvaros in Pannonia Inferior. He died after serving for 28 years and was buried by his wife Samosata Aurelia Cansuana.[44] Lucius Caecilius Urbanus was an optio valetudinarii with Legio III Augusta at Lambaesis, where he set up a dedication to the divine imperial house 'of the Augusti', describing himself as 'curator of armament works'.[45] There were also specialists in animal treatment. [A]llio Quartionus served as a *medicus veterinarius* with Cohors I Praetoria, though all we know about him is that he died in Rome aged eighty-five.[46] Remarkably, an Alio is listed as a veterinarius at Vindolanda at the end of the first century in a list of payments and receipts; he may have been the same man.[47] Lucius Cliternius was a veterinarius with an unspecified legion in Pannonia Superior.[48]

Roman military doctors had plenty of work, though most of what they had to deal with were probably infections from disease, or injuries caused by accidents or in training. There were manuals available, including the section on medicine in a general encyclopaedia compiled by Celsus in the early first century AD. He provided a useful section on how to extract missiles while in the process avoiding severing arteries and veins. Arrows, for example, were best extracted by making a counter-opening on the other side of the body and pulling out the projectile in the direction of travel, but larger weapons had to be withdrawn with special equipment which he goes on to describe. There were also special precautions for dealing with poisoned missiles.[49]

Other problems arose from the unhealthy conditions in which soldiers might find themselves. In 69 some of Vitellius' troops chose an unsuitable location on the west side of the Tiber at Rome in which to camp:

A large proportion camped in the unhealthy districts of the Vatican, which resulted in many deaths among the common soldiery; and the Tiber being close by, the inability of the Gauls and Germans to bear the heat and the consequent greed with which they drank from the stream weakened their bodies, which were already an easy prey to disease.[50]

Of thirty-one Tungrian soldiers reported sick at Vindolanda in about 90, ten were suffering from 'inflammation of the eyes'.[51] The thirty-one sick amounted to more than 10 per cent of the unit present at the fort on the day the strength report was drawn up.

Soldiers were of course susceptible to any epidemic that affected the rest of the population, especially if they were concentrated together in forts or camps. It can be difficult now to work out which diseases were being referred to in ancient sources. 'Tertian agues' (four-day fevers) were often reported by people living close to the marshy banks of the Thames in the seventeenth century. Now usually identified as malaria, the disease was far more widespread along river valleys in northern Europe in pre-modern times than is usually realized. A long-lost altar found at Risingham, an outpost fort beyond Hadrian's Wall, was set up to a goddess by Aelia Timo. A 1789 reading names the goddess as Diana, but readings made at earlier dates called her Tertiania. If this is correct, then Aelia Timo was worshipping the personification of a tertian ague.[52]

During Severus' campaign in Caledonia, 'the water caused great suffering to the Romans'.[53] Whether this means disease or the practical problems of tackling river crossings and swamps is unknown. During the reign of Marcus Aurelius an unidentified 'pestilence' affected the whole Empire, killing many soldiers as well as civilians.

It is unidentifiable now, though smallpox and measles have been suggested.[54] A rumour at the time had it that a soldier in Babylonia had accidentally cut open a golden casket dedicated to Apollo. From this a noxious vapour allegedly escaped across Parthia and then the known world; it was carried there by the returning army of Lucius Verus, Aurelius' co-emperor, who died from it. The theory reflected the belief, common well into the 1800s, that disease was carried through foul air (miasma).[55]

One solution to illness was for a soldier to visit a spa, a type of facility that the Roman army seems to have had involvement with in a number of places. Aufidius Eutuches was an obsequious freedman, probably of Greek origin, who left two altars to Sulis at Bath in the hope that the welfare and safety of his patron and former master Aufidius Maximus, the centurion of Legio VI Victrix, would be protected.[56] If the offerings were occasioned because Maximus was sick, we do not know what happened to him. Julius Vitalis was a *fabriciensis* (armourer) with Legio XX, based in Chester, when he arrived at the spa in the late first or early second century, probably because he was seriously ill. He was from Gallia Belgica, having joined the legion at the age of twenty. Nine years later he came to Bath where he died. His armoury colleagues funded the funeral out of their subscriptions.[57] The tombstones of other soldiers who died at Bath indicate that they had served in any one of various different legions around the province, suggesting that they came to the spa for treatment.

## MEMENTOS OF THE FRONTIER

Roman military works are awe-inspiring today and have become popular tourist destinations. In antiquity they were yet more impressive, and they were also visited by tourists. Hadrian's Wall in northern Britain was unique. No other Roman military frontier matched its sophisticated combination of wall, ditches, roads, forts, milecastles and watchtowers. Stretching over 73 miles (117 km) from one side of the province to the other and supplemented with outpost forts, supply bases and coastal installations, it must have been as much a source of fascination then as it is today.

Some of the enterprising tradesmen, or perhaps military metal-workers themselves, who helped service the needs of soldiers and their communities along the Wall came up with the idea of producing souvenirs of the frontier. Several small enamelled bronze pans with handles have been found that have the names of forts along the Wall as part of their design. Two of these, the 'Amiens Patera' and the 'Rudge Cup', found in France and in Wiltshire in Britain respectively, have a design representing the Wall and its crenellations (or forts) beneath the inscription. The 'Ilam Pan', found in Staffordshire, bears beneath the fort names a frieze made up of so-called 'Celtic' curvilinear decoration.

The pans are all different, yet were clearly designed with the same purpose in mind. They record the names of some of the forts in the Wall's western sector but not the same number. The most interesting is the Ilam Pan because it uniquely includes the names both of the Wall and of either the manufacturer or the customer who commissioned it, RIGORE VALI AELI DRACONIS. *Rigore val(l)i* means 'strictly in the order of the frontier', using a Roman military surveying term. AELI probably belongs to *val(l)i*, making the meaning 'strictly in the order of the Aelian frontier'; Aelius was Hadrian's family name (Publius Aelius Hadrianus), and thus supplies the name by which the Wall was known, just as the Hadrianic fort and bridge over the River Tyne where the Wall originally started was known as Pons Aelius . As it happens, an altar was found at Kirksteads, north of the Wall, set up to an unknown god by Lucius Junius Victorinus, legate of Legio VI Victrix, 'because of successful achievements *trans vallum* (beyond the frontier)'.[58] As for Draconis, '[by the hand/the possession] of Draco', he may have been either the maker or buyer.

None of the three pans was found anywhere near the Wall. That one was found in France only makes it more likely that these were personal keepsakes of a visit to the frontier, or perhaps even of military service on it. Fragments of similar vessels are known from elsewhere, but without any surviving references to the Wall. Those that do refer to the Wall probably belong to a regional variant type, adapted to make bespoke versions for the souvenir trade in a province where military installations were more concentrated than anywhere else in the Empire.[59]

Up to this point it has probably seemed as if Roman soldiers spent their lives for the most part exclusively in the company of their compatriots. In a legal sense that was supposed to be the case; under the emperors soldiers were barred from legally marrying until the early third century. In reality it was inevitable that soldiers would seek out, or be sought by, female company. The presence of unofficial army wives and children is found in evidence ranging from tombstones to artefacts, and reflects another aspect of the Roman army reminding us that soldiers shared many of the same concerns that we do.

# FOURTEEN

# WIVES AND LOVERS

## FAMILY LIFE ON THE FRONTIER

*To the spirits of the departed, Lucius Calpurnius Valens, optio*
*of Cohors I Lepidiana equitata, Roman citizens, century of*
*Ponticus, lived forty years, served 18. Calpurnia Leda, his*
*wife, erected this.*

A military wife commemorates her
dead soldier husband in Asia[1]

Around the time of the Second Punic War, the Roman comic playwright Plautus' *Miles Gloriosus*, 'The Swaggering Soldier', first appeared. The principal character is Pyrgopolinices, who spends his time boasting about his military accomplishments, including killing 7,000 men. Pyrgopolinices is followed around by an obsequious acolyte called Artotrogus who constantly flatters Pyrgopolinices' vanity, describing all the women who are in love with Pyrgopolinices and who likened him to Achilles, desperate for Artotrogus to let them see him.[2] Plautus was poking fun at Pyrgopolinices (whose name incidentally means something like a 'city fortified with walls and towers'), but the joke must have had some basis in reality.

316

Lucius Cornelius Scipio Africanus (236–183 BC), the celebrated hero of the Second Punic War after his defeat of Hannibal at Zama (202 BC). He was widely admired for centuries after his death, his military exploits and anecdotes about him being constantly cited by later authorities. A bronze bust found at the Villa of the Papyri, Herculaneum. Photo: Miguel Hermoso Cuesta.

Gaius Julius Caesar (100–44 BC), the celebrated statesman, general, and dictator of Rome. Caesar's military leadership was primarily commemorated in his Commentaries on the Gallic Wars. These accounts became a highly influential textbook on military leadership in the Roman period and long afterwards. The loyalty of his men became legendary. From a portrait on a silver denarius struck at Rome, February–March 44 BC, shortly before his assassination.

Arch of Septimius Severus, Rome. One of the best-preserved of all Roman triumphal arches, Severus' arch still dominates the Forum in Rome and commemorated both his rise to power and his Parthian war. Constructed in 203.

Marcus Aurelius (161–80) and the Praetorian Guard. Marcus Aurelius is greeted by the Praetorian Guard at Rome. The armour of the praetorians is idealized. From his triumphal arch (demolished).

Plautus' play was set in Ephesus and involved largely Greek characters. Perhaps Roman women too preferred soldiers to civilian men, though there is not a single surviving instance of any testimony by a Roman woman to that effect. In an uncertain and often insecure world, however, especially in frontier provinces, the prospect of securing a soldier as a partner and the potential father of her children might well have offered a woman an attractive and reliable prospect. If so, this was to be far more the case under the emperors, when the army had become a steady career.

The Roman army was far from being only a man's world, though it hardly needs to be said that women did not serve.[3] Nor were soldiers under any emperor from Augustus to Septimius Severus apparently allowed to marry until they were discharged. The reasons were obvious. Wives and children offered distractions, were financial liabilities, and gave soldiers reasons to avoid taking risks.[4] Nonetheless, the state seems to have taken no action to prevent soldiers acquiring wives. They might be local women, but most seem to have been drawn from Romanized communities in the region and perhaps were the daughters of other soldiers, including veterans. No soldier that we know of was punished for marrying, and it is quite clear from tombstones, dedications and even documents that plenty had wives. The growth in unofficial marriages and families belonged largely to the second century AD. In part this must have been because such unions were likely to produce potential recruits who were already familiar with army life. Hadrian's decision to end conquest and to consolidate the Empire's frontiers meant that the forts and fortresses became settled establishments. The opportunity to establish a long-term relationship thus became much more common, and since this is an entirely natural inclination it is hardly surprising that there is a great deal of evidence for the presence of women and children in military areas.

Roman armies on campaign usually travelled with vast baggage trains, accompanied by families and camp followers of all sorts. At forts that remained in existence for any substantial period of time, it was in the vici or canabae that soldiers kept their unofficial wives and families. Some soldiers also owned slaves, who formed part of their personal *familia* and were sometimes freed so the soldier had a loyal freedman who would take care of his interests. These settlements are often the most productive sources of evidence for life in the army, especially with regard to religion and families.

Of course any military marriage or liaison was liable to sudden disruption if the legion was ordered to move on. Legio XIIII was transferred from Britain permanently in 70 and after occupying a series of different bases eventually ended up at Petronell (Carnuntum) on the Danube. Legio III Cyrenaica was at Alexandria when it was ordered to relocate to Bostra in Arabia Petraea in 106; it returned to Egypt in 119 before being sent back to Arabia in the early 140s.[5] What happened to the women and children of the soldiers is unknown, but some must have been affected, either having to follow the legion or waiting in the hope of its return.

By far and away the most famous Roman military wife and mother was Agrippina the Elder. She was Augustus' granddaughter through her mother Julia the Elder, Augustus' only child. Her father was Marcus Vipsanius Agrippa, Augustus' principal supporter and general during the civil war of the 30s BC. She was married to Germanicus, the brilliant young general and golden boy of the Julio-Claudian dynasty. He was the grandson of Augustus' empress Livia through her first marriage and his mother was Antonia Minor, daughter of Augustus' sister Octavia. Germanicus and Agrippina therefore had a brilliant and formidable pedigree in the imperial family, which meant their six children promised a great future. Agrippina exemplified the importance of a Roman woman's honour and purity to her husband's professional success.

Agrippina's fame among the army came about because she unhesitatingly accompanied Germanicus on campaign, together with some or all of their children. In AD 14, during Germanicus' campaign in Germany, she was not only with him, but also pregnant and accompanied by her infant son Caligula. With the Roman army on the point of mutinying Germanicus ordered Agrippina and Caligula to go home to safety. She was disgusted by the suggestion that a descendant of Augustus should commit such a cowardly act, but he convinced her to go. As she departed the soldiers were horrified at the shame of having been the cause and pleaded 'She must come back, she must stay!' After a speech from Germanicus the mutiny collapsed, the ringleaders being denounced and executed.

The following year the Roman forces came under threat from a major German advance. With the tribes threatening to cross the Rhine, the simplest solution would have been to destroy the Roman bridge over the river, but according to Tacitus that would have amounted to a humiliating act of cowardice. It was Agrippina herself who stepped in to stop the bridge being demolished. Tacitus said she was a 'woman of great spirit' and added that she now meant more to the army than any military commander.[6] It was a dangerous development in a society that excluded women from formal power. After Germanicus' untimely death in 19, Agrippina was systematically pushed out of the limelight, exiled and murdered (see Chapter 9).

Agrippina was an unusually prominent military wife. She was, after all, a major member of the imperial family. Other women too were present with the army even on campaign, but it is rare for us to hear about them. When Varus and his three legions were wiped out in the Teutoburg Forest in AD 9 (see Chapter 7), the sources say his baggage train included women, but no individual woman is specified in the extant accounts.

One of the few empresses to accompany her husband on campaign was Julia Domna, wife of Septimius Severus, when he set out to conquer Caledonia (Scotland) and thereby toughen up their sons Caracalla and Geta. As early as 193 she had been awarded the title *mater castrorum* ('Mother of the Camps' or 'Mother of the Army'). This honour placed her in a quasi-divine role as a protective female figure.[7] The designation was not new: Marcus Aurelius had endowed it on his wife Faustina the Younger in 174 after she travelled with him on a campaign against the Quadi.[8] But Domna made the most of her new title. A small bronze plate found at Rome, and once fixed to a carriage belonging to her, pronounced that the vehicle and its goods were on the business of Julia Augusta Domna, Mother of the Camps, and were therefore immune from liability for payments or duties.[9]

In 208 Julia Domna arrived with Severus and their sons Caracalla and Geta in Britain, where they remained until Severus' death at York in 211. Domna's presence was unsurprising. She was a woman of high intelligence and came from an aristocratic Syrian family with a well-defined sense of status and entitlement. Domna does not seem to have involved herself with the Roman army in the way Agrippina did, but she did participate in imperial diplomacy.[10] She was commemorated in this capacity during the reign of her son Caracalla, throughout which she presided as the dowager empress, and became an important figure in military observances.[11]

Agrippina and Julia Domna were of course women of high status. The position was completely different for ordinary soldiers. In the Republic, soldiers were often married because they had been summoned from their everyday lives to fulfil the obligations to military service for which every freeborn Roman citizen male was liable. If they were lucky they returned home to their wives and families. Only men of senatorial or equestrian status serving as commanding officers were unaffected by a ban on soldiers marrying, which Augustus is believed to have introduced and which remained in force until the end of the second century. The law was only concerned with the status of an unofficial union between a soldier and a woman, and therefore denied the wife and the children any of the legal privileges normally granted to the families of Roman citizens. The children, for example, were classified as illegitimate, and were therefore unable to inherit their fathers' estates unless they were explicitly named as heirs.[12] This could cause serious problems for women and their offspring if they did not act. When, for example,

Isidorus of Alexandria married a woman called Chrotis, they were both civilians. But Isidorus then joined the army, serving in Cohors I Theborum under the name Julius Martialis. While he was in service Chrotis bore him a son called Theodorus, but overlooked registering his birth. Julius named Theodorus as his heir but seems then to have died. Chrotis had to petition on her son's behalf, which she did in 115. The judgement was that although Martialis could not have a legitimate son while he was a soldier, it was possible for him to have made his son his legal heir.[13]

In 119 Hadrian issued a clever letter which circumvented the law formulated by his predecessors, and helped soldiers' families. He ordered that children born during the period of a man's military service, who would otherwise be excluded from inheriting their father's estate on the grounds of illegitimacy, could claim the property on the legal basis that they were kinsmen by birth of the deceased. Hadrian also ordered that his letter be made public so that soldiers and veterans could take advantage of the privilege. The copy that survives was originally displayed at the winter camp at Alexandria of III Cyrenaica and XXII Deiotariana.[14]

Conversely, the soldier Octavius Valens (his unit is unknown) and his wife Cassia Secunda were doubtless dismayed by the judgement made in 142 by the prefect of Egypt, Valerius Eudaemon. He declared that their son – who was born during his military service – could be neither Valens' legitimate son nor an Alexandrian citizen. The judgement was predicated on the fact that the two were not legally married. Valens tried to argue his case but the prefect was adamant and refused to change his mind.[15]

Whatever the legal problems some soldiers encountered with their unofficial wives and children it seems that such families were not only commonplace but also an essential part in the psychological and emotional well-being of the troops. Tacitus described how comfortable the garrison in Syria had become by 69. It is quite likely that some of the soldiers were married and made no secret of the fact:

For in fact, the provincials were accustomed to soldiers and delighted in cohabiting with them. Many of them were joined to the soldiers through friendship and kinship, and through their long years of service the familiar camp was loved by the soldiers as their home.[16]

This cut both ways. Retired soldiers, if they were forced to relocate to colonies in unfamiliar places, were known to abandon their wives and children there and sneak back to their old bases (see Chapter 15).

Praetorians were also prohibited from marrying while in service, and this had been the position since the inception of the Guard under Augustus. In practice unofficial unions did take place and it is apparent from a number of sources that during the second century such arrangements were sometimes accepted by the authorities, right up to and including the emperor. Veteran praetorians were offered improved help in starting families in January 168. In order to help these veterans acquire wives, any sons born of such marriages would now count for the veterans' fathers-in-law when it came to seeking claims for intestate property or claiming exemption from *tutela* (legal guardianship). The intention was clearly to encourage the prospective fathers-in-law to regard marrying their daughters to praetorian veterans as an asset. Previously, the legal privileges had only been given to such men through sons born to their sons.[17] Legal judgements made in the third century show how military wives living with their husbands while they served the state enjoyed a certain amount of protection. For example, Diocletian and Maximianus ordered that a wife who was away with her soldier husband and discovered that her home had been sold during her absence was entitled to get the house back, if she could prove it was hers.[18]

Some of the most memorable and moving evidence for Roman military life involves the soldiers' families and children. In 44 one of Claudius' measures was to allow soldiers to enjoy the privileges enjoyed by married men, in order to compensate them for being unable to marry legally.[19] On the face of it, the offer was a strange one. The privileges included control of a wife's property – if there was no wife it was meaningless. It is more likely that Claudius had found a way to allow soldiers with unofficial wives to benefit in the same way as other men. A petition made in 63 to the prefect of Egypt, Caius Caecina Tuscus, by legionary veterans claiming that the rights of their children, born during their service, were inferior to those of auxiliaries has already been mentioned (see Chapter 1). It illustrated how complicated the arrangements were for different parts of the armed services. In 117 a later prefect of Egypt, Marcus Rutilius Rufus, denied a wife called Lucia Macrina the right to a claim on the estate of her deceased soldier husband, Antonius Germanus, on the grounds that 'a soldier is not permitted to marry'. The papyrus recording this and other related cases is the principal evidence for the existence of the ban on soldiers marrying mentioned above.[20] When Septimius Severus granted soldiers 'the right to live at home with their wives' in 197, he was probably acknowledging reality rather than significantly altering practice.[21] Whether the permission to cohabit was the same as being legally able to marry in military service has long remained a matter of debate, though in general it is assumed now that it did.[22]

Severus seems to have been bowing to the reality of how things had been for generations. Even his predecessors seem to have been prepared to overlook breaches of regulations. Gaius Acilius Relatus was a veteran of Legio VIII. The inscription of around 71–100 from Aquileia that names him records that he had been discharged from service, perhaps early, and now had a five-year-old son called Gnaeus Acilius Saturninus. The boy was legitimate because he was born after his father's discharge. The inscription also records that he had an 'illegitimate daughter' called Caesia Procula, evidently born while he was in military service. The formula used was SP.F., for *spurii filia*, 'daughter of illegitimacy'; well known from other inscriptions. It seems to have been merely a statement of fact, and not an indicator of shame.[23] Publius Accius Aquila was a centurion with Cohors VI, a part-mounted cohort in Bithynia and Pontus around 111–13, during the reign of Trajan. He wrote to the governor, Pliny the Younger, about the legal status of his daughter, requesting that she be given Roman citizenship. Pliny reported to Trajan, 'I considered it difficult to say no, seeing that I know how much you are accustomed to offer patience and humanity with requests from soldiers.' In Trajan's reply he explicitly granted the soldier's daughter *civitatem Romanam*, 'Roman citizenship', and sent confirmation in writing.[24]

Accius Aquila was a Roman citizen; though whether he was a legionary detached to command auxiliary cavalry or a citizen recruited directly into an auxiliary centurionate in his own right is not clear. But it does not really matter. The fact that his daughter was of unsettled status demonstrates that Aquila and the mother were not joined in a marriage or liaison of equal rank. The daughter was therefore either illegitimate or perhaps had been born to a wife of intermediate status, such as a Latin citizen (see Glossary), or a provincial, *peregrina*. Clearly, Accius Aquila had no reservation about publicizing the existence of his family to the governor and even to the emperor. If there were any potential recriminations, they were clearly gathering dust from lack of use. Clearly it was normal for Trajan to accede to such requests. Here then, in one letter, we have evidence that soldiers families' might be of lesser status than themselves while in service, and that the 'system' allowed for routine requests to have the offspring elevated to equal status.

For commanding officers the position concerning wives was straightforward. They were allowed to marry, the only proviso being that they should not marry a woman from the province in which they were stationed. The reason was probably that such an association might discourage the officer from fulfilling his duties if they involved having to attack his wife's tribe, for example. The most famous letter of all found at Vindolanda is the one written by Claudia Severa to her friend Sulpicia Lepidina. Both women were the wives of the prefects of auxiliary units. Claudia was the wife of Aelius Brocchus, who commanded an unknown unit at an unknown place called Briga, presumably somewhere not very far from Vindolanda; Sulpicia was the wife of Flavius Cerealis, prefect of Cohors VIIII Batavorum at Vindolanda. Another document found at Vindolanda shows Cerialis was in post in 104, during the reign of Trajan. Claudia writes asking Sulpicia to attend her birthday party and sends greetings from herself, her husband and 'my little son' (whose name was probably Aelius too). The letter is brief and appears to have been written by a scribe, but there is a postscript in another hand, probably Claudia's own. If that is so, this is the earliest known Latin inscription by a woman.[25] The letter paints a touching picture of two young military wives – they may well have still been only in their late teens or early twenties – trying to maintain a social life at the furthest reaches of the Empire. Claudia's reference to her son proves that at least one officer's child lived at Briga. The presence of others is indicated by a child-sized leather shoe found at Vindolanda. More recently, Vindolanda's waterlogged soil has given up a miniature sword made of wood, from a level dated around 120. The model weapon has a functional and crude look about it, as if it had been made to serve as a toy.[26]

Claudia Severa's and her son's fates are unknown. Within a few years her husband Aelius Brocchus had been made prefect of Ala I Contariorum (cavalry armed with lances) in Pannonia Superior.[27] If she and the child had survived then like so many military wives over time they must have followed Brocchus on the laborious journey by sea and land across Europe to a new posting on another bleak frontier, though it is possible they were closer to home. Either way, after their brief emergence from the mists of the past the family disappeared and are never heard of again.

Children of course needed to be educated in some way, though boys were always prioritized. A writing exercise also found at Vindolanda includes the first four words of another line from the *Aeneid* – 'Meanwhile, winged Victory flittering through the trembling town' – and a tiny fragment containing parts of the first two lines of the poem.[28] Both are likely to be connected with teaching children, probably those of officers, and must again reflect a determination to maintain a cultured existence even on this remote military frontier. In Egypt a soldier called Julius Terentianus placed his children and his other private affairs in the hands of his sister Apollonous in Karanis. In 99 she wrote to him to say 'do not worry about the children. They are in good health and are kept busy by a teacher.'[29] Terentianus had perhaps been widowed, but another possibility is that this was an example of a marriage between brother and sister. Such unions were a long-established practice among the elite in pharaonic Egypt but became much more common among ordinary people in the Roman period, possibly as a strategy to hold on to inheritances, and might involve either biological or adoptive siblings.[30]

Numerous tombstones and monuments attest to other women who shared their husbands' military lives, many out on the frontiers. When Petronius Fortunatus set up a monument recording his military career as a centurion at Cillium in Africa in the third century he was eighty. He added a note that his 'beloved wife' Claudia Marcia Capitolina was sixty-five, and that the two of them had erected the monument as a memorial to their son, who had long predeceased them.[31]

Tiberius Memmius Ulpianus was prefect of a Cohors II equitata of Roman citizens and became tribune of Legio III Augusta at Lambaesis. There, with his wife Veratia Athenaide, who was probably Greek, and their daughter Memmia Macrina he made a dedication to the eastern sky god Jupiter Optimus Maximus Dolichenus.[32] An unusual example of marital collaborative repair work was organized by Lucius Antonius Sabinianus, a cornicularius in Legio I Adiutrix, and 'his wife Aurelia Aeliana' at Gyor in Pannonia Superior. They rebuilt another temple that had collapsed through age, dedicated to the Capitoline Triad – Neptune, Father Liber and Diana – during the reign of Severus Alexander.[33] The two seem to have acted autonomously rather than on behalf of the legion, which would have been odd anyway given Antonius Sabinianus' rank. The temple was then perhaps a modest affair, but the inscription made no secret of the fact that the couple were married.

The centurion Publicius Proculinus was based at, or passing through, the Wall fort at Housesteads. At Carrawburgh, the next fort to the east, a temple of Mithras (known as a mithraeum) is known to have existed and can be visited today. In 252 Proculinus set up an altar to Mithras at Housesteads, incidentally telling us that it was also erected in the name of his son Proculus.[34] Proculinus' family probably lived in the civilian settlement attached to this fort or another. In due course, if he had not already done so, Proculus would have followed his father into the army.

It is rarely clear where a soldier's wife had come from. Generally we have only her name to go on. Gaius Maesius Tertius was a trooper with Ala I Hamiorum Sagittariorum, a Syrian unit, when he was discharged in 109 during the reign of Trajan. His discharge diploma, which was found in North Africa in the province of Mauretania Tingitana, also names his wife Julia Deisata, who is specifically described as a Syrian. She must have come with him when he joined the army or followed him out there at a later date. By the time her husband left the army they already had children.[35] Julia is described as 'daughter of Julius', making it likely she was freeborn. Soldiers could also acquire their wives in the provinces where they were stationed, sometimes as slaves.

## SLAVE WIVES

Soldiers sometimes picked their wives by buying them as slaves. Owning slave girls might have been one way of providing themselves with what they wanted without having to turn to prostitutes. Gaius Julius Longinus was a veteran of Legio VIII Augusta who came from Philippi in Macedonia but retired to Riati in Italy. He built his tomb sometime after 79 for his family which included his wife, named also as his freedwoman. Helpis had been his slave-woman when he was a soldier, as is proved by the fact that their son was born a slave. As a freedwoman she became Julia Helpis, the boy as a freedman of his father became Gaius Julius Felix.[36] In 150 the trader Aeschines Flavianus of Miletus sold an experienced slave girl to a soldier of the Classis Ravennatis called Titus Memmius Montanus. Although written in Ravenna, the tablet that records the transaction was found in the Fayum in Egypt where Montanus must have taken it, probably because that was where he came from in the first place. The cost was 625 denarii, a substantial sum of money and around twice his official annual income as a soldier.[37] It was a lot to pay for a servant, and therefore it is likely he had other plans for her.

The poet Martial in the late first century counted among his friends a centurion called Aulus Pudens from Umbria, who appears to have campaigned in Domitian's Dacian War and may have reached the rank of primus pilus. Pudens married a woman called Claudia Peregrina ('Claudia the Provincial'), also known by Martial as Claudia Rufina. Martial tells us that Claudia was very fertile and that she had 'sprung from the woad-stained Britons'; he was particularly fascinated by how successful the marriage was. Where Pudens met Claudia it is now impossible to say. One possibility is that she was enslaved in Britain and bought by Pudens from a slave trader who had brought her to wherever Pudens was garrisoned at the time.[38] She might alternatively have been the daughter of a Briton serving in the Roman army.

Gaius Longinus Cassius was a veteran of the fleet at Misenum. He probably came from Egypt, because in the late second century he retired to the Fayum, where he drew up his will in 189. He made provision for freeing his two female slaves, Marcella and Cleopatra, who were then aged over thirty, and Cleopatra's daughter Sarapias, who was also his slave. He does not say so but Sarapias was probably his daughter, because he left her two plots of land as well as a one-third share in two houses and a palm grove. Marcella and Cleopatra seem to have been his only heirs, despite the fact that he left to a single kinsman a bequest of 4,000 sestertii, since he shared the rest of his estate equally between the women.[39] Longinus Cassius appears to have had a convivial retirement in the company of two women of whom he was clearly extremely fond but had purchased originally as slaves. If either woman had died by the time the will was opened, then he made provision for Marcella's share to go to three named males, and Cleopatra's to one named male. These were presumably the women's other children, though they were not identified as such and nor were they described as slaves.

Regina ('Queen') did not live long enough to benefit from her husband and former master's will. She was a tribeswoman of the Catuvellauni in Britain who lived in the early third century AD. Her name is Latin and was probably given to her when she was enslaved. She had quite possibly been enslaved by Britons servicing the Roman military market, unless she was born into servitude in a Romanized household. Provincials are also attested dealing in slaves.[40] Regina was bought by a man of Syrian origin called Barates, who came from Palmyra. Evidently he had taken, or took, a liking to her, because he freed her and married her, probably when she was in her mid-teens or early twenties. The couple lived in the civilian settlement outside the frontier supply base fort of South Shields at the mouth of the river Tyne, some way east of the end of Hadrian's Wall. Barates was almost certainly a soldier based at the fort or on the northern frontier nearby. In the event, their marriage was cut short, as so many were in those days, when Regina died at the age of thirty. We do not know why: childbirth complications or disease are only two of the possibilities. Barates was evidently greatly distressed, because he commissioned an extravagant tombstone for his 'freedwoman and wife' in the florid eastern classical style of his homeland. It must have been made for him by a specialized sculptor who lived close to the fort. Regina is portrayed seated spinning and with her jewellery box. Beneath the Latin inscription an extra line in Palmyrene script was added saying 'Regina, freedwoman of Barates, alas'. Barates may be the same man attested at the military base of Corbridge further west, where he was described as a flag bearer who had died aged sixty-eight.[41] The presence of the Palmyrene text probably means that Barates continued to use his native language privately, even if like other soldiers in the Western Empire he was familiar with Latin on an everyday basis.

Around the time that Regina lived at South Shields, a case was brought before the emperor Caracalla on 17 February 214 by a soldier called Marcus. It concerned a slave that Marcus had bought for his mistress. Caracalla's judgement was:

> If you prove before the governor of the province that the female slave in question was purchased with your money, even though it was stated in the bill of sale that she was destined as a gift to your concubine, he must order her to be restored to you. For although this donation may be valid where matrimony does not exist, I am still unwilling that my soldiers should, by means of perfidious blandishments, be plundered in this way by their concubines.[42]

It is a pity we do not know where Marcus was based or what sort of soldier he was, but evidently Caracalla believed Marcus was being taken advantage of, even if technically it was legal for him to make the gift.

There is no doubt about the origins of Claudia Rufina and Regina, but the wife of Titus Flavius Virilis is more of a mystery. Flavius Virilis did well in the army in the third century. He was promoted to centurion, in which capacity he served in Britain in Legiones II, XX, VI and again in XX, before being moved off to III Augusta at Lambaesis in North Africa; he also did service with III Parthica Severiana, probably somewhere in Mesopotamia. He had reached the age of seventy when he died. His wife Lollia Bodicca was among those who contributed 1,200 sestertii to pay for his tombstone.[43] Lollia Bodicca's name is almost certainly a British one and, given her husband's career, it is likely (though not certain) not only that he met her in Britain but also that she was a native Briton, and perhaps originally a slave. The prefix Bod- is well known in British contexts and those of the north-western provinces and is obviously linked to the name Boudica, the tribal warrior leader of the rebellion of 60–1.

Regina and Lollia Bodicca are interesting precisely because they were so unusual. The vast majority of the known wives of soldiers and veterans have so-called 'Roman' names in the form of a nomen and cognomen, such as 'Severia Secundina', rather than indigenous ones. This might mean that soldiers predominantly found wives whose origins were similar to their own, or who came from local families that had been Romanized through enfranchisement or through being freed by a Roman citizen.[44] Lollia Bodicca had retained a native name, but only because it had been adapted into a Romanized form. Regina's native name, if indeed she ever had one, is lost. Although she was a Briton and a slave, her Latin name was a mark of her absorption into Roman culture. The bias towards 'Roman' women may thus be partly an illusion, with native names being dropped or marginalized by some as part of the social climbing opportunities afforded to women who found security and a future in being married to a Roman soldier. Indeed, remarkably, a law in Roman Egypt was created to address, by ordering their detention, the problem of non-citizen Egyptian wives of veterans who passed themselves off as Romans.[45] Either way the Roman identity of soldiers' wives, whether genuine or assumed, reflects an environment in which army communities tried to distinguish themselves from the indigenous populations.

## MOTHERS AND CHILDREN

Children often lived in the vicinity of forts and fortresses, many probably having been born there. The best-known child attested on a Roman military campaign was Caligula, or 'Little Boots', on account of the fact that his mother Agrippina the Elder dressed him in military costume with a miniature version of a soldier's footwear known as *caligae*. As an adult he bitterly resented the nickname, which he regarded as a 'reproach and disgrace'. A legionary centurion 'of the first maniple' once got into trouble for using it, probably during Caligula's campaign in Germany while emperor.[46] One account of Caligula's birth in 12, certainly fictitious, had it that he was born in the legionary winter quarters on the Rhine frontier. Suetonius, who reported the rumour, doubted it was true while recording some popular lines that circulated at the time:

> He who was born in the camp and reared amongst his
> nation's arms,
> Gave at the outset a sign that he was destined to rule.[47]

Claudia Severa's little boy (mentioned above) was the son of an equestrian prefect so his status was legitimate – unlike that of the children of unofficial unions, of whom there must have been many. The sad truth is that most of the children whose presence is attested in a military context are known only from the tombstones which recorded their premature deaths. Infant mortality was of course extremely high, despite the better diets and conditions available to military families. Mascellus, a trumpeter (*cornicen*) with Legio II Traiana Fortis Germanica at Alexandria, commemorated his 'sweetest daughter', whose name is lost but which we can guess was Mascella. She lived three years and twenty days. The child's mother was not included, either because she was dead herself or perhaps because Mascellus was keen to gloss over the fact that he had an unofficial wife.[48] Bruttius Primus, another soldier of the legion, buried his 'dearest daughter Bruttia Rogatina' who died aged eleven. Her mother also goes unmentioned.[49]

Simplicia Florentina, 'of the most innocent spirit', had lived ten months when buried by her father Felicius Simplex, a centurion of Legio VI Victrix.[50] Septimius Licinius, an *immunis librarius* with Legio II Parthica at Albano Laziale in Italy, buried his 'dear son Septimius Licinianus' who had died aged three years, four months and twenty-four days.[51] Blaesus was a boy of ten who died at Birdoswald, and was commemorated on a tombstone set up by his brother.[52] Julius Simplex, a soldier at Maryport, buried his ten-year-old son Ingenuus,[53] while at Lambaesis, Quintus Aemilius Dativus had the tragic task of burying two of his children. One was his nineteen-year-old son Quintus Aemilius Felix, a soldier of Legio III Augusta who must have only recently signed on with the legion. The other was his daughter Aemilia Januaria, who died aged five years and twelve days. Both were commemorated on the same stone.[54] Aemilius Dativus tells us nothing about himself, but if he was a legionary at Lambaesis he had probably retired and had his son legitimized.

Mothers were not always omitted from the memorials of soldiers' children. When Aurelia Heraclita died at Dunaujvaros in Pannonia Inferior aged seventeen years, seven months and twenty-two days, she was buried by her parents Marcus Aurelius Heraclitus, a spear-thrower (*hastatus*) with Legio VII Claudia, and her mother Pia. The name Pia is a female version of *pius*, which means faithful and right-eous; its brevity makes it certain she did not have the status of a Roman citizen and was probably a local woman, like Regina, who had been given or adopted a Latin name.[55] Aelius Vitalis, 'who lived more or less thirteen years', was commemorated by his parents Aurelius Januarius, a soldier with Legio II (or III) Italica, and Quirilla 'his mother' with a dedication to their 'unhappy and sad son' at Solin in Dalmatia at some point after 165, when the two legions were founded.[56] Her name and the vagueness about their son's age suggest an informal liaison, albeit a long-standing one.

Julia Paterna seems to have been a widow when she buried her son Julius Amandus, the thirty-year-old *librarius consularis* and soldier of Legio III Italica at Augsburg.[57] No husband is mentioned but it is inconceivable that he had not been a soldier too, dying either in service or in retirement. Julia Paterna must have been at least in her mid-forties when she set up the gravestone. If her husband had waited until discharge to marry her, then he might easily have been twenty to thirty years her senior. As such, and as a widow in a military community, Julia Paterna can have been far from alone.

Atilia Vera was the dutiful daughter of Lucius Aemilius Paternus. Her father had had a remarkable career, serving as a centurion in several legions including VII Claudia, I Minervia, XIIII Gemina and II Augusta, as well as Cohors IIII Urbana and Cohors IIII Praetoria, and she wanted to honour his achievements. During his time he was decorated by Trajan for his exploits in Dacia and Parthia, receiving the *corona vallaris* twice. Without his daughter's efforts to commemorate his achievements on an inscription we would know nothing about him. The inscription was found at Perolet in Spain, which was perhaps where he had retired.[58]

## Extended families

Just because a soldier was based at a fort in a remote province did not mean he was necessarily far from members of his extended family. There is some evidence that siblings and others might follow a soldier to his posting. Chrauttius was a soldier at Vindolanda in northern Britain in the late first century AD. He wrote to his friend and former comrade Veldedeius at the provincial capital of London, Veldedeius having evidently been sent there on detachment to serve on the governor's bodyguard as a groom (*equisio*). The letter concerned trivial business, including Chrauttius' hopes that a pair of shears he wanted from a veterinary doctor called Virilis could be sent to him.[59] He also wanted his greetings sent to his sister Thuttena, who therefore must have lived in London; not something one might have expected had Chrauttius been a recruit who had signed up with the Roman army on the Continent, leaving his family behind when he was posted to Britain. Instead it looks more as if he came with his family in tow, perhaps not even joining the army until he was in Britain. Why the letter was at Vindolanda, if it was originally sent to London, can only be explained if Veldedeius was subsequently restored to his old unit and returned to Vindolanda bringing the letter back amongst his effects, unless it was a copy.

Unlike Thuttena, Togia Faventina appears to have lived close to her brother Togius Statutus. Both names show a mixture of native and Roman forms. Togius was a soldier who served for 19 years with the Numerus Exploratorum Divitiesium Antoniniana, an irregular auxiliary division of scouts from Deutz by Cologne in the third century.[60] He was buried at Mainz where he must have been stationed, but it was his sister who took care of his funeral.

In another instance it is clear that a whole family was settled near the military base. At Mainz a retired soldier of Legio XXI Rapax called Quintus Marcius Balbus died and was buried along with his son Celer. No explanation is provided in the inscription about whether they died together or at different times, and nor is any information supplied about their ages. An oddity about the text is that Balbus was referred to as a *missicius*, a rare word which means 'discharged' without the usual formula for an honourable discharge, *honesta missio*.[61] According to his monument at the headquarters of Legio II Augusta at Caerleon in south Wales, Tadius Exuperatus 'died on the German expedition' at the age of thirty-seven. The monument commemorated both Tadius and his mother Tadia Vallaunius, and was set up by his sister Tadia Exuperata.[62] Vallaunius is a native female name – which, along with the fact that the soldier had a family at Caerleon, makes it likely that Tadius was a son of a legionary who had taken a local woman as his wife, joining Legio II Augusta himself when he was old enough. His father had probably died some time before, explaining why he is not mentioned. Neither the stone nor the *Expeditione Germanica* can be dated with any certainty, but in 213 Caracalla led a campaign against the Alamanni, a tribe of the Upper Rhine, and this may be where Tadius met his end.

Julius Valens was buried at Caerleon by his wife Julia Secundina and his son Julius Martinus. Julius Valens had reached the epic age of one hundred (or thereabouts), a fact stated with no elaboration or ceremony as if it was of not the slightest importance, although the figure is probably an approximation rather than an accurate measure. He had probably served as a legionary with Legio II Augusta for at least 25 years, which means that he must have lived in and around the fortress for another half century or more once he retired. Julia may well have been a second wife. She was to die later at the comparatively modest age of seventy-five, and was buried by their son close to her husband. [63]

Aurelius Marius, an 'optio of the standard-bearers' with Legio XIIII Gemina when it was based at Alba Julia in Dacia, made a dedication to the goddess Nemesis (Fate) along with his wife Severia Secundina and their freedwoman Mariana Bonosa and freedman Marinianus. The relationship between the last two is not specified, but it looks as if the soldier and his wife perhaps lived with or in close association with two former slaves who were now married. [64]

However, most soldiers, like the Egyptian recruit Apion (see Chapter 2), were based a long way from the places where they had grown up but were anxious to maintain ties. The letter Apion sent home to his father shows his anxiety to keep in touch and that he had the means to do so. Aurelius Poleion had left his home in a village called Tebtynis in the Fayum in Egypt and gone to serve with Legio II Adiutrix at Budapest in Pannonia Inferior during the third century. Poleion wrote home to his brother Heron, his sister Ploutou and his mother Seinouphis (a bread seller), and itemised other relatives. He posted the letter via a legionary veteran in Egypt who was supposed to forward it to his family. He used bad Greek and other linguistic peculiarities that suggest he had written it himself. Poleion was clearly upset that despite having already written home a number of times, he had had no reply, 'I do not cease writing to you, but you do not have me in mind', he said. There is a hint that he might have joined the army as a result of falling out with his family and says that as soon as he learns they are thinking of him he will request leave from the legate of the legion and come home for a visit. His anxiety may in part have been out of fear for his inheritance rights (see the case of Julius Terentianus earlier in this chapter). What became of this plea we do not know, or whether Poleion ever saw his family again. [65]

## MARITAL STRIFE

Not all Roman soldiers had long, stable and faithful unions with women, regardless of the latter's status. Marriages could break down. Julius Antiochus was a soldier with a vexillation of Legio IIII Scythica at Dura-Europos when he divorced his wife Aurelia Amimma, a woman 'of Dura', in 254. This is recorded in a document found at Dura-Europos which was compiled after the divorce and dealt with the distribution of property and possessions. Under its terms Antiochus agreed to hand back everything he had received from her and accepted her right to form a union with someone else, while Amimma made the same undertakings to her former husband. A breach on either part was on pain of a fine of 1,000 denarii. Antiochus signed on his own behalf but a male civilian signed on Amimma's because she was illiterate, itself a noteworthy but not unsurprising facet of female inequality at the time.[66]

## ADULTERERS

Officers and soldiers alike were of course often confronted with the opportunity to cheat on their wives. Inevitably some soldiers embarked on adulterous trysts, even though men who had been found guilty of the crime were not allowed to enlist in the first place. If as soldiers they were convicted of the offence they were regarded as being released from their oaths, and thus were thrown out of the army.[67] As so often in Roman society, different standards applied to the elite. Scipio Africanus, so often referred to by later Romans as a model of virtue and restraint, was also respected for the example he set when resisting taking advantage of women who had been captured. Such women were liable to be treated as chattels and subjected to sexual abuse and exploitation by a conquering Roman army. Ammianus Marcellinus, writing almost six centuries after the Second Punic War, commented on how the emperor Julian (360–3) followed

Scipio in avoiding conduct that might result in being 'unnerved by passion'.[68] The story had its origins in Polybius, who recounted how Scipio had turned down the offer of a young woman after the capture of Cartagena in Spain in 210 BC, his self-control 'earning the respect of his men'.[69] Scipio was however not as restrained as he was made out to be. The ideal Roman military wife was a woman who unquestioningly supported her husband and made sure her reputation enhanced his. Scipio's wife, Tertia Aemilia, was hugely admired in later years for various reasons, one of which was that she turned a blind eye although she knew her husband was sleeping with a slave girl. After Africanus died she married the girl off to a freedman of hers rather than seek revenge. To have acknowledged her husband's actions would have been openly to accuse him of lacking self-control, thereby denting his public image.[70] For a man of his standing, evidently an affair with a slave girl simply did not 'count'. Sleeping with the wife of another senator, however, or indeed anyone of freeborn status, might have had different consequences. Had his wife been unfaithful though, Scipio would undoubtedly have been seriously compromised.

During Trajan's reign, Pliny the Younger was called to Città Vecchia in Etruria, Italy, to act as the emperor's adviser during a number of court cases. The second one involved a military tribune's wife, a woman called Gallitta. Her husband, a man of senatorial rank, was on the cusp of entering civilian public office but his career now hung on a knife edge, for Gallitta had been having an affair with a centurion. The disgrace had the potential to destroy the tribune's career. If a wife was unfaithful or compromised her reputation in any way it reflected badly on the husband for failing to keep her under control. In order to distance himself, Gallitta's husband had reported to the governor – presumably of an unidentified province. This makes it likely that the centurion was serving in one of the frontier legions. The fact that Gallitta's indiscretion had been with a man of lower social rank made it far more serious, and the governor had passed the case on to Trajan. The emperor looked at all the evidence, 'cashiered the centurion and banished him'. Gallitta's husband did not want her punished but was criticized for effectively condoning her behaviour because he had allowed her to remain in the family home after reporting the case. He was forced to stand up in court and speak against Gallitta, but did so only reluctantly. She was condemned under the *lex Julia de adulteriis* (the Julian Law on adultery, brought in by Augustus in 17 BC) to lose half her dowry and one-third of her estate, and was banished to an island. Trajan took the opportunity to name the centurion in court (Pliny did not record it) and spoke about military discipline, saying he did not wish similar cases all to come to him. He must therefore have been well aware that Gallitta's case was far from unusual.[71]

The emperor Macrinus was notoriously vicious when it came to disciplining troops. When some of his soldiers had sexual relations with their host's female servant (the host was probably the owner of a house where they had been billeted), one of his frumentarii spies discovered what had been going on and told Macrinus. He interviewed the culprits and, when he had determined the story was true, ordered a peculiarly disgusting punishment. Two exceptionally large oxen were brought and cut open while still alive. The soldiers were then rammed into the dying animals' guts with only their heads protruding, so they could still talk to each other. It was an entirely novel punishment with no precedent. They might have ruminated on their good fortune, given that on other occasions Macrinus allegedly had adulterers tied together and burned alive.[72]

Before becoming emperor in 270, Aurelian was a commander with a reputation for being a ruthless disciplinarian. One story was that one of his soldiers had been caught having had an affair with the wife of a man in whose house he had been billeted. His punishment, ordered by Aurelian, was to be tied by his feet to the tops of two trees that had been bent down to the ground. When the trees were released, they flew upwards so quickly that the man was torn in two. There is no means of substantiating the anecdote, but if it circulated during Aurelian's time it may well have helped reinforce his authority as a man to be feared.[73]

## PROSTITUTION

References to prostitutes were sometimes quite oblique and tied up with a more general intolerance of adultery or immorality rather than of prostitution itself. The Latin word for the best class of prostitute was *meretrix*, which meant a 'female wage earner'.[74] The more general word, *scortum*, was gender neutral, a reminder that both men and women were involved in selling their bodies. Prostitution was a well-established phenomenon in Roman culture and the women concerned made their profession obvious in the way they dressed, wearing bright colours.[75] Among them was Novellia Primigenia, who according to a scrawled graffito found in Pompeii made her living at Nocera near 'the Roman gate in the district of Venus' (obviously a euphemism for the prostitutes' district). There was a praetorian base near to Nocera.[76] It is not hard to imagine similar commerce going on at most forts in some shape or form.

Prostitution and the use of prostitutes were both legal activities and it is certain that soldiers were among their clients, though finding specific evidence for this is difficult. Soldiers stationed in forts were perfectly capable of seeking prostitutes in local settlements, and under such circumstances we would be unlikely to know about it. In 90 in Egypt 'women for prostitution' were charged 108 drachmas as the toll for a journey on the Red Sea whereas women classified as 'soldier's women' were charged only 20 drachmas.[77] If the larger fee was for a single prostitute, its level must have reflected their ability to pay. Attice at Pompeii charged four sestertii (one denarius) at a time when a legionary was paid only around two-thirds of that a day before stoppages.[78] Caligula had felt able to levy a tax on prostitutes equal to the amount they received from a single client;[79] it was one of the taxes he ordered his Praetorian Guard tribunes and centurions to collect, doubtless in part because, being among the patrons of the brothels, they were bound to know where the prostitutes lived and worked. The state therefore had a vested interest in allowing prostitution to continue.[80]

A papyrus from Egypt preserves a judge's decision to send a soldier, who had been having sexual relations with both a mother and her daughter, into exile. Hadrian determined that a woman 'on whom a disgraceful suspicion' fell (implying she was a prostitute) could not benefit from a soldier's will.[81] A woman who lived with a soldier 'as his concubine' was legally prohibited from benefiting should the soldier die, or so the jurist Papinian believed on the basis of military law in the late second or early third century. If he died in service, or even within a year of being discharged, and had made a will naming her as his beneficiary she was disallowed from receiving his estate, which would be appropriated by the state.[82] Hadrian's and Papinian's views were probably based on an effort to ensure that in general, soldiers' wives and children were not disadvantaged by prostitutes or adulteresses. Ensuring they could not benefit from soldiers' wills was a way of discouraging them from having relations with soldiers who already had partners, even if in Hadrian's time those relationships were themselves still technically illegal.

In reality it is possible that some, perhaps many, of the unofficial wives of soldiers had formerly been prostitutes, or perhaps remained so, especially if they were separated from their husbands for long periods. While it would not do to cast unnecessary aspersions, it is easy to see that the arrival of an army unit and its sustained presence automatically presented the local population with customers who had not only a range of needs but also the means to pay for them. Whether a commanding officer was prepared to put up with prostitutes was another matter altogether.

Fabius Maximus Verrucosus, who fought against Hannibal in the Second Punic War, was once faced with two problem soldiers. One, an infantryman from Nola, was exceptionally brave but was suspected of having Carthaginian sympathies. The other, a Lucanian cavalryman, was a good soldier but was having an affair with a prostitute. Keen to keep both men on rather than punish them, Fabius Maximus made a point of praising the Nolan infantryman and decorating him, so that the man abandoned his Carthaginian leanings. As for the other, Fabius enabled him to buy his lover out of prostitution, with the result that the cavalryman was far happier to fight.[83]

That occasion had a better outcome for the prostitute than those who followed Scipio Aemilianus' army at Numantia in 134–133 BC. Among other reforms and disciplinary measures, Aemilianus threw out of the camp 2,000 prostitutes (not necessarily all female) who had followed his men. Generally invisible in the sources, prostitutes must then have been a constant part of Roman army life; if the surviving section of this book of Livy's is correct, they were a physical presence within the Roman encampment at that date. In fact, if the figure is right, they approximated to about one-sixtieth of the strength of Aemilianus' army.[84] Around 150 years later Valerius Maximus was still frothing with disgust at the thought of the 'disgraceful and shameful bilge-water' Scipio Aemilianus had had to empty out 'of our army'.[85]

Despite the large numbers of forts known, and their associated civilian settlements, evidence for buildings or facilities that can be associated with prostitutes is virtually unknown. The camp at Dura-Europos is a rare exception, perhaps the only one. The walls of a house found there within the military compound bore graffiti suggesting that female entertainers had lived there around 250. This example is, however, so unusual that it is impossible to know whether it is a chance survival of a common arrangement or a private enterprise. Certainly no senior officer who wanted to live up to the reputation of an exalted forebear like Scipio Aemilianus would have tolerated anything so blatant.[86]

## HOMOSEXUAL AND HOMOEROTIC RELATIONSHIPS

Roman attitudes towards homosexual or homoerotic relationships in the army varied across time. In antiquity the concept of being either homosexual or heterosexual did not really exist. Instead there was a general acceptance of more fluid preferences, especially where they involved older married men and their indulgence in relationships with younger male slaves.[87] In Plautus' comedy *Pseudolus* the pimp Ballio mocks an officer's slave called Harpax, 'at night, when the Captain was going on guard, and when you were going with him, did the sword of the officer fit your scabbard?'[88] The line was clearly designed to resonate with a popular view of what some officers expected from their slaves, men who were in no position to fight back. The great soldier-emperor Trajan was well known for his inclinations. Cassius Dio said that 'in his relations with boys he [Trajan] harmed no one', adding this observation to various comments about the emperor's personality.[89]

To what extent consensual same-sex relationships existed between soldiers of similar status is much harder to unravel.[90]

Nonetheless, most of the evidence indicates that homosexuality in the Republican army was not officially tolerated. Polybius included homosexual acts as one of several offences, along with stealing from the camp or giving false evidence, for which a soldier could be clubbed to death; this punishment applied to both the active and the passive partner, so long as the latter were acting voluntarily.[91] Many years earlier, during the Third Samnite War, the military tribune Marcus Laetorius Mergus was sent for trial, accused of trying to seduce his *cornicularius*. Unable to bear the shame, Laetorius first absconded and then committed suicide. That was not enough for the Roman people, in whose name the deceased tribune was still convicted of the crime of *impudicitia* ('unchastity'). Laetorius, said Valerius Maximus, had been pursued to the Underworld; the fact that his story was still circulating three centuries later was a mark of how notorious it had remained.[92] That the trial had been conducted in the civil courts shows that the offence was not treated purely as a military affair.

Marius perhaps bore the Laetorius episode in mind when he was confronted with an incident involving a young soldier called Trebonius. Having subjected Trebonius to constant sexual harassment, to no avail, Marius' nephew Gaius Lusius, who held a command in Marius' army, had finally summoned the young man to his tent, determined to get what he wanted in a classic example of abuse of power. Trebonius had no choice but to present himself, but had no intention of giving in. Lusius evidently tried to use force to make him submit, so the soldier pulled out his sword and killed Lusius. Trebonius was sent for trial and spoke in his own defence, explaining how Lusius had previously solicited him with presents. Marius exonerated Trebonius for his conduct and decorated him with a crown.[93]

Another story from around 280 BC concerned the decorated centurion and four-times primus pilus Gaius Cornelius. Accused of having had sexual relations with a 'freeborn youth', Cornelius tried to defend himself on the basis that the young man had been an active prostitute, but did not deny the offence. The tribunes threw him into prison anyway and left him to die there. The story demonstrates that the army was followed by both male and female prostitutes, because Cornelius' defence would otherwise have been nonsensical.[94]

In the early third century AD, at the supply base fort of South Shields, a young man called Victor was buried and this text placed on his tombstone:

> To the spirits of the departed (and) of Victor, a Moorish tribes-man, aged twenty, freedman of Numerianus, trooper of Ala I Asturum, who most devotedly conducted him to the tomb.[95]

The text seems quite neutral. However, the tombstone is unusually elaborate and must have been a costly commission. It also has some important symbolism in the composition. Victor is shown in a long-sleeved tunic and robe lounging on a couch. Below him a boy stands holding up a cup towards the deceased. The figures are contained within panels framed with architectural features of Palmyrene design originating in Syria and which must have been carved by the same man or workshop that produced Regina's tombstone. Whether the men shared a sexual relationship can only be conjectured, but the unusually affectionate nature of the piece suggests that the possibility is a real one. The wearing of a long-sleeved tunic (*chiridota tunica*) by a man was explicitly associated with a preference for male partners. Scipio Africanus had definitely disapproved of such attire. He once described how 'a young man who with a lover has reclined (at meals) in a long-sleeved tunic on the inside of a couch, and is not only partial to wine, but also to men; does anyone doubt that he does what sodomites are accustomed to doing?'[96] Victor's tombstone amounts almost to a visual realization of Scipio's words, while replacing condemnation with veneration. It must mean that by then Victor's relationship with Numerianus was conducted openly and in safety. On the other hand, Victor was a former slave and freedman and in this context the relationship may have been more acceptable, especially if Numerianus was somewhat older.

In war no one was safe, though provoking outrage by routinely including rape in a litany of stock abominations was part of writing Roman history, as Sallust did when he threw 'the rape of maidens and boys' into a list of 'the horrors of war'.[97] At the Second Battle of Bedriacum in 69, the army of Antonius Primus burst into the city and reportedly started fighting with one another over the best-looking young women and youths.[98] During the same civil war corrupt soldiers, probably centurions, were engaged in pressing Batavian provincials into the Roman auxiliary forces to support Vitellius. Among their techniques was to call up the elderly and sick who they knew would pay bribes to be exempted. They also had other plans. According to Tacitus, they earmarked 'the most striking-looking boys to be dragged into debauchery', much to the outrage of the Batavians who then rebelled.[99]

One strange variant on this topic took place in 89 after the rebellion of Lucius Antonius Saturninus, governor of Germania Superior. Antonius Saturninus had planned to use Legiones XIIII Gemina and XXI Rapax as the basis of his rebellion. He was defeated by another general, Lucius Maximus, and in the aftermath Domitian initiated a brutal purge, executing a large number of people. Julius Calvaster, a military tribune of senatorial rank on Saturninus' staff, was in a particularly susceptible position. Evidence had been found that he had frequently attended meetings with Saturninus on his own, and so he was accused of having been a co-conspirator. Calvaster had nothing whatsoever to hand with which to defend himself and it looked as if he was bound to be found guilty. However, he came up with an imaginative solution. He claimed merely to have been having a sexual relationship with Saturninus. Since Calvaster was good-looking his claims seemed entirely credible. He was acquitted.[100]

Conversely, the emperor Elagabalus' sexual inclinations seem to have had a wholly negative effect on the soldiers in the East. The cousin of Caracalla, he had only recently been declared emperor after the brief and violent reign of Macrinus and represented a restoration of the Severan dynasty. According to his biographer, though, his 'depraved manner' and habit of engaging in 'unnatural vice with men' soon led the soldiers to start planning to depose him and place another cousin, Severus Alexander, on the throne.[101] Elagabalus' choice of men for senior positions only worsened his standing. Selecting a dancer to be prefect of the Praetorian Guard was just one of numerous unsuitable appointments, his main criterion allegedly being the size of the men's sexual organs.[102]

Wives and children were part of a plan for the long term. Soldiers who invested in families were looking to the future. Many found it hard to separate themselves from the forts and fortresses where they had spent much of their lives, but they also hoped that they would be able to enjoy retirement and sometimes a second career. Veterans, and often their families too, turn up all over the Empire. Some could not bring themselves to leave the army at all.

# FIFTEEN

# VETERANS

## THE EMPEROR'S DIE-HARDS

*Publius Tutilius, son of Publius, of the Olufentina tribe and*
*veteran aquilifer of Legio V, overseer of the veterans, was*
*twice-rewarded by the emperor. He was born in [43 BC] and*
*died in [AD 29]. For himself and for his son Publius Atecinx*
*and his daughter Deminca, and for Andoblato and Gnata, son*
*and daughter of Publius by his will he ordered this done.*

The tombstone from Milan, dated to
AD 29, of a legionary veteran who led
other veterans into his seventies[1]

From the reign of Augustus on, if a Roman soldier lived long
enough to retire he was able to enjoy a remarkable package
of privileges and rewards so long as he was honourably
discharged at the end of his term of service (*honesta missio*). Men
who had reached this point were discharged on 7 January, but
this was probably arranged only on alternate years because of the
logistics involved. However, a wounded soldier who was no longer
able to serve could still be awarded another form of honourable

discharge (*missio causaria*), but from Caracalla's reign on (211–17) the benefits were reduced. If he had disgraced himself he was dishonourably discharged and lost any rights to retirement grants (*missio ignominiosa*).[2]

Many soldiers did not make it to retirement at all. One estimate suggests that around a third of men who enlisted at twenty had died by the age of forty-five, another that as many as half had expired by then, but such figures are hardly surprising in the general context of any pre-modern society.[3] The principal difference is that a Roman soldier in the days of the emperors had better reasons than most to look forward to surviving his term of service. A soldier who survived to retirement, however, might receive not only a gift of money or land, the *praemium militia*, but also various entitlements called together an *emeritum*, which made him an *emeritus*. Emeritus meant someone who had earned his status through merit. They included for example exemption from any obligation to civic duties and tolls. Domitian proclaimed in 94 that veterans be free from liability for 'all public taxes and tolls [*vectigalia*]'. A Latin copy of the proclamation, written on a wooden tablet on the occasion of the discharge of veterans from Legio X Fretensis, was found at the Fayum in Egypt. It had been made by one of the veterans, Marcus Valerius Quadratus, who noted where the original stone inscription was displayed in Alexandria.[4]

In Egypt in 103 the veteran Lucius Cornelius Antas produced his evidence of service and honourable discharge to a government official so that his right to exemption from the poll tax could be recorded.[5] Diocletian and Maximianus confirmed exemption from public duties for veterans who had been honourably discharged.[6] Taken together, all these awards were supposed to set a retired soldier up for civilian life, though exactly what he received depended on how long he had served and in which part of the army, as well as the date he retired, since some emperors added extra privileges.[7] However, veterans were not exempt from taxes on inheritance, or from property taxes, and were also obliged to contribute to the upkeep of roads.[8]

Veteran soldiers in general, however, proved to be one of the most valuable resources, not only for the Roman army but also for all of Roman society.

## VETERANS DURING THE REPUBLIC

The position had been very different in the Republic when there was no standing army. In those days, soldiers were ordinary citizens fulfilling their duty to the state perhaps only for one campaign. In theory they went home to their farms and businesses after the war had finished, hoping to find their affairs as they had left them. That often turned out not to be the case, but the state had no further obligation. Some carried on in the army, fighting for example in the disastrous Battle of Lake Trasimene. When the consul Gaius Flaminius was killed by the Insubrian horseman Ducarius at the height of the battle, it was veterans who gathered round his body and prevented Ducarius from despoiling it by blocking him with their shields.[9]

One of the risks was that veterans would turn into dangerous bands of bitter landless men once the fighting was over. This had been a principal concern of the reforming tribune of the plebs Tiberius Gracchus in 133 BC. He had seen how veterans returned home to discover their land had been stolen by wealthy senators and absorbed into their vast estates, and was badly worried by the destabilizing effect this was having on Roman society (see Chapter 2).

A veteran of the army in the Republic was therefore usually thrown back on his own resources, despite the initiative taken after 107 BC by Gaius Marius to set aside some funds for his men when they retired. We know little about Republican veterans as individuals because in those days the habit of producing funerary inscriptions was far less well-established (at least, very few survive). However, there are some interesting cases of men who overcame the challenges of being discharged and became successful, or notorious. In 63 BC a former centurion called Gaius Manlius became involved in the senator Sergius Catalina's conspiracy to topple the consuls. Manlius had served under Sulla and gained much military experience, but he was corrupt and had made a great deal of money out of his time in the army. By 63 BC he had spent it all and was eager for an opportunity to make more.[10] Manlius was an extreme example, but he illustrated well the importance of providing for veterans if they were not to become outlaws.

Another rare instance of a known veteran from the Republic was Lucius Orbilius Pupillus. He came from Benevento in Campania and was born about 113 BC. His name preserves his origins; *pupillus* means 'orphan'. His parents were murdered by family enemies, though the reason is unknown. His first job was as a public servant, assisting the magistrates, a post that shows he was educated and literate. Around the time of the Social War, when Rome fought its Italian allies, he joined the army, serving as a cornicularius in Macedonia, and then moved on to a cavalry unit. As a cornicularius Pupillus was working in a supervisory administrative role, and clearly using his education.

Pupillus served out his term with the army and then retired to Benevento where he had to forge a living. He took the opportunity to go back to studying, something he had forgone since childhood. Pupillus worked a teacher from then on, moving to Rome in 63 BC to continue the job. He made little money but built up quite a reputation for being bad-tempered and beating the children he taught, earning the nickname 'The Flogger'. On one occasion, when giving evidence in court, he was asked by the lawyer Varro Murena, a hunchback, what he did for a living. He answered, 'I move hunchbacks from the sun into the shadows.' The phrase seems to have been a metaphor for suggesting he moved mediocrities out of the limelight, presumably by exposing their shortcomings.

Orbilius Pupillus seems to have been disgruntled about his teaching experience. He wrote a book about the unpleasantness visited on teachers 'by indifferent or selfish parents'. He was commemorated in Benevento with a statue near the capitol, which depicted him accompanied by two book boxes and dressed in the manner of a Greek.[11]

That he had served in the army was an important part of the esteem in which Pupillus was held but he had had to make his own way afterwards. Gaius Nasennius was 'first centurion with the eighth cohort' of an unspecified legion during the war fought in Crete in 68–67 BC. Crete had been supplying mercenaries to support Mithridates VI against Rome, and also serving as a pirate base. Nasennius returned to his private affairs after the war. He became wealthy in the city of Suessa (modern Sessa Aurunca in Campania), possibly with the assistance of booty or some sort of ad hoc grant when he left the army unless like Manlius (above), he had made money as a soldier. Nasennius subsequently supported the tyrannicides Brutus and Cassius following the assassination of Caesar in March 44 BC. In the early summer of 43 BC he asked Cicero to recommend him to Brutus for a position of some sort (Nasennius was surely too old to fight by then). We have no idea why Nasennius felt able to approach Cicero, but he was not alone in doing so. Cicero wrote Brutus a letter and made much of Nasennius' personal qualities and wealth. His military service was a key part of Cicero's endorsement.[12] However, late in 43 BC Cicero was executed on Antony and Octavian's orders as they pursued their own ambitions and revenge for Caesar's death, completely defeating Brutus and Cassius at Philippi in 42 BC. Suessa became a military colony at some point in the next few years, settled by veterans from Octavian's army (see below). Nasennius had backed the wrong side, though nothing is known of the personal consequences for him or his family.

Of course, one solution for an experienced soldier was to re-enlist. In the late Republic there were plenty of opportunities. In the latter stages of Caesar's war in Gaul, Legio XI had proved itself to be an exceptionally promising unit after eight campaigning seasons. But Caesar's VII, VIII and VIIII legions, made up of veterans, still outclassed them when it came to courage and experience.[13] One veteran of those days returned to his hometown and set his family on a path to history. In 48 BC, when Caesar defeated his arch-rival Pompey at the Battle of Pharsalus, among Pompey's soldiers was a man called Titus Flavius Petro from Rieti in Perugia, about 70 miles (112 km) from Rome, who had served either as a centurion or as a volunteer veteran, an *evocatus*. After he fled from the battlefield and made his way home, Caesar had the magnanimity to offer men like him a pardon and an honourable discharge. Flavius Petro thereafter became a 'collector of monies'. His son Sabinus also became a tax collector, working in Asia. Sabinus in turn had two sons, Sabinus and Vespasian; the latter was to become emperor in 69.

Quintus Annaeus Balbus was fifty-three when he died, his tombstone describing him as having been a 'soldier of Legio V' who had been decorated twice. By the time of his death he was a *duumvir* at Thuburnica in what is now northern Tunisia, a position that must mean he had been given money and opportunity when discharged as a veteran. The form of the text suggests a late Republican date and thus he may have been one of Caesar's recruits when the legion was raised from non-citizens in Transalpine Gaul in 51 BC. The legion was serving in Africa by 47 BC by which time the men had all been made Roman citizens, and was called *veterana legio quinta* 'the veteran Fifth legion'.[14]

## AUGUSTUS' PROVISIONS FOR VETERANS

The prospects for veterans started to change and become more reliable once the wars of the late Republic came to an end. The answer to dealing with veterans for a Roman general was to settle conquered territory and money on them. Octavian, after he became emperor as Augustus, bragged that he had ended up with about 'half a million Roman citizens who had sworn the military oath to me'. Following his victory over Antony at Actium in 31 BC he had no need of an army of such size. Indeed, he had to get rid of most of it so that he could claim to have restored the Republic and brought peace. He said that he had settled 300,000 veterans 'in colonies or sent them back to their home towns ... and to all of them I gave land or money as a reward for their military service'. Augustus did not do this all at once. The first 120,000 men were paid off in 29 BC, the others at later dates in his reign.[15]

Perhaps among them were two men of Legio XI, a legion founded by Julius Caesar and which served Octavian throughout the period 42–31 BC. The two were exceptional instances of soldiers who had fought for Octavian at the Battle of Actium. One, Marcus Billienus, was so proud of being 'in the naval battle' that he took the cognomen Actiacus, becoming Marcus Billienus Actiacus, as if a British or American veteran had added 'Trafalgar' or 'Midway' to his name. After Billienus retired from military service he settled in the colony of Este in north-eastern Italy, where he rose to be elected as a decurion in the town council. He was not the only man to adopt the name: Quintus Coelius Actiacus, who also served in Legio XI, must have been at Actium too, because like Billienus he too was settled at Este.[16]

Augustus paid out 600 million sestertii to buy land for retired soldiers in Italy, and a further 260 million for land in the provinces. He boasted that 'in the recollection of contemporaries I was the first and only person to have done this' – meaning that his predecessors who had founded colonies had not actually paid out any money to do so. He added a further 400 million sestertii to the amount allocated to soldiers who were discharged. In 29 BC Augustus paid out 1,000 sestertii to every veteran already settled in a colony, using the war booty he had amassed to cover the cost.[17]

Dealing with veterans was obviously an expensive commitment, but an essential one.

It took time to develop and regularize such a system of payments, insofar as anything was ever regularized in the Roman army. There was also a vested interest for the state in delaying discharge, since a shortage of troops at a crucial moment would be hard to make good with new recruits. However, one of the reasons the Rhine garrison mutinied in AD 14 was because the veterans had not been released and had been forced to stay on. Some of the older soldiers were still serving more than 30 years after enlisting. Promises had to be made that anyone who had served 20 years would be discharged.[18]

Tiberius, Augustus' stepson and successor, was notoriously mean, although his reputation was in large part created by Roman historians who took any opportunity to criticize him. Suetonius said that Tiberius 'carried out discharges of veterans rarely, waiting to seize the money when they expired, dying of old age'.[19] Nero, on the other hand, established a colony for praetorian veterans at Anzio, forcing the richest of the *primi pili* centurions to move there too. However, later in his reign his extravagance had reached such astronomical heights that he had to suspend retirement grants to veterans, as well as pay to the serving soldiers.[20] This was a disastrous decision which benefited his rival Galba.

Nonetheless, in the centuries that followed hundreds of thousands of retired Roman soldiers went back to their communities, stayed near the fort where they had been stationed, or settled in new lands. They took with them their retirement grants and the skills they had gained. So long, that is, as they had been honourably discharged. Soldiers who had been dishonourably discharged, or had been discharged on medical grounds, were not entitled to any of these privileges.[21]

Not all soldiers wanted to leave the army. Some like Publius Tutilius of Legio V, who appears at the start of this chapter, stayed on in the army, providing invaluable experience – or at least he would certainly have thought so. He stayed perhaps 30 or more extra years, finishing up under Tiberius as a *curator* in command of other veterans still with the legion. They were probably organized into a wing of their own attached to Legio V.

## DIPLOMAS

Retired auxiliaries and members of the fleets had, in relative terms, a great deal more to look forward to than legionaries. For the most part auxiliaries had had to wait until their term of service was up to become Roman citizens. This changed when Caracalla made all free men of the Roman world into Roman citizens in 212. Until then the award of citizenship had been one of the greatest incentives for the majority of auxiliaries to serve. Some auxiliary units were made Roman citizens as a special honour to reward achievement, although some of the more irregular units had no such entitlement.

When an auxiliary soldier was honourably discharged, usually as one of a group, his name and details were inscribed on a bronze tablet in Rome, a place many had probably never visited and probably never would visit, on a wall at the back of the temple of the deified Augustus. (The temple is long lost, but is thought to have been in a valley below the Capitoline Hill at the north-west end of the forum.) However, there seems to have been a specialist industry that would supply copies of the discharge as personal souvenirs, along with the details of the emperor, the units, the province involved and the date. Made on bronze plates, many have survived though they are often badly damaged and incomplete. These records of soldiers' discharge reflect the importance of written evidence in the Roman military world, providing the proof of a soldier's legal status and that of his family. Today the plates are known as diplomas (Latin plural is *diplomata*), but the ancient name for them is lost. An oddity is that they are only known for praetorians and auxiliaries, including members of the fleet. Legionaries neither received nor commissioned such copies.

Praetorian discharges were also commemorated in inscriptions displayed on the wall at the temple of the deified Augustus, close to a statue of Minerva. Surviving diplomas record that a distinctive formula was applied to praetorians. No such diploma from a date before the mid-70s is known, but it is unlikely that the special form of address had changed. On 2 December 76 a member of the Praetorian Guard called Lucius Ennius Ferox was given his honourable discharge by Vespasian. In the recorded text, found at Tomi in Moesia Inferior, the emperor addressed the praetorian directly, referring to his 'courageous and loyal performance of military service' in 'my

Praetorian Guard'. He was commemorated for having done his duty. This emphasized the close personal relationship that was supposed to exist between the emperor and his praetorians, and reflected the oath these soldiers had taken.

The praetorian was granted the right of marriage to any woman, and regardless of her status their children would be Roman citizens. This privilege was only conferred once; subsequent wives and children would not be eligible. In another praetorian discharge diploma also found in Moesia Inferior and dated 7 January 228, in the reign of Severus Alexander, Marcus Aurelius Secundus of Cohors I Praetoria was commended for his loyal service and awarded these marriage rights. It is odd that the existing wives of auxiliaries were allowed the same rights as the wives of praetorians, but the praetorian entitlement may have been based on the assumption that such men would only have been living with Italian women who were Roman citizens.[22] Until *c.* 140 existing children were included, but not thereafter. The terms for veterans of the fleet were much the same, except that their existing children were still admitted to the citizenship, though they had eventually to prove their wives were the mothers of the children.

On 19 January 103, in the reign of Trajan, auxiliary soldiers who had served 25 years were discharged in Britain from four cavalry regiments and eleven cohorts during the governorship of Lucius Neratius Marcellus. The auxiliaries (apart from one cavalry regiment whose men had already received the award) were awarded citizenship for themselves, their children and their descendants. If they were already married those marriages became legal, and a first marriage after discharge would also be legal. All this information is recorded on a diploma made for Reburrus, son of Severus, who had risen to the position of decurion in Ala I Pannoniorum Tampiana. The diploma added the key certification at the end which read 'copied from the bronze tablet set up at Rome behind the temple of the deified Augustus, near (the statue of) Minerva'.[23] Reburrus' diploma, which is well preserved, must have been an extremely important possession which validated his status in retirement. However, it was found in a field near the village of Malpas in Cheshire, Britain, in 1812, about 12 miles (20 km) south of the legionary fortress of Legio XX at Chester. The find spot makes it possible Reburrus had settled somewhere in the vicinity of a legion to which the regiment had once been attached, quite possibly during Agricola's campaign into Scotland between *c.* 78 and 84.[24]

Diplomas also turn up in such incongruous locations that Reburrus may have had no connection with Chester at all, though he was undoubtedly in Britain in 103. Another possibility, but one that is impossible to prove, is that there was a market for militaria such as diplomas both in antiquity and in more recent times, in the same way that medals from the First and Second World Wars are enthusiastically traded today. Moreover, the metal was recyclable in antiquity. One fragment that turned up in Cirencester had been cut down from a diploma into a small circle of bronze. It probably came from a local antiquarian's collection, but the Roman metalworker who used it to make a disc had probably acquired it as scrap bronze. Another diploma, found at the city of Volubilis (capital of Mauretania) in Morocco, had also been cut down to serve as a lid.[25]

Why legionaries' formal discharge was not recorded in the same way is unknown, unless there was no legal aspect of their status or rights that needed confirming. Not all legionaries were satisfied with this. On 22 January 150 the legate of Syria, Villius Cadus, was sent a special request by 22 Egyptian members of Legio X Fretensis. They had begun their military careers in the fleet at Misenum in Italy but Hadrian had transferred them to the legion, a move that would have required making them Roman citizens. Their careers over, they wanted to return home to Egypt with written proof that they were legionary veterans. Lucius Petronius Saturninus composed the petition on behalf of his fellows, and it was written out by a man called Pomponius. When Villius Cadus received it he endorsed it as requested, including the lines:

Legionary veterans do not normally receive a written document. However, you want it to be made known to the prefect of Egypt that you have been discharged from your military oath by me on the orders of our emperor [Antoninus Pius]. I will give you your bonus and written document.[26]

## Veterans in the civilian world

Veterans might return to their old homes, settle near the forts where they had served or move to a colony – unless of course they chose to re-enlist. For a state that had virtually no other means of asserting and exerting its power over the general population other than through the army, colonies of veterans were a vital resource. They created settlements where thousands of trained and experienced men, loyal to the Roman state, could act as a military reserve. A rebellion by the Salassi tribe in north-western Italy provided Augustus with the opportunity to do just that. Once the revolt had been crushed, land was seized to settle veterans from his Praetorian Guard. The new city was named Augusta Praetoria Salassorum (Aosta), and local men of military age were cleared out by being sold into slavery.[27] In Britain following the invasion of 43 a new colony at Colchester was established in 47 after Legio XX was sent out to campaign in the west, showing that it was an early priority.[28] The archaeological remains have revealed that the colonists made use of the fortress's lay-out and some of its buildings as they converted the site into the beginnings of a fully-fledged Roman town in a remote province.

Lucius Poblicius of the Roman Teretina tribe was a veteran of Legio V Alaudae, based at Xanten, which disappeared from history by the late first or early second century AD. His exceptionally impressive first-century tomb showcased the status he must have reached in the veteran community in the city he knew as Colonia Claudia Ara Agrippensium (Cologne), where he was buried. A substantial podium with pilasters contained Poblicius' cremated remains and those of his children, his wives and his freedmen, while statues of Poblicius and family members stood on the podium between four Corinthian columns which supported an elegant tapering roof flanked by sphinxes.[29]

Other veterans attained similarly high status in the civilian world. Lucius Silius Maximus was a veteran of Legio I Adiutrix at Alba Julia in Dacia. After retiring he moved to the canabae of the fortress where he served as a magistrate in the civilian settlement, 'the first' said to have done so.[30] His new job represented an early stage in the settlement's history. It would later earn formal incorporation under Marcus Aurelius. A dedication by Marcus Lucilius Philoctemon records that he was a senior civic magistrate, *duumvir*, of Alba Julia.[31] Having been honourably discharged from service with Legio V Macedonica, Gaius Valerius Pudens likewise moved a short distance to the legion's fortress canabae at Iglitza on the Danube, where in 138 he was serving as one of the settlement's magistrates alongside colleagues who were civilians but may also have been veterans.[32] Pudens must have done well to have fulfilled the necessary property qualification for civic office – whatever that was in this location. The text recorded an unspecified building given by Pudens, his senior magistrate colleague Marcus Ulpius Leontius, and the settlement's aedile Aelius Tucce to the veterans and citizens of the canabae. The structure was dedicated to Hadrian, sick at the time and approaching death, but who of course had also served in Legio V Macedonica as tribune during its time in Moesia Inferior. Perhaps Pudens and the emperor were known to each other because Hadrian, who was said to have had a remarkable memory, 'even knew the names of the veterans whom he had discharged at various times'.[33]

At Capua a veteran called Lucius Antistius Campanus was honoured by the magistrates of the city when he died. He was 'an excellent man', they said, who had completed his military service 'during hard and dangerous campaigns' in their testimonial. Exactly when he retired is unknown but he seems to have earned the admiration of both Caesar and Augustus; most likely he served under Caesar in the mid-40s BC and under Augustus into the 20s BC. He was then settled by Augustus at Capua and perhaps lived into his eighties, since one restoration of the text suggests that his death came after 14 when Augustus died and was deified. Antistius Campanus, said the dedication, had generously spent his own money on various local causes 'for everyone's benefit', rather than only that of his own family, as well as working hard for the community even though he was getting on in years. He almost certainly served in all the civic magistracies. The senior civic duoviri magistrates voted that his body be carried from the forum to the place of cremation, and that a gilt statue be erected to commemorate him and record the decree by the town council in his favour. Finally, they voted that he be buried in a suitable plot by the Via Appia, the road that led to Rome from Capua.[34] Antistius Campanus of course did benefit his family. In 13 BC his son of the same name was serving as one of the duoviri at Capua and participating in the dedication of a temple of Jupiter and the Lares.[35]

Many veterans established families and futures like Antistius Campanus. Others were not so lucky. Gaius Aeresius Saenus, a veteran of Legio VI Victrix at York, commissioned a tombstone for his family and, in anticipation of his own demise, himself. This grieving man had said farewell to his wife Flavia Augustina, who expired aged thirty-nine years, seven months and eleven days, his son Saenius Augustinus who had survived only one year and three days, and a daughter whose name is lost but who lived for one year, nine months and five days.[36] Flavia Augustina, like so many mothers of that time, had experienced the tragedy of losing two infant children. The tombstone, however, was not a bespoke piece. The family is depicted with the parents standing behind their children, but the latter figures are more appropriate to a son and daughter of around eight and six years respectively. It was clearly purchased 'off the shelf' and an appropriate inscription inserted in the blank space.

Not all soldiers were impressed by the prospect of retiring to a colony in the provinces. Some found they were not even allowed to. According to the mutineer Percennius, who had worked up the troops over their grievances in Pannonia after the death of Augustus in AD 14, there were still men serving after 30 or even 40 years. Even if such a man managed to get out of the army, Percennius said, he was liable to find himself being offered a parcel of land in 'waterlogged swamps' or on 'uncultivated mountain-sides'.[37] This cannot have been entirely true. Inscriptions from all round the Empire show that retired legionaries could and did retire to colonies in Italy, or in pleasant locations out in the provinces. These provincial colonies were often in towns close to their former forts, or which had replaced the fort when the legion moved on. Not surprisingly the settled veterans often took up with local women, especially as once discharged they could contract a legally valid union that gave their wives and the children all the normal privileges that went with marriage to a Roman citizen. Tacitus describes the German inhabitants of the colony at Cologne, referring to how they had intermarried with the first colonists and become 'allied' with them.[38]

One veteran who settled far from home was Gaius Julius Calenus. A veteran of Legio VI Victrix, based at York. He came from Lyon in Gaul but evidently decided not to return there when he was discharged, probably because he had a family in Britain. Julius Calenus settled in the colony at Lincoln, 80 miles (130 km) to the south and founded in the late first century AD, where he must have been given a grant of land. His daughter Julia Sempronia set up her father's tombstone there.[39] Lincoln's modern name of course preserves the settlement's Roman name Lindum Colonia. Gaius Cornelius Verus came from Tortona in Italy. By the time he was discharged from the army he was a legionary with Legio II Adiutrix, which was based in Budapest. He decided to settle in Colonia Ulpia Traiana Poetovio (Ptuj), a colony in Pannonia established by Trajan in 103, where he had been awarded a 'double grant of land', perhaps because his job in the legion already earned him double pay. His tombstone says that he was serving as clerk to the governor of the province. However, he did not live long enough to enjoy his retirement. He died at the age of fifty and was buried at Ptuj.[40]

The trades and skills acquired in military service probably, and sometimes definitely, lay behind a veteran's choice of post-army career. Vitalinus Felix, a veteran of Legio I Minervia, lived long enough to retire and chose to settle, not in or near the legion's base in Bonn, but in Lyon in Gaul with his wife Julia Nice and their son Vitalinius Felicissimus. Described as a 'wise and faithful man', Vitalinus Felix set up in business as 'a trader in the pottery craft'. He died aged fifty-eight years, five months and five days.[41] His choice of trade is an interesting one. Traders in pottery (*negotiatores cretarii*, literally 'traders of pots') were often, as so many men involved in commercial enterprises were freedmen. Vitalinus Felix had perhaps learned to manufacture or deal with pottery while serving in the legion – the army had vast needs for ceramics. On the other hand, Lyon was a major Roman city and trading centre. It lies around 90 miles (143 km) from the massive red-slip samian-ware potteries of Lezoux, whose products dominated tableware in the Western Empire in the second century. Vitalinus Felix may therefore have been an entrepreneur who shipped local pottery to markets further afield.

Gaius Gentilius Victor definitely stuck with what he knew. After retirement from Legio XXII Primigenia he worked during the reign of Commodus in the late second century as a dealer in swords (*negotatior gladiarius*) at Mainz, the legion's base.[42]

Nothing else resembling either of these inscriptions is known, but it is certain that many other veterans built second careers. Julius Demetrius once served with Cohors III Augusta Thracum and settled in a village near the unit's base at Sachare in Mesopotamia. He is recorded there in 227 as having purchased for 175 denarii a plot of land elsewhere in the area with fruit trees and 600 vine-stumps by a vineyard, perhaps both for pleasure and as a source of income. The transaction was witnessed by several serving soldiers. Another veteran had to act on behalf of the vendor, Otarnaeus, who was illiterate, illustrating how much better educated soldiers and ex-soldiers could be than the civilian population.[43]

More often, veteran praetorians would return to their homes. This must have been even more likely to happen once praetorians were recruited from the whole army, as happened under Severus from 193. Sextus Quinctilius Seneca was a veteran of Cohors III Praetoria when he made a dedication to Jupiter on the island of Rab in Dalmatia (Croatia).[44] Gaius Terentius Mercator, veteran of Cohors III Praetoria, expired at Como, where he was buried according to the instructions in his will.[45] Gaius Carantius Verecundus, a veteran of Cohors VII Praetoria in Flavian times, was buried by his freedmen at Riati in Italy.[46]

The Roman military veteran Nonius Datus continued to work as an engineer after his military service, attached to Legio III Augusta in North Africa. When the townsfolk of Saldae in the province of Mauretania Caesariensis got into difficulty with the tunnel they had been digging to bring a mountain water supply to their city – the two digging parties, which consisted of military personnel, had made a mistake in surveying the hill and ended up missing each other – Nonius Datus was called in to help them solve the problem. It was not at all surprising to find an old soldier stepping in like this; indeed he had to be called back several times to keep the project on track.[47] Roman military veterans pop up all over the Roman Empire in numerous contexts that reflect the skills they had learned when in the army.

The unusually named Eltaominus had served as an architectus with Ala Vettonum, a cavalry unit sufficiently distinguished to have been awarded Roman citizenship in service. The Ala is attested at various locations in Britain, including Binchester where Eltaominus' altar was found. Eltaominus described himself as an *emeritus* ('veteran') on a dedication to Fortuna Redux ('the Home-Bringer'). He had perhaps been away on a trip and had returned to the fort, where he presumably lived in the vicus.[48]

In the Fayum in Egypt in the early second century AD a veteran called Lucius Bellienus Gemellus made a living as a farmer in retirement. Thanks to the survival of some of his personal archive, a surprising amount is known about his everyday life. Gemellus was a networker who knew the value of keeping local officials on side, so he made sure to send them gifts. He also participated in religious ceremonies. On 6 November 110, for example, he wrote to an agent with instructions to 'buy us some presents for the Isis festival for the persons we are accustomed to send [them] to, especially the strategoi'.[49] Another document of 110, written to his son Sabinus, records that he wanted olives and fish sent to a man called Elouras who had recently been made deputy to the strategos Erasus, and cabbages sent to Erasus who was about to celebrate the festival of Harpocrates.[50] These and other documents from the archive paint a picture of a veteran thoroughly involved with local life, managing it all almost as if he was still in the army with its love of bureaucracy and records.

Gemellus seems to have been happy, if busy, but a former soldier could all too easily turn into a disgruntled old man. Gaius Julius Apollinarius (not the ambitious young soldier of the same name mentioned in Chapter 3) lived at Karanis in Egypt between 169 and 172 after his military service in Cohors I Apamenorum. He had established himself as a well-to-do landowner, buying and selling land there even while he was still in the army, as well as acquiring an unofficial wife. But in retirement he was furious that the five-year exemption from compulsory public services (such as acting as a tax collector) for veterans had been overturned when a demand came for him to perform those services after only two years. This was not even supposed to happen to the native population, or so the furious Apollinarius moaned; he expected the rules to be even more strictly observed when it came to someone like himself. In the year 172 he sent off his petition to the strategos, explaining that he wanted to be able to take care of his property 'since I am an old man on my own'.[51]

## SIGNING ON AGAIN

Some soldiers were unable to resist the temptation to sign up again. When they did so they were called *evocati Augusti*, 'men recalled to arms by the emperor'. Such men were absolutely essential to the Roman army's ability to fulfil its duties. While he was governor of Germania Superior, Galba made as much use of veterans as he did of soldiers still in service. By training them all at the same pace he pushed them into such good condition that he was able to hold back any barbarian incursions.[52]

Gaius Vedennius Moderatus came from Anzio in Italy. He served in Legio XVI Gallica for ten years, doing well enough to be rewarded with a transfer to Cohors VIIII Praetoria probably in 69. He retired having clocked up a further eight years' service in the Guard, but was promptly recalled to sign on for another 23 years under Vespasian and Domitian as a reservist *architectus armamentarius* with specialist artillery skills, probably because he was regarded as indispensable.[53] No doubt Moderatus had acquired his invaluable experience with artillery after Legio XVI Gallica threw in its lot with Vitellius during the Civil War, so his transfer to the Praetorian Guard may have come about as a result of Vitellius' policy of allowing his soldiers to choose which section of the army they wished to serve in. Soldiers who so wished were allowed to join the Guard 'however worthless' they were, but in Moderatus' case he may already have proved his worth.[54]

Moderatus' military service thus extended to over 40 years, during which time he was decorated by both Vespasian and Domitian. He must have been well into his sixties when he died, early in the reign of Trajan. His tombstone advertises his special skills by, for example, depicting on one side a ballista.[55] Moderatus was unusual. He stayed with the Guard for an exceptionally long time and presumably reached the point where the idea of leaving was beyond his ability to contemplate. As an interesting aside, a bronze plate has been found from a catapult used in one or other of the battles at Bedriacum in 69. The embossed plate was fitted to the front of the catapult with a hole for the bolt to pass through. Above the hole is the name of the legion, in this case Legio IIII Macedonica, while on either side are a pair of military standards bearing bulls, the legion's emblem, and the names of the consuls for 45 in the reign of Claudius, the governor of Germania Superior, Gaius Vibius Rufinus, and the centurion Gaius Horatius, who was princeps praetorii, 'centurion in charge of the headquarters'.[56] The veteran catapult was thus well over two decades old at the time of the battle and was still in service.

Remaining in the army was potentially dangerous. Aulus Sentius, from Arrezo, was a veteran of Legio XI sometime in the first century AD. He 'was killed in the territory of the Varvarini in a small field by the river Titus at Long Rock [in Dalmatia]', evidently while still serving in the front line.[57]

Veterans of course had skills that could be useful in other contexts than the battlefield. When Otho wanted his dead rival Galba's praetorian prefect Cornelius Laco killed – the hapless Laco had been under the impression that he was to be exiled to an island – he commissioned a loyal veteran to go and murder him.[58] Nymphidius Lupus on the other hand was used in a more positive way. He rose to the heights of being *primus pilus* in Legio III Gallica, where he knew Pliny the Younger around the year 81 when the latter was a military tribune in the same legion. 'From then on I began to develop a special regard for his friendship', wrote Pliny to Trajan, and the two evidently stayed on excellent terms thereafter. When Pliny became governor of Bithynia he wrote to Lupus and asked him to give up his retirement and join him as a sort of adviser. When Lupus agreed, Pliny wrote to Trajan to ask for promotion for Lupus' son by way of repayment.[59]

## PROMOTION BY HEROISM

Tiberius Claudius Maximus did well during his time in the army in the late first and early second century, but apparently not so well that anyone else could be bothered to record the fact. Therefore he commemorated himself. As a veteran he commissioned an inscription found at Philippi that documented his time in Legio VII Claudia. Having been a *quaestor equitum*, a title which is otherwise unknown and probably meant he was in charge of the treasury of the legion's cavalry, he then served in the legate's mounted bodyguard before being made a vexillarius of the cavalry in the legion. Decorated for his feats in Domitian's Dacian War, he fought in Trajan's Dacian and Parthian wars where he served as a duplicarius in Ala II Pannoniorum before being promoted to decurio for – so he claimed – capturing the Dacian king Decebalus and bringing his head to Trajan. The claim is spurious, because Decebalus had already taken his own life; but it is plausible that Claudius Maximus found the body, severed the head and took it to the emperor.[60] He continued to serve after his official term of service was up, and finally received his honourable discharge as a *voluntarius* in the army of Mesopotamia under Terentius Scaurianus.[61]

## BAD LUCK

Not all soldiers retired from the army to enjoy life. At some point before the mid-first century BC one veteran, whose name is unknown, came home to find that his father had been falsely told that he had died. The father had named other heirs in his will as a result and had himself then expired. The soldier therefore found he had nothing, while his father's friends, who had been the beneficiaries, shut him out in spite of the personal sacrifice he had made fighting for the Roman state. Forced to pursue a legal campaign in the forum at Rome, he eventually won his case by a unanimous vote and was restored to his family estate.[62]

Another veteran whose retirement went disastrously wrong was Claudius Pacatus. Pacatus had risen to the ranks of the centurionate by the time he was discharged in the late first century. Unfortunately, it was discovered in 93 that he was a slave and had evidently managed to conceal the fact when he enlisted. Domitian, acting in his capacity as *censor*, ordered that Pacatus be returned to his former master.[63]

Even those who made it into retirement did not necessarily live long enough to make the most of their freedom from duty. Gaius Julius Decuminus, who served with Legio II Augusta at Caerleon, died when he was only forty-five. He was by then a veteran of the legion and had remained close to its base but evidently did not live long to enjoy the privileges of his status. His anonymous wife was left to organize his funeral.[64]

Wherever a Roman soldier came from, and wherever he served, the experience stayed with him for life. For some men it proved impossible to let the army go, especially if retirement meant unfamiliarity and a loss of contact with old comrades. Normally, veterans were settled in colonies drawn from their old units. In 61 under Nero, however, the colonies at Tarento and Anzio in Italy were peopled by veterans drawn randomly from bases all over the Empire. The idea was that they would make good a decline in the local populations; but the scheme fell flat when most of the veterans, faced with other men they had never seen before, slunk off back to the frontier provinces where they had served. Some even abandoned wives and children in the colonies, preferring a retirement near their old bases where the sight, sound and smell of the Roman army made them feel at home.[65] Once a soldier, always a soldier.

# SIXTEEN

# JUPITER'S MEN

## RELIGION AND SUPERSTITION

*To the Invincible Sun God Mithras, Everlasting Lord, the
centurion Publius Proculinus, on his own behalf and that of his
son Proculus, willingly and deservedly fulfilled his vow, in the
consulship of Gallus and Volusianus.*

Dedication to Mithras in 252, found at
Housesteads fort on Hadrian's Wall[1]

The single most conspicuous aspect of the Roman Republic
that made it superior to all others, said Polybius, was its religious belief. It was, he declared, what held the Roman state
together, pervading public and private life.[2] He was not talking about
the Roman army, but in fact the army perfectly reflected that aspect
of Roman society. Indeed, it is the army that probably provides us with
the bulk of our evidence for Roman religion. The vast majority of individual soldiers we know about have been identified from their religious
dedications or their tombstones. They worshipped classical deities,
local native cults from their homelands or their postings, or exotic
eastern mystery cults like Mithraism, which was especially popular

among soldiers. They also carried these cults around the Empire with them. Individual religious belief, moreover, was only one part of the story. The army also participated in observation of the official state cult, according to a strict calendar and cycle that formed an essential part of the army's allegiance to the Roman state and the emperors.

Apart from the mystery cults, Roman religion was primarily transactional. A soldier, or a civilian, typically wanted a service or support from a deity and in return promised to fulfil a vow in the form of a sacrifice or gift recorded in a dedication. The same relationship operated between the gods and the state or the army unit. The initial request rarely survives, other than occasionally at shrine sites where messages to the deity were deposited, for example the sacred spring at Bath in Britain; even then it is unusual for anything about the individual involved to be identifiable, largely because they did not usually name themselves. Instead, we have the vow fulfilment records, in the form of stone altars set up by a grateful worshipper. Some of these altars are roughly carved and crudely inscribed, making it more than likely they were made by the soldiers concerned, although the very presence of inscriptions reflects the higher levels of literacy found in the Roman military community. The main records of state cults meanwhile are honorific dedications in the form of altars and statues in the expectation of or in thanks for the deity's protection.

There was an obvious reason why soldiers placed so much emphasis on the goodwill of the gods. They desired protection in warfare and also to win victories, both as individuals and as an army. The narrative that the great pantheon of divinities, especially the Capitoline Triad of Jupiter, Juno and Minerva, as well as Mars, had backed Rome's rise to greatness was engrained in Roman culture at all levels and in all contexts, but nowhere more so than in the army.

No wonder then that Publius Aelius and Marcus Aurelius Severinus made a dedication in 217, almost certainly to Jupiter Optimus Maximus, fulfilling vows they had previously made. They were beneficiarii on the staff of the commanding officer of Legio II Adiutrix which had returned to the legion's base at Belgrade from Caracalla's second Parthian campaign. They had more than one good reason to make an offering. Not only had they survived the war, they had also survived the murderous reign of Caracalla, who had been assassinated during the campaign in April 217.[3] Lucius Septimius Veranus, a veteran of Legio II Adiutrix, was another survivor. He made a dedication at Székesfehérár in Pannonia to Jupiter Optimus Maximus in 218, shortly after the accession of Elagabalus, in fulfilment of vows he had made when taking part 'in the Parthian expedition'.[4]

## SUPERSTITION

Roman soldiers were notoriously superstitious, and hypersensitive to the interpretation of omens or auspices. They could refuse to fight as a result of the slightest upset, whereas a good omen could inspire them to achieve extraordinary military feats. They looked on in awe at natural phenomena and wondered whether they portended good or ill.

The superstitious tradition went back a long way indeed, into the days of the Republic when the Roman army first flexed its muscles. When in 340 BC the consuls Titus Manlius and Publius Decius offered animal sacrifices before leading their soldiers into battle against the Latins, the soothsayer (*haruspex*) was concerned that the liver of Decius' sacrificial victim was damaged on one side, but since all the other signs were acceptable and Manlius' sacrifice had been favourable Decius was happy to go ahead with the battle. After a sound start the Roman forces were in trouble, so Decius called a priest to guide him through an emergency prayer on the battlefield to Janus, Jupiter, Mars and other gods 'that you prosper the might and victory of the Roman people'. Decius threw himself into the battle and died in the fray, but the day was won.[5]

In 293 BC the consul Lucius Papirius Cursor led a Roman army during the Third Samnite War against the Samnite people of central Italy, who opposed the expansion of Roman power. Worked up with excitement about fighting, his men were infuriated when a battle at Agnone was postponed by a day. The mood even affected one of the *pullarii*, the men in charge of the sacred chickens that accompanied a Roman army. He eagerly 'reported to the consul that the birds had eaten so greedily that the corn dropped out of their mouths onto the ground', recounted the historian Livy. It was a lie – in reality the birds had refused to eat, a bad omen which the *pullarius* wanted to cover up. Unaware of the truth, Papirius Cursor announced with delight that the auspices for battle were excellent and that they would be fighting with the gods on their side. With preparations already under way, some of the cavalry heard that the story about the chickens might have been untrue. The news reached Papirius, who decided to continue on the basis that the gods would punish the pullarius if he had been lying. In the event, a freak Samnite spear flew through the air immediately before the battle started, striking down the dishonest pullarius. Papirius was satisfied that a guilty man had been punished by the gods, when at the same moment a crow near him 'gave a loud and distinct caw'. Papirius announced that it was a sign from the gods, proving beyond doubt that they were behind the Romans. He was right. The Romans won, and followed up their victory by seizing the city of Agnone itself. They carted off so much gold and silver from the war that by 290 BC it was said the public buildings of Rome and its allies could be decorated with it.[6]

Just how serious an issue the sacred chickens could be was amply demonstrated by a disastrous incident in 249 BC, during the First Punic War. Publius Claudius Pulcher, leading the Roman fleet off Sicily, decided to take the auspices by seeing what the sacred chickens were doing. They refused to eat, which was not the sign he was hoping for. He had them thrown overboard, declaring that if they wouldn't eat they could drink instead. His fleet was defeated in the Battle of Drepana that followed, serving him right for ignoring the signs. No wonder the Roman armed forces, collectively and individually, learned to set so much store by the sacred chickens and other sources of omens, good or ill.[7]

The arrogance and impatience of the Roman consul Gaius Flaminius before the catastrophic Roman defeat at Trasimene in 217 BC is illustrated by his readiness to dismiss portents. Ignoring advice to wait for the forces of his fellow consul, Servilius Geminus, Gaius Flaminius was determined to confront Hannibal. He told his army to march and 'vaulted upon his horse', unsettling the animal so much that it stumbled forwards and flung Flaminius over its head. So terrible a portent caused 'dismay' among everyone present, but there was more to come. Despite using all his strength the signifer was unable to pull the standard out of the ground. Furious, Flaminius yelled, 'Are you bringing a dispatch from the Senate too, forbidding me to fight?' He insisted that if the men's hands were so 'numb from fear' that they could not pull the standard out of the ground then it should be dug up. When the column moved off the officers were 'terrified by the double prodigy'. The calamitous defeat justified their apprehension.[8]

In 209 BC, before he attacked the Punic coastal city of Cartagena in Spain, Scipio came up with an idea to inspire his army – or perhaps he really believed the dream he claimed to have had. Having outlined to the Roman troops and naval force his plan of attack, which involved cutting the city off on the landward side, he explained that Neptune had come to him in his sleep and told him what to do. Together with the promise of golden crowns for the first men to scale the wall, the combination of a clearly thought out plan of attack with divine backing was the perfect way to inspire the Roman forces.[9] The assault was no walkover because the Carthaginians defended themselves fiercely, but Scipio prevailed with the use of his land forces and the navy.*

Scipio also understood the value of being able to come up with an optimistic take on the spur of the moment, thereby deflecting a bad omen. When he landed in Africa with his forces he stumbled

---

* See Chapter 9 for the brutal treatment of the inhabitants on Scipio's orders.

as he disembarked. That could have been alarming to the watching soldiers but he had the presence of mind to turn round to them and say, 'Congratulate me, soldiers, I have hit Africa hard!'[10] He may have unwittingly inspired the Roman rebel senator Sertorius, an ingenious and resourceful general who fought a number of wars in the early first century BC. In his final war he seized control of most of the Iberian peninsula, fighting off the Roman armies sent to defeat him. Sertorius was especially skilled in using superstition as a tool to manipulate his own men. He invented dreams and omens to inspire the native troops he had recruited to his cause.[11] In the end his rebellion only ended because he was assassinated.

Pompey's end, which came when he was assassinated at Pelusium in the Nile Delta in Egypt on 28 September 48 BC, was said to have been foretold before by bad omens of a kind that would have been particularly alarming to Roman troops. In 49 BC, when he landed at Dyrrachium (Durrës, now in Albania), some of Pompey's soldiers were killed by thunderbolts, and even more mysteriously 'spiders' allegedly occupied the army's standards. Other omens were seen in Rome 'that year and a short time previously', including, as Dio reported, a total eclipse of the sun. There was a temptation always to maximize the significance of such signs, and that year's eclipse is no exception. An annular eclipse in 50 BC did indeed pass across southern Egypt and Arabia on 21 August, but the sun was only about 30 per cent obscured in Rome,[12] while another annular eclipse on 7 March 51 BC had passed through northern Italy and was about 80 per cent in Rome.[13] Dio of course had to use records that were around 250 years old when he wrote. Perhaps either he or his sources had jumbled the events and exaggerated them into a single total eclipse event to make a better story.[14]

The extraordinary case of the so-called 'rain miracle' took place in 174, during the reign of Marcus Aurelius. The emperor was fighting a war against the Quadi tribe. A battle was going badly because a vastly superior enemy force had surrounded the Romans. It was a baking hot day so the Quadi backed off, confidently believing that the heat and lack of water would force the Romans to capitulate. They posted guards around their foes so that no one could get past to fetch water. The Romans were 'scorched by the heat', some of them wounded, but they had managed to retain their position when suddenly clouds formed, followed by a torrential downpour. A story was thereafter put about that an Egyptian magician called Arnuphis, a friend of the emperor's, had used his spells to encourage several gods – including Mercury, here treated as 'god of the air' – to bring down the rain.

The story was also originally recorded by Cassius Dio, but his text at this point is lost. All we have are excerpts and a summary created in the eleventh century by Xiphilinus, who added a comment that he was sure Dio was wrong. The legion involved was XII Fulminata ('Thunderbolt'). Xiphilinus provided his own explanation, which was that Marcus Aurelius had a legion made up of Christians from Melitene. He was told by an officer amid the impending disaster that Christians could achieve anything by prayer. The emperor asked them to pray and it was the Christian God who obliged. It was for this that he was said to have called the legion Fulminata.[15] However, since the legion is attested with that name at earlier dates, the tale was clearly apocryphal.

Whether these stories had any truth to them is not important. The real point was that – Roman historians being especially fond of such tales – they were firmly fixed in the canon of Roman military history. It is likely that many soldiers had heard all or some of them, even if only anecdotally. Stories of Roman defeat following ominous warning signs that were ignored or challenged are bound to have preyed on their minds.

One anonymous soldier at Haltonchesters on Hadrian's Wall was clearly very affected by seeing a lightning bolt strike the ground about half a mile (800 m) south-west of the fort from where he must have watched the storm. He set up a stone to mark the spot, neatly inscribed FVLGVR DIVO(RV)M, 'the lightning of the gods', leaving no doubt about whom he thought responsible.[16]

## Imperial and state cults

Roman troops swore allegiance to the state and the gods through the person of the emperor, a process established under Augustus. It was essential to reinforce the harmony (known as *concordia*) between the emperor and the army in the common interest. That ritual was continued thereafter and formed part of an annual cycle of religious observations. Given that the army was not only widely dispersed but also included huge numbers of non-citizen auxiliary troops, the cycle played an important role in reinforcing a sense of common allegiance and cohesive identity. Every fort headquarters contained a shrine, the *sacellum*. Here the unit's standards were stored, and official rituals observed which included the veneration of the standards as religious objects. Soldiers from the resident unit could participate as priests, and the commanding officer presided over the rites as the senior man or men did in all Roman communities. In the Roman world priesthoods were not exclusive professions. The Castra Praetoria in Rome had its own temple of Mars where Titus Aelius Malchus, a mounted praetorian of Cohors III Praetoria, also served as a priest in the cult of Mars. This kind of dual role was commonplace and gave men like Malchus prestige and status in their units.[17]

Trajan wrote to Pliny to acknowledge the news that the appropriate vows had been made in Bithynia on 3 January: 'With pleasure, my dear Pliny, I learned from your letter that the soldiers and provincials have fulfilled their vows to the immortal gods for my safety, and have pronounced them for the future, with general delight.' Twenty-five days later Pliny wrote again to tell Trajan that the day of the emperor's accession had also been observed.[18]

State cults reinforced the religious foundation of the state through the protection of traditional Roman gods, principally Jupiter Optimus Maximus ('Jupiter, the Best and Greatest'). Roman military units worked according to a calendar of official dates and anniversaries concerned only with Roman divinities. Soldiers were free to observe any other cults as a matter of personal conscience, unless their beliefs conflicted with the authority of the Roman state and its cults. For this reason Christianity (see below) was susceptible to persecution until 313.

One state religious calendar, the *Feriale Duranum* ('the Festivals of Dura'), has survived. Found in the ruins of the temple of Artemis Azzanathkona at Dura-Europos, originally built in AD 12–13 but by the third century apparently the fort's administrative headquarters, it was the official religious calendar for the period *c.* 225–7 of the mounted Cohors XX Palmyrenorum based at Dura-Europos. The anniversaries and festivals it listed represented the blurring of politics and religion in a cycle that reinforced the relationship between the emperor and his soldiers. Written in Latin, even though Syria was in the Greek-speaking part of the Roman world, the document was probably produced for general distribution to military bases all over the Empire and regularly updated. For example, on 3 January the unit made offerings to a variety of official gods such as Jupiter and Minerva in honour of the welfare of the current emperor, Severus Alexander. As the year progressed further observations followed, including the birthdays of past deified emperors and empresses such as Hadrian on 24 January and Matidia on 4 July, Claudius on 1 August, and the birth of Rome. Great imperial military victories were also listed, for example Septimius Severus' defeat of Parthia on 28 January. Needless to say, emperors whose reigns had ended in ignominy go unmentioned, and the content is restricted to state cults. The focus was entirely on the emperors and the imperial family. The only Republican figure included was Julius Caesar on 12 July, from whom Augustus claimed descent as his adoptive son (he was in fact Caesar's great-nephew). This also explains the inclusion of Germanicus, who though not an emperor remained one of the most celebrated members of the Julio-Claudian family and military leaders in imperial history.[19]

In order to work properly, the state religious calendar needed to be synchronized with other sites where the same anniversaries were being celebrated, and in particular Rome, to make sure they were all being observed simultaneously. A bronze fragment of what seems to have been a sophisticated water clock and calendar has been found in the remains of the third-century fort at Vindolanda. Referring to the month of September the piece of metal has holes for pegs which evidently used to mark the date as the year progressed. Similar devices must have been used in forts and other official places across the Empire.[20]

The religious rites were conducted in front of the whole company, or at least those who were currently at Dura, to remind them not only of their fealty to the imperial house but also of the benefits they enjoyed from the emperor's protection and leadership. That this form of state religious propaganda did not always work was obvious from the number of occasions, especially in the third century, when soldiers toppled the emperor in favour of their own candidate, who had usually offered them more money. By 235 Severus Alexander had been emperor for 13 years – a long time for the soldiers, as Herodian observed; they found serving the young emperor 'unprofitable' because it had been so long since his accession handout.[21] The emperor was on campaign and the soldiers had been greatly unimpressed by his performance, regardless of the religious ceremonies they had participated in on his behalf. They much preferred the strongman soldier Maximinus, who had been promoted to the Praetorian Guard by Septimius Severus and whom Severus Alexander had placed in charge of military recruitment and training.[22] They regarded Maximinus as a fellow soldier and made him emperor. Maximinus promptly 'doubled their pay, promised an enormous bonus of money and in kind, and cancelled all punishments'.[23] Shortly afterwards he sent a tribune and centurions to kill Severus Alexander and his entourage in his tent: 'They burst into the tent and slaughtered the emperor, his mother [Julia Mamaea] and all those believed to be friends or favourites', said Herodian, who lived through those times himself.[24]

No matter that the day of Severus Alexander's acclamation as emperor (26 June) and the birthday of his mother Mamaea had been celebrated in the annual military calendar at Dura-Europos; the whole occasion made a mockery of the state religious calendar and exposed the real basis of military loyalty. Three years later Maximinus, who had faced rebellions and a Senate that appointed its own emperors, was killed on campaign by the disgruntled men of Legio II Parthica with the assistance of praetorians. Instability was to continue into the fourth century, but each emperor in turn remained the subject of the annual cycle of observances conducted on the same days at every fort throughout the empire where his authority held sway.[25]

A cache of altars dating to the second century were dedicated to Jupiter Optimus Maximus at the fort of Maryport in Britain. Set up on the orders of a succession of commanding officers of Cohors I Hispanorum milliaria, probably following the same or a similar calendar to the one from Dura-Europos, these altars are so well preserved that it is likely a new one was carved each year in preparation for an annual ceremony to renew the vows, and the old one buried. Lucius Antistius Lupus Verianus was one of the prefects whose names appear on the altars. Unlike most of the other men named, he added where he came from: the city of Sicca Veneria, a colony in the province of Numidia.[26] The majority of the altars abbreviate the god's name to I.O.M., reflecting the formulaic nature of the ritual, which was conducted in the name of the commanding officer on behalf of and before the whole unit. The dedication would have taken place on the parade ground, a feature known at some forts including Hardknott, not far from Maryport (see Chapter 13). It is certain that exactly the same sort of dedications were made at every other fort. At the temple of the Palmyrene Gods at Dura-Europos, a wall-painting depicts the tribune Gaius Julius Terentius 'in action' making a sacrifice to three Palmyrene deities and the tutelary spirits of Dura and Palmyra. Terentius is accompanied in the fresco by 21 other soldiers, among them a signifer. Terentius seems to have been killed later during his time as tribune, possibly in an attack in 239 by the Persians, and was commemorated in another inscription by his wife Aurelia Arria.[27]

A similar religious ritual appears on a distance slab of Legio II Augusta from Bridgeness on the Antonine Wall, dating to 142–3. It commemorated the completion of 4,652 paces-length of the Wall in the name of Antoninus Pius. On one side is a panel showing a Roman cavalryman trampling on four enemy warriors. The other illustrates six men involved in the sacrifice of three animals (a pig, a sheep and a calf) at an altar. One of the men, dressed in a toga, seems to be leading the ceremony, while the sixth, a *victimarius*, is clearly in charge of enticing the animals to come forward. No one is named apart from the emperor in the main text, but the ceremony may be a depiction of a senior officer, perhaps even the then legate of Legio II Augusta, presiding over a *suovetaurilia*. This was a purification land ritual involving the sacrifice of these three animals, and entailing prayers to Jupiter, Janus and Mars. It was supposed to ward off 'sickness, barrenness and destruction, ruin and unseasonable influence'. The use of the ritual on a military frontier was perhaps intended to prepare the men for the building work ahead.[28]

The observances were not always properly maintained. Aurelius Attianus was a prefect of Cohors II Gallorum at Old Penrith in north-western Britain, probably in the third century. He restored a temple to the divine imperial house and Jupiter Optimus Maximus Dolichenus, explaining that it had 'collapsed through old age'. The phrase may be rhetorical in the sense that he was trying to amplify his achievement, but the temple must have been in a significant state of disrepair to warrant the work.[29] Attianus was not alone. In 211 Priscus, a centurion of Legio XXX Ulpia, was sent to the military colony at Cologne to take charge of restoring a temple also dedicated to 'the divine imperial house and Jupiter Optimus Maximus Dolichenus'. The same reason was given – collapse through old age – the inscription adding that the temple had to be rebuilt from the ground up.[30]

One of the most remarkable of all Roman military shrines is that in the Temple of Luxor in Egypt. The temple was begun around the end of the fifteenth century BC and went through a number of modifications right up to and including the time of the Ptolemaic pharaohs, shortly before the Roman conquest. In the Roman period the temple was still substantially intact. By the late third century AD it had become the centre of Roman administration in the area, and had also been incorporated into a Roman fortress, sufficiently large to house a vexillation, which was built in and around the temple. The layout of the fortress, not surprisingly, bore little resemblance to that of most military installations, since far and away the most dominant features were the vast hypostyle halls and pylons of the Eighteenth and Nineteenth Dynasty pharaohs, by then well over fourteen centuries old. The Roman structures included an apsidal shrine built into the Egyptian temple and dedicated to the Tetrarchy emperors of the late third and early fourth centuries. Some of the shrine's frescoes have survived; they show Roman officials worshipping at the feet of the emperors, as well as soldiers of the garrison with spears and shields and their horses.[31]

## LOCAL AND REGIONAL GODS

At the Second Battle of Bedriacum in Italy in 69, Legio III Gallica, based in Syria from not long after the Battle of Actium a century earlier, greeted the rising sun in Syrian fashion, a custom the legion had developed during its time in the East.[32] Roman soldiers were especially attracted to the power of local cults in the lands where they were garrisoned, or to the idiosyncratic attractions of exotic eastern cults like Persian Mithras and Egyptian Serapis. This provides us with a wealth of evidence for individual soldiers' practices and their concerns about protecting their interests and the afterlife. Dealing with a native deity was based on the straightforward premise that its power could either protect an arriving Roman soldier or be turned against him. Honouring that god with offerings and dedications was a means of ensuring, for example, that an auxiliary cohort stationed in a remote frontier fort was protected rather than threatened. The process usually involved creating a Romanized version of the cult. This meant depicting the deity in Roman form and sometimes identifying a Roman equivalent in order to create a hybrid cult, especially in the remoter north-west provinces.

In Britain, on Hadrian's Wall, at Carrawburgh a spring still rises a few yards to the west of the fort in a little valley, bubbling amongst the moss and grass. It was once sacred to a water nymph known as Coventina, who is unknown anywhere else and therefore seems to have been a native British deity. In 1876 it was discovered that the spring was filled with Roman offerings to the nymph, including coins, brooches, altars and other dedications. Titus Cosconianus, a prefect of Cohors I Batavorum stationed at the fort, not only made his own dedication in the form of an inscription but also added a relief carving of Coventina, giving us some idea of how the Romans perceived her. His Coventina is depicted as a typical Roman water goddess; she lies on her back, facing the onlooker and holding a sprig of leaves. Yet, once carved and dedicated, the offering was never to be seen by Roman eyes again.[33] Where Cosconianus obtained the carving from, we do not know. It might have been manufactured to order in the fort vicus, or perhaps he commissioned it from a sculptor working in the military town at Corbridge, not far to the south.

Sometimes soldiers brought their local gods with them, with the result that a native deity was incongruously transported to an entirely new location. The Germanic deity Garmangabis was worshipped in Britain by a detachment from an irregular unit of Suebians stationed at Lanchester between 238 and 244, during the reign of Gordian III. Suebians included members of a number of different Germanic tribes (the name survives in the modern German region of Swabia). The Suebian soldiers made their dedication to both Garmangabis and the 'divinity of the emperor' as a fulfilment of a vow. The text was elaborately and stylishly carved on an altar found not far from the north-west corner of the fort, perhaps at a location once set up as a shrine.

The soldiers of Legio I Minervia at Bonn were fond of worshipping the mother goddesses known as the Matronae Aufaniae, to the extent that a third of the surviving dedications made by soldiers there were addressed to the cult.[34] Marcus Clodius Marcellinus, a soldier of the legion, was typical with his simple dedication in fulfilment of his vow to them, but he was in good company. Even legates of the legion participated in the cult. Lucius Calpurnius Proclus honoured the goddesses, and so did his wife Domitia Regina in her own right.[35] Visitors to the fortress joined in too, such as Publius Prosius Celer, praefectus castrorum of Legio VIII Augusta, then based at Strasbourg.[36] In 205, during the reign of Septimius Severus, a beneficiarius on the governor's staff called Titus Flavius Severus and a woman who must have been his wife, Successinia Tita, made a joint dedication.[37]

On the other side of the Empire Aurelius Diphilianus set up an altar at Dura-Europos to Zeus Baetylos, inscribing it: 'To the god of (his) nation Zeus Ba(e)tylos, of those who live along the Orontes, Aurelius Diphilianus, soldier of Legio IIII Scythica Antoniniana, has dedicated this'. The 'Antoniniana' component of the legion's name dates the dedication to the reign of Caracalla or Elagabalus in the early third century, while the Orontes is the river that flowed through the city of Antioch in Syria. Diphilianus evidently came from somewhere along its banks and wished to honour the god of his homeland.[38] Gaius Longinus was a legionary standard-bearer serving with Legio V Macedonica in the second century, perhaps on one of its eastern campaigns. He commissioned what must have been an expensive dedication in Greek to the god Apollo of Nisyra, his homeland. Appropriately enough, Apollo is depicted on horseback carrying a double axe, in the manner of an Anatolian sky-god.[39]

## EASTERN MYSTERY CULTS

By far and away the most popular eastern cult among soldiers from the late second century onwards was Mithraism, inspired by the

Persian cult of Mithra, a deity of light and truth. Roman Mithraism emphasized valour and fortitude in the face of adversity, but very much developed its own beliefs, character and iconography. Like a number of other contemporary eastern cults, Mithraism involved admission only by initiation into sacred mysteries, and promised its ecstatic adherents a joyous afterlife. It spread into the Roman Empire, carried by traders and soldiers as they moved about the Roman world. Open only to men, Mithraism appealed to soldiers because of its emphasis on bravery and endurance, symbolized by Mithras as the bringer of light (good) that triumphed over the forces of darkness (evil). Although Mithras was also identified with Sol Invictus, the Undefeated Sun God, the religion was essentially monotheistic and reflected the emergence in the Roman Empire of a number of eastern saviour cults, such as Christianity and Isis worship, focused on a single deity. Mithraism's exclusive and elitist nature helps explain why it faltered in the face of Christianity, which accepted all comers including women. Mithras, however, was most conspicuously successful in the army, for a variety of reasons.

The central element of the Mithras story was that he had been engaged in a fight to the death with a bull created at the dawn of time, though this key component appears only to have evolved in the Roman version of the cult. Mithras killed the bull in a cave, thus releasing the blood that contained the essence of life, and celebrated the fact in a feast with the Sun God. Little detail is known today about the beliefs because Mithraism was very secretive, and so successfully later suppressed by Christians. However, the bull's death seems to have been the crucial moment of redemption when the trials of life on earth could be escaped and a higher, better life attained.

Mithraism so far as we know had no 'home', nor any sort of empire-wide hierarchical authority as Christianity was developing. The result was considerable variation in practices but some elements recur. A temple to Mithras, known as a mithraeum, was basilican in form, with a nave and aisles, but it had no windows in order to recreate the mystery and symbolism of the cave. They were also small and could accommodate no more than 30–40 adherents. Many have been found near Roman forts from Hadrian's Wall to Syria, port towns such as London and Ostia, and also Rome itself, reflecting how the cult had been transmitted across the Roman Empire by soldiers and traders. Mithraism was a congregational cult in which the participants faced a large relief depicting the sacred killing, the 'tauroctony', where they participated in the feast that formed a pivotal role in the rites by celebrating the bull's death. Theatrical props enhanced the sense of being in a special place; there were, for example, cosmic symbols like the signs of the zodiac which encircled a symbolic entrance to the cave, and perforated altars through which lamps cast eerie pools of light and shadow across the congregation. Statues of Mithras's associates Cautes and Cautopates stood in the nave. It is these features that make mithraea so easy to identify, even if there are no inscriptions.

Mithraism was an exclusive religion and only accepted men who were willing and able to endure extreme physical tests. In this way they could demonstrate that they were of the same calibre as the cult's eponymous hero. The faithful progressed through a cycle of seven challenging initiation rites that led them up the ranks of the cult's hierarchy. Initiates were also expected to be able to utter the correct liturgical responses to a series of questions that recounted Mithraic beliefs.

Christians were revolted by Mithraism, angered by what they saw as resemblances to their own practices. Ironically, their observations have helped us understand better what was going on during the ritual. 'The monstrous images there by which [Mithraic] worshippers were initiated as Raven, Bridegroom, Soldier, Lion, Perseus, Sun Runner, and Father' were mentioned by the Christian priest Jerome, recounting a hierarchy that increased its appeal to the Roman army.[40] Two centuries earlier, the Christian writer Tertullian described the ritual in more detail, using military terminology as he did for Christianity with his references to 'fellow soldiers of Christ':

Deep in a cave, in the camp of darkness, a crown is presented to the candidate at the point of a sword, as if in mimicry of martyrdom, and placed upon his head. Then, he is admonished to resist and throw it off, perhaps slipping it on the shoulder of the god, saying 'Mithras is my crown'. He is immediately acknowledged as a soldier of Mithras if he throws the crown away, saying that in his god he has his crown. Thereafter he never places a crown on his head, and uses that to identify himself if anywhere he is tested on his oath of initiation.[41]

The mithraeum at Dura-Europos is the most easterly known in the Roman Empire. Rather than being built as a standalone or underground structure, it was constructed in a room in a private house, close to the fort curtain wall between two towers. The layout was the normal basilican form. It was intended for use by members of the garrison, Cohors XX Palmyrenorum, but the evidence here and elsewhere suggests that the cult was mainly followed by officers. Ordinary soldiers may even have been deliberately excluded in some cases. However, when Secundinius Amantius, a 'cornicularius of the prefect' of Legio XXII, set up a dedication to 'the Undefeated God Mithras and of Mars' at Mainz, he added an unusual phrase, 'by (his Father) Primulus' permission'.[42] Primulus was apparently one of the worshippers who had reached the highest grade (Father) and was therefore in a position to admit new adherents who had passed the tests. The text suggests Secundinius was being given a special privilege, perhaps because he was not an officer.

The Dura mithraeum was originally constructed in 168–9. It was subsequently demolished and rebuilt in 209–11, with further modifications c. 240, only to be destroyed around 16 years later when the military compound's defences were remodelled.[43] Under the upper of the two tauroctonies displayed was a Palmyrene inscription that records the shrine's original construction by the cohort's commanding officer: 'A good memorial; made by Ethpeni the *strategos*, son of Zabde`a, who is in command of the Archers who are in Dura. (This was set up) in the month Adar of the year 480 (168)'.[44] Another inscription followed in 209–11, erected by the procurator Minicius Martialis and the centurion Antonius Valentinus, the latter described as *princeps praepositus* ('first in command'), in association with two vexillations of Legio IIII Scythica and Legio XVI Flavia Firma. The text recorded their restoration of the shrine.[45] The vexillations were presumably passing through, but their work shows that the worship of Mithras was not a fringe activity.

About as far as it was possible to be from Dura-Europos while still remaining within the Roman Empire, one of several mithraea built on Hadrian's Wall was constructed close to the south-east corner of the fort of Carrawburgh near the shrine of Coventina. Carrawburgh was built in the early 130s, a few years after most of the forts attached to the Wall, but the mithraeum did not follow for around another seventy years. Modified several times over its life, it survived until it was destroyed in the early fourth century, its ruins abandoned in an area that became a late Roman rubbish dump. Still visible today in a dank and windswept hollow beside the fort's south-west corner, the excavated mithraeum once contained several altars, all dedicated 'To the Undefeated Mithras' by the equestrian commanding officers of the fort. These were Lucius Antonius Proculus and Aulus Cluentius Habitus, who both describe themselves as prefects of Cohors I Batavorum, while a third, Marcus Simplicius Simplex, says merely that he was prefect. His name is known to have been a Romanized form of native names in the north-west provinces, for example in the Rhineland. Unlike the other two altars, his features a portrait of the god which when found still bore traces of red paint. Piercings in the stone allowed a lamp placed behind it to shine through, the rays emanating from his head in the manner of a solar deity.[46]

Antonius Proculus is the only one of the prefects whose later career seems to have left evidence. A Lucius Antonius Proculus turns up in Egypt as the *epistrategos* of the Thebaid region. Aulus Cluentius Habitus meanwhile came from Larino in eastern central Italy, which he called Colonia Septimia Aurelia Larinum. The name shows that the town had been made a colony during the reign of Septimius Severus, so the altar must be contemporary with or after that time.[47]

Mithraea are attested at a number of other forts in Britain, but the best known of all was found in London not long after the Second World War. Mithraea are well-known in ports and major cities where the numerous traders shared the military interest in a saviour deity associated with endurance and heroism. However, the most evocative relic from the London mithraeum was commissioned by a soldier.

Ulpius Silvanus was a veteran of Legio II Augusta. His name suggests that a forebear had become a citizen during the reign of Trajan. He seems to have joined the legion at its colony in Gaul at Orange, but therefore will have spent his career in Britain where Legio II Augusta was permanently stationed after the invasion of 43. Silvanus paid for a small but elegantly carved marble depiction of the tauroctony, the central event in the cult. The stone came from the Luna quarries in Italy and would have been an expensive investment.[48] The mithraeum lurked on the dank and waterlogged banks of the Walbrook stream, a tributary of the Thames. At 59 ft (18 m) long and 26 ft (8 m) wide it was relatively large for a mithraeum. We do not know if Silvanus lived in London, if perhaps he had been a member of the governor's bodyguard, whether he was passing through, or whether the sculpture was in fact acquired by a third party and brought to Britain. Although the style of the lettering is more appropriate to the late second century, the mithraeum was built around the middle of the third; it would have been frequented by soldiers because of its proximity to the fort of the governor's garrison in London.

Military officers commemorated Mithras in many places throughout the Empire. Marcus Valerius Maximianus, legate of Legio XIII Gemina, made a dedication to 'the Undefeated Sun-God' at the legion's base at Alba Julia in Dacia.[49] But the impact of the Mithras cult and others was not by any means universal. Mithraism is virtually unknown in Germania Inferior and in Egypt, even among the military.

Although the mithraeum at Dura-Europos was destroyed when the defences were rebuilt, mithraea in Britain seem in several cases to have been deliberately desecrated and destroyed. The Carrawburgh building was torn down around the time of the reign of Constantine the Great, who legitimized Christianity in 313. There is no evidence to prove who was responsible, but the most likely candidate is Christians who, buoyed up with confidence in their new status, were bent on wiping out a cult that seemed dangerously close to Christianity in concept and nature. In the military zone these iconoclasts were probably Christian soldiers (see below). The official end of Mithraism came under the Christian emperor Theodosius I (379–95), who in 393 banned all pagan worship.

Another mystery cult favoured by soldiers was that of the old Hittite sky god Dolichenus, who was conflated with Jupiter. Ulpius Amandianus served with Legio XIIII Gemina at Petronell in Pannonia Superior during the reign of Maximinus I. His dedication to Jupiter Dolichenus reads like a curriculum vitae, which is appropriate enough because it was clearly made in the hope of promotion. He reeled off his various jobs which included librarius (clerk), armourer, signifer, optio of the second centurion in Cohors VIII of the legion, and finally candidate for the centurionate. He made the dedication in association with Ulpius Amandus, a veteran who was probably his father.[50] Although Jupiter Dolichenus is well attested in military locations in the Western Empire, the cult is not, for example, known in Egypt.

The worship of Serapis (also known as Sarapis), a Graeco-Roman hybrid Egyptian god whose name is a conflation of Osiris and Apis, was spreading in Italy in the first century BC, as was the cult of Isis. Isis' role in the rebirth of Osiris made the cult attractive to those interested in the notion of a saviour cult that promised rebirth. The army was no exception. Gaius Julius Antigonus, a centurion with Legio V Macedonica during its time at Turda in Dacia between c. 166 and 274, made a dedication to both Isis and Serapis along with his wife Flavia Apollinaria. There was a temple to the same two deities at Lambaesis, where the governor of Numidia, Lucius Matuccius Fuscinus, 'finished and decorated the temple, which his predecessors had started, with a porch and columns with his money in association with his wife Volteia Cornificia and their daughter Matuccia Fuscina' in the mid-second century.[51] The temple was located in the vicus outside the legionary fortress of Legio III Augusta, which had been incorporated as a *municipium* by the late 160s. It would have been frequented by soldiers, veterans and their families. The ambitious Gaius Julius Apollinarius of Legio III Cyrenaica, mentioned in Chapter 3, wrote to his father Julius Sabinus at Karanis in Egypt on 26 March 107, acknowledging the protection of 'Sarapis' (*sic*) during a journey as the fleet recruit Apion had done. Whether this represented genuine belief or was a colloquial superstitious formula, expressed without even thinking about it, is hard to say. In another letter concerning a journey to Rome from Alexandria via Syria, Asia and Achaea (Greece), he referred to acknowledging 'the favours of the lord Sarapis'.[52]

The cult even found its way to northern Britain. Claudius Hieronymianus, commander of Legion VI Victrix, built a new temple to Serapis at the legion's base at York. The temple itself has never been found but the dedication slab has survived.[53] The Greek name Hieronymianus is as unusual for Britain as the god of his new temple; Claudius Hieronymianus had quite possibly spent some of his time to date in the East acquiring an exotic taste in religion. If so, he was well and truly in touch with the Roman zeitgeist in the 190s. Septimius Severus was not only a North African but was also beguiled by the exotic superstitions and mysteries of the East. Indeed, the temple may even have been built to please him, perhaps during his British campaign of 208–11. In any event, before long Hieronymianus had moved on to become governor of Cappadocia.

The worship of Isis, increasingly popular among civilians under the emperors, also appeared in the army; but its known distribution is patchy, and even then it was often associated with that of Serapis. Like Mithraism, access was through a process of initiation. At Potaissa, the base of Legio V Macedonica and also a military colony, there was a college of worshippers of 'Isis of the Countless Names'. A centurion of the legion, Gaius Julius Antigonus, was one of those who along with his wife made a dedication to both Isis and Serapis.[54] At Micia in Dacia, Ala I Hispanorum Campagonum and their prefect Marcus Plautus Rufus made a dedication 'to the Queen Goddess Isis', while at Čačak in Moesia Superior Gnaeus Pompeius Politianus, tribune of Cohors II Delmatarum milliaria, set up an altar to Isis and Serapis.[55] Instances of whole units making such dedications to Isis are extremely rare.

Evidence for the presence of Christians in the army before the legalization of Christian worship is limited. A centurion from Caesarea called Cornelius, who served in a unit styled Italica, is described in the Acts of the Apostles as having had visions and an encounter with St Peter.[56] In 295 Maximilianus was aged twenty-one and the son of Victor, a Christian soldier serving in Numidia. The law then obliged the son of a soldier to follow his father into the army. Although Maximilianus fulfilled the physical require-ments, he refused to become a soldier on the grounds of his faith in an early example of conscientious objection. He was executed on Diocletian's orders, became a martyr and was canonized as a saint.[57] In 298 Marcellus, a centurion in the first cohort of an unspecified unit, threw down his belt and sword and refused to swear an oath of allegiance. When challenged, he said, 'It is not proper for a Christian man, one who fears the Lord Christ, to engage in earthly military service.' He too was executed and later canonized.[58]

Once the Empire had become Christianized, these problems diminished. Constantine had issued his Edict of Milan in 313 tol-erating Christianity after a vision before the Battle of the Milvian Bridge in 312 when he defeated his rival Maxentius. He allegedly saw a Christian cross and the inscription 'conquer with this'.[59] It was the beginning of the end for the great Roman gods credited with Rome's military might over past centuries. Perhaps it was no coincidence that by then the Roman army was changing out of recognition from the one that Caesar, Germanicus or Hadrian would have recognized.

# EPILOGUE

itus Cissonius apparently served in the army with no distinction and was similarly insignificant in civilian life as a veteran living at Antioch in Pisidia. Nevertheless, he and his brother appear to have been men who enjoyed a good time, and they immortalized themselves as such. Some soldiers, perhaps the majority, were never promoted or did anything worthy of note but stolidly continued with military careers, sometimes long after they could have legitimately retired.

Lucius Caesius Bassus served in Legio VII Claudia Pia Fidelis for a remarkable 33 years. He died between 42 and 57, aged fifty-three, so he must have enlisted under Augustus or Tiberius when he was twenty. He was still in the army when he expired at Split in Illyricum. His epitaph says all this but adds nothing more.[2] His nameless heir, who could have been a fellow legionary or a freedman, organized the tomb. There is no mention of wars, decorations, promotion, an unofficial wife or children. During the time he served there was little fighting apart from the invasion of Britain, in which VII Claudia was not involved. Entitled to leave the army when he was forty, or forty-five at the latest, he must have elected to stay on with the legion he knew. He was perhaps by then institutionalized and chose the security of what he knew over the uncertainty of life outside the barracks. Cissonius and Caesius Bassus stand for the countless ordinary soldiers who made no recorded mark on history and achieved nothing of note, but whose presence in the ranks made the Roman army's achievements possible.

Over millennia almost all lines of human descent eventually die out. Infant mortality, disease, accidents, war and infertility all play their part in bringing to an end hereditary lineage in countless different ways and at myriad different times. Of course, some lines escape and continue, meaning that the further we go back in time there is a smaller and smaller number of people from whom we are all descended. Anyone with ancestry in the lands surrounding the Mediterranean, and in parts of north-western Europe, has ancient forebears who lived in the Roman Empire. It is now impossible to trace any such lineage that far back, but if it could be done then it is certain that those remote ancestors would include members of the Roman army.

For several years, rumours have circulated that certain villagers in Liqian in western China, close to the Gobi Desert, are descended from legionaries who fled from the disastrous defeat at Carrhae in 53 BC. Recently, studies of the DNA of those villagers have suggested that more than half of it is Caucasian in origin.[3] Of course it is impossible to demonstrate a connection, since the ethnic origin of the legionaries concerned is unknown. There is also the possibility that others of European origin, such as traders, settled there – if only one or two of them fathered children, they would have tens of thousands or more descendants many centuries later. Nonetheless, the sheer romance of the tale makes it a beguiling one.

Men who served in the Roman army were conscious that they

were part of a long and venerated tradition, and were aware too of the way emperors appealed, often optimistically, to the army's identity. Philip I produced his version of an established coin type with military standards and the legend FIDES EXERCITUS ('the faithful army'). In 249 some of the 'faithful army' were serving under his general Decius and decided they preferred him. Shortly afterwards Decius met Philip in battle at Verona. Philip's army was comprehensively defeated and both he and his son were killed. Decius was declared emperor as Trajan Decius. He issued a sestertius with the legend GENIVS EXERCITVS ILLVRICIANI ('Genius of the Illyricum Army'), commemorating the garrison in the Illyricum region by honouring its divine personification, but was killed in battle against the Goths in 251.[4]

The emperor Numerian issued an unusual coin during his brief and turbulent reign in 283–4, at a time when the Empire had become more militarized than ever before. The reverse depicted the emperor himself standing between two seated and bound captives and bore the legend VNDIQVE VICTORES, 'everywhere victorious'.[5] Clearly aimed at the army, the type was unprecedented, though a few later emperors copied it. *Undique* is an older Latin word used often in literature, frequently in military contexts: it features, for example, 36 times in Virgil, and was also used by Caesar and Livy,[6] while Numerian was said to have been far more interested in literary matters than anything to do with the army. The origin of the paired words seems to be Livy, who describes Scipio's forces at the Battle of Ilipa of 206 BC, in the Second Punic War, as *victores se undique inveherent*: 'everywhere victorious they charged' or 'everywhere the victors charged'.[7] The grammatical usage is different from that on the coin, but there is no other surviving instance of the two words being associated in Latin literature. The legend on Numerian's new coin was surely echoing Livy's account of an extremely famous battle involving one of Rome's most celebrated and legendary generals. If so, the allusion was made by a regime trying to associate itself with a remote and mighty military past, however indirectly.

Livy's contribution to perpetuating the memory of famous past wars and generals was moreover discussed by the author of the life of Probus.[8] These references were part of a grand contemporary tradition in the late Empire. The Gallic panegyricists who trumpeted the achievements of the Tetrarchy emperors, such as Diocletian and Maximianus, over the next few years favoured references to great men of the Roman Republic, usually military figures and mainly of the Second Punic War.[9]

Within a generation, in the early fourth century, Constantine I issued another unprecedented coin type. This time the reverse showed two soldiers with spears and shields, accompanied by the legend GLORIA EXERCITVS, 'the Glorious Army'. By that time it was a different organization from the one that had conquered the Empire in past centuries. Constantine disbanded the Praetorian Guard after it backed his enemy, Maxentius, in a civil war and was defeated by him at the Battle of the Milvian Bridge in 312. The emphasis now was on defence of the frontiers against barbarian incursions and a reliance on fast-moving smaller forces, with an emphasis on cavalry. We know from a late Roman document, the *Notitia Dignitatum*, that the names of legions still existed, but they either contained far fewer men than they used to or were more or less permanently split up. Legio II Augusta, for example, had been moved from its long-time base at Caerleon in Wales to the far smaller coastal fortified compound at Richborough, on the east coast of Kent in south-east England. Part of Legio X Gemina was still in Vienna, where it had been since Trajan's reign, but the rest had been converted into a mobile field unit.[10] Beyond that we know little about how the legions were organized at this late date, or about the individual men.

Despite the changes, the great old traditions were far from forgotten. Vegetius' book about Roman military institutions was written by the early fifth century and drew heavily on earlier works and traditions to describe the Roman army in a mixture of facts and anecdotes. It was thus a rather idealized, even nostalgic, piece of writing and therefore not necessarily reliable when it comes to detail. Vegetius variously cited stories about Scipio Aemilianus and Cato the Elder, and the works of Sallust, among others, relating for example how Aemilianus had to knock ill-disciplined troops into shape at Numantia, even though this had happened 500 years before he wrote.[11]

## TOMBS AND GRAVESTONES

Funerary inscriptions have been cited throughout this book. Their texts are often perfunctory and the stones frequently damaged but no other class of evidence has told us so much about the men who served in the Roman army. The right of a soldier to decide how his life was commemorated was enshrined in law as the emperor Severus Alexander ordered in a judicial ruling made between 222 and 235 concerning the obligations of a soldier's parents:

> *The Emperor [Severus] Alexander to the soldier Cassius*
> A father and a mother who are the heirs of their son, who was a soldier, should not fail to comply with his will, in which he provided for the erection of a monument to himself, for although all complaints on this ground have been abolished by former constitutions, still the parents cannot avoid experiencing regret, and being conscious that they have neglected their duty by failing to comply with the last will of the deceased.[12]

If Cassius ever had a funerary monument it has never been found. This is hardly surprising, since the overwhelming majority of Roman funerary inscriptions have been lost or are illegible. Nonetheless, those that do survive communicate above all else the sense of self that was such a powerful characteristic of the Roman era. The desire, even the need, to commemorate oneself, or to be memorialized by one's family members, was a product of a period when for the first time in western history ordinary people appear in the record in sufficient numbers to tell us who they were, where they came from and what they did. The information is often incomplete, tantalizingly abbreviated or ambiguous, but for the most part Roman tombstones provide information that would be impossible to garner from any other source. Soldiers dominate the record, especially in the frontier provinces where so much of the army was stationed, because they and their families were more likely to commission tombstones than any other single class of person. They were keen to commemorate their military identity in death. The habit was most common between the middle of the first century AD and the late third century, though the majority of surviving monuments belong to the first and second centuries. The only other group of people in the Roman world whose lives are so well recorded is the civic elite in Rome and other cities.[13]

A tombstone like that of Lucius Antonius Quadratus, the decorated soldier of Legio XX, found at Brescia had an agenda that went way beyond recording his feats. The monument honoured his family past, present, and yet to come. Not only were his descendants supposed to be inspired by his military achievements, they would find their own standing enhanced by association, while in turn Antonius Quadratus would have felt that he had been inspired by his own forebears. The cult of ancestors was a powerful and fundamental one in Roman culture. The historian Sallust remembered that he had often heard how Scipio Africanus and others frequently used to say 'how greatly' the sight of the wax masks of their ancestors had 'set them on fire' to aspire to manly courage, valour and strength, qualities summed up in the concept of virtus which was so important to the Roman army.[14]

These memorials were originally erected in cemeteries along roads outside the forts, colonies, towns or settlements where the men had died, whether in service or in retirement. The descendants of those lucky enough to be remembered in this way would gather by the tombs or cenotaphs at the annual feast of the Parentalia (or Feralia) on 18–21 February to recall the achievements of their heroic forebear and bask in the honour he had brought them and their own descendants. Passers-by were supposed to look up admiringly and be inspired too.[15]

Caesar always celebrated the efforts of his bravest men, whether they died or lived to tell the tale. Just before the Battle of Pharsalus in 48 BC he asked Crassinius, one of his centurions, how he thought the battle would go. Crassinius answered, 'We shall conquer, O Caesar, and you will thank me, living or dead.' Crassinius covered himself in glory that day but was killed. Caesar gave the centurion's body full military honours and had a tomb built specially for Crassinius alone, close to the mass burial mound where the rest of the dead were interred.[16]

Relatively few soldiers who fell on the battlefield, however, were commemorated with tombstones. There was no fixed protocol for dealing with the Romans' military dead. War cemeteries as we understand them did not exist in the sense that they were visible. In 90 BC the Senate decreed that those killed in war were to be interred where they fell in order to avoid the public spectacle of their remains coming back to Rome and putting off others from fighting. This instruction followed the death during the Social War of the consul Rutilius Lupus, whose body and those of other aristocrats were brought to Rome for burial, making a 'piteous spectacle'. However, even after cremation became preferable, expressly because it became known that bodies buried 'in wars abroad were dug up again',[17] bodies from the Varian disaster in AD 9 were placed in mass graves, some of which have since been found. Germanicus presided over the interment of the bones, laying the first turf for the burial mound himself.[18]

The manpower and inclination needed to bury the dead was not always available. After the First Battle of Bedriacum in 69 a few of the corpses were buried by family members who came to collect them. The vast majority of the dead were left where they were to rot, as Vitellius was able see for himself almost six weeks later, 'a dreadful and revolting sight'.[19] The tombstone of Gaius Cesennius Senecio, who died in Severus' war in Caledonia, recording how his body was brought to Rome is exceptional.[20] Of those whose bodies were not recovered (the vast majority), some were commemorated by their families with cenotaphs such as that of Marcus Caelius, who was killed in the Varian disaster (see Chapter 7). But such monuments are rarely found. One need only consider the levels of organization, acquisition of land and sustained maintenance exercised by the Commonwealth War Graves Commission and the US Government's American Battle Monuments Commission since the First and Second World Wars to see how much effort is needed to institute and sustain any form of reliable policy for dealing with military dead.

Physical remains identifiably those of Roman soldiers are rare. Even the so-called soldier or praetorian at Herculaneum (Chapter 3) is only identified as such on the balance of probabilities; nothing is known about the man's exact age, origins or unit, unlike the detailed information that can appear on a tombstone. The main reason bodies are so rarely found is because for most of the period in which Roman military tombstones were being made, cremation was the preferred form of burial. Even then it is rare to be able to associate cremated remains with a tombstone. Marcus Favonius Facilis, the centurion of Legio XX buried at Colchester not long after the invasion of Britain, is a rare exception. Cremated remains are useless for assessing age at death or physiological condition, and in Facilis' case his perfunctory memorial inscription tells us nothing about his age: we are told only that he served with Legio XX, was from the Roman Pollio voting tribe, and had two freedmen called Verecundus and Novicius.[21] It is possible that the two were freed in Facilis' will. Severus Alexander, almost two centuries later, confirmed in a ruling on 20 December 226 that a special privilege awarded soldiers was the right to free a slave by stating the fact in his will with the words 'I give and bequeath to my freedman . . .'[22] As such, these freedmen and their descendants thereafter carried on their former masters' names; in Favonius Facilis' case they would have been named Marcus Favonius Verecundus and Marcus Favonius Novicius.

An extraordinary survival from Thebes in Egypt is a painted burial shroud bearing a life-size picture of a soldier called Tyras, apparently named after a Greek city on the Black Sea coast. Tyras wears a military tunic with cloak, there is a ring on the little finger of his left hand, and he holds a sword, while staring out with a piercing gaze at the onlooker, his head slightly tilted. He is flanked by Osiris and Anubis, symbols of the Egyptian afterlife. We know nothing else of him apart from his name, but his physique and powerful expression form a remarkable record of a man who served in the late second or early third century and whose appearance in life as a soldier was preserved this way.[23] It was not at all unusual for people of Graeco-Roman origin to adopt Egyptian burial customs, albeit with a highly classicized tone. Tyras' portrait is realistic and full frontal, and his name is written in Greek.

The Roman army, and indeed the wider population, seem to have been relatively indifferent to the names and achievements of individual soldiers from the past. The time and trouble the heirs of deceased soldiers went to in order to set up monuments counted for nothing as wars were forgotten and even the great legions dwindled into rumps of their former selves. Lucius Valerius Geminus came from a place called Forum Germanorum, now identified as San Lorenzo di Caraglio in north-western Italy. He served in Legio II Augusta during the early conquest of Britain, but died around the age of fifty in the late first century at Alchester. Perhaps one of the legion's former bases, Alchester later developed into a normal small Roman town; when it was eventually walled, the builders found Valerius Geminus' tombstone and hacked it up into hard-core to use in the foundations.[24]

A large number of tombstones have survived at the fortress of Legio XX at Chester in Britain, mainly because when in later years repairs to the fortress wall were needed, a legionary cemetery was pillaged for its tombstones, which were summarily heaved out of the ground and carted off to use in the wall filling, presumably no relatives or descendants being still around to raise any objections. The tombstone of the veteran Gaius Aeresius Saenus and his family at York was removed from the cemetery there at a later date, probably in the late third or fourth century, and reused as the lid of a coffin for a male burial.[25] We can assume that even if Aeresius Saenus had any descendants – given the early deaths of the two children mentioned, he probably did not – they either knew nothing about this outrage or did not care. The tombstone of Flavinus, a signifer with Ala Petriana, was dug out of the remains of a Roman military cemetery on Britain's northern frontier and carted off for use in Hexham Abbey, perhaps in the seventh century when the first abbey was constructed. The tombstone finally ended up in the foundations of the south transept porch in the twelfth-century rebuild of the abbey. Flavinus was twenty-five years old, had served seven years and was in the turma of Candidus.[26] Without the builders of the medieval abbey we would probably know nothing about him. The same practices went on across the Empire; ironically the reuse of a tombstone was often the best way of ensuring its survival.

Lucius Pompeius Marcellinus never had the chance to earn himself a moment in the Roman army's annals of infamy and glory. But for his tombstone, found in an unexpected place, we would know nothing about this young equestrian Roman embarking on a professional life. Marcellinus, who came from Rome, was made a tribune commanding the auxiliary Cohors I Ligurum during the second century AD. Evidently en route to join his unit, Marcellinus died at the great city of Ephesus in Asia at the age of twenty-three, probably from disease. His grieving mother Flavia Marcellina and sister Pompeia Catullina commissioned an elaborate tombstone to be set up at Ephesus depicting the family hero on horseback, his spear held aloft, in a military career that had scarcely started.[27] His record shows how random our knowledge of the men of the Roman army can so often be.

The Roman army never came to a formal end. It was modified in many different ways and gradually adapted to various types of fighting in different historical contexts. By the fourth century the emphasis had switched to static frontier garrisons and crack mobile field units. Eventually the army evolved into that of the Byzantine emperors, and changed out of recognition. The names of the legions gradually disappeared from the record. As early as the fourth century the post of praetorian prefect had become a high office of state in its own right, even though the Praetorian Guard had been disbanded by Constantine. As forts and fortresses were given up most fell slowly into ruin, collapsed and were buried, the stone of their walls and cemeteries robbed for reuse in countless different buildings throughout the Middle Ages and into early modern times. Some survive in recognizable shape, while others are completely buried or demolished and cleared away. Archaeological excavation has revealed huge quantities of evidence for the way they were designed and their histories, and also of the lives of the soldiers who lived in them. Aerial photography, especially in the Near and Middle East, has revealed some of the most remarkable evidence for Roman military strong-holds in the region. One thing is certain though. The names of the mighty legions will last forever now, enshrined in books and the remains of their fortresses, and revived by re-enactors.

No single soldier, no single event and no single campaign could ever possibly sum up the Roman army and do it justice. Perhaps the last words should be left to a pair of soldiers, one of them anonymous. Both were immensely proud of their time in the mighty Roman army. Their epitaphs – which seem to have been composed by the men themselves – stand for those of countless comrades who lie today in unmarked, unnamed and unknown graves from the outpost forts beyond the Euphrates to the windswept hills of northern Britain across which the shattered remains of Hadrian's Wall still stand.

The emperor Hadrian was lucky enough to witness an outstanding example of military training and prowess when he watched a display given by an anonymous member of a milliaria auxiliary unit. The soldier concerned was so pleased with his performance that he appears to have dictated his own triumphantly smug epitaph before his death, unfortunately omitting his name and where he came from:

> I am that man who was once famous on the Pannonian shore and first in bravery among 1,000 Batavian men. With Hadrian as judge I succeeded in swimming in full armour across the vast waters of the mighty Danube, and with a second arrow I transfixed and broke the arrow which I had shot from my bow, while it was still suspended in the air and falling back. No Roman or barbarian soldier was ever able to outdo me in throwing the javelin, no Parthian in firing the arrow. Here I lie and here I have sanctified my deeds on this memorial stone. Let people see if anyone can emulate my feats after me. By my own example, I am the first person to have performed such things as these.[28]

Titus Flaminius' memorial, meanwhile, is different from the countless others that baldly recount a man's name, age, length of service, unit and perhaps something of his family. Titus Flaminius came from Faenza in north-eastern Italy. He served for 22 years with Legio XIIII Gemina, having signed up at the age of twenty-three. The tombstone is badly damaged, and apart from his feet the figure that represented him is lost. But the text is intact. Sometime around the middle of the first century AD, around 70 years before Hadrian witnessed the anonymous soldier's feats, Titus Flaminius found himself in Britain on the Welsh Marches in the legion's new fortress at Wroxeter. The climactic experience of his life must have been taking part in the invasion of Britain not long before in 43, serving with a legion founded by Julius Caesar exactly a century earlier, though he probably had no chance to take part in the legion's finest hour when it fought Boudica (the tombstone does not mention the titles Martia Victrix, which the legion won for that campaign). Instead, Titus Flaminius spent the last few years of his life with Legio XIIII as it forged its way across the middle of a new province by fighting, laying out roads and building a fortress. In the event he died at the age of forty-five, probably around the years 50–60. He never returned home to enjoy the fruits of retirement. Perhaps retirement would have been an anti-climax. Titus Flaminius considered himself fortunate to have been a Roman soldier, as he claimed in his unusual and unique epitaph.

I served as a soldier, and now here I am. Read this, and be happy – more or less – in your lifetime. (May) the gods keep you from the wine-grape, and water, when you enter Tartarus. Live honourably while your star gives you life.[29]

# ROME'S PRINCIPAL WARS

(Selected – some dates are approximate)

First Samnite War 343–341 BC
War against the Latin Revolt 340–338 BC
Second Samnite War 326–304 BC
Third Samnite War 298–290 BC
Pyrrhic War 280–275 BC
First Punic War 264–241 BC
Second Punic War (Hannibalic War) 218–201 BC
First Macedonian War 214–205 BC
Second Macedonian War 200–196 BC
Seleucid War (Antiochus III) 192–188 BC
Third Macedonian War 172–168 BC
Fourth Macedonian War 150–148 BC
Third Punic War 149–146 BC
Conquest of Greece 146 BC
Numantine (Celtiberian) War 143–133 BC
Jugurthine War 118–116 BC
Marius' war against the Cimbri and Teutones 104–101 BC
Social War against Roman allies 91–88 BC
First Mithridatic War in Asia 88–84 BC
Second Mithridatic War in Asia 83–81 BC

Civil War in Italy 82 BC

War against Sertorius in Spain 76–73 BC

Spartacus (Servile) War 73–71 BC

Third Mithridatic War in Asia 73–63 BC

Cretan War 68–67 BC

Pompey's War against the Cilician pirates 67 BC

Caesar's Gallic Wars 58–50 BC

Parthian campaign under Crassus 54–53 BC

Civil War 44–31 BC

Pannonia Campaign 13–12 BC

Tiberius' German Campaign 8 BC

Varus' campaign against Arminius 9

Germanicus in Germany 15–17

Claudian Conquest of Britain 43 (continues intermittently
   thereafter for over 160 years)

Corbulo in Armenia 55–60

Boudican War in Britain 60–1

First Jewish War 66–73 (Vespasian and Titus' triumph in 71)

Civil War 68–9

Chatti War 82–3

Dacian War 85–8

Trajan's First Dacian War 101–2

Trajan's Second Dacian War 105–6

Parthian War 113–17

Second Jewish War 132–5 (also known as the Bar
   Kokbha Revolt)

Dacian War 157–8

Marcomannic Wars 166–75

Marcus Aurelius' Eastern Campaign 175–6

Civil War 193–7

Septimius Severus' Parthian Campaign 197–8

Septimius Severus' British War 208–11

Caracalla's Alamanni Campaign 213

Caracalla's Eastern Campaign 215–17

Severus Alexander's Persian Campaign 231–3
Severus Alexander's Alamanni Campaign 234–5
Gordian III's Persian Campaign 243–4

From this time on, the remainder of the third century and then the fourth century were characterized by an endless series of frontier wars, civil wars, and coups of varying importance.

# EMPERORS FROM AUGUSTUS TO VALENTINIAN I AND VALENS

Augustus (27 BC–AD 14)

Tiberius 14–37

Caligula 37–41

Claudius 41–54

Nero 54–68

Galba 68–9

Otho 69

Vitellius 69

Vespasian 69–79

Titus 79–81

Domitian 81–96

Nerva 96–8

Trajan 98–117

Hadrian 117–38

Antoninus Pius 138–61

Marcus Aurelius 161–80 (with Lucius Verus 161–9; with Commodus 177–80)

Commodus 180–92

Pertinax 193

Didius Julianus 193

Septimius Severus 193–211 (with Caracalla 198–211, and also
   with Geta 209–11)

Caracalla 211–17 (with Geta 211–12)

Macrinus 217–18

Elagabalus 218–22

Severus Alexander 222–35

Maximinus I 235–8

Gordian I 238

Gordian II 238

Balbinus and Pupienus 238

Gordian III 238–44

Philip I 244–9 (with Philip II 247–9)

Trajan Decius 249–51 (with Herennius Etruscus and
   Hostilian 251)

Trebonianus Gallus 251–3 (with Volusian 251–3)

Valerian I 253–60

Gallienus 253–68

Claudius II 268–70

Aurelian 270–5

Tacitus 275–6

Probus 276–82

Carus 282–3

Carinus 283–5

Numerian 283–4

The Tetrarchy (rule of Four Emperors):
   Diocletian 284–305 (with Galerius 293–305)
   Maximianus 286–306 (with Constantius I 293–305)
   (followed immediately by a confusing period of abdications
      and overlapping reigns)

Maximianus (2nd reign) 306–8

Constantius I 305–6

Galerius 305–11

Severus II 306–7
Maximinus II 309–13
Maxentius 306–12
Licinius I 308–24
Constantine I 307–37
Constantine II 337–40, Constans 337–50, Constantius II 337–61
Julian 'the Apostate' 361–63
Jovian 363–4
Valentinian I 364–75, Valens 364–78

## 'The Gallic Empire'

Postumus 259–68
Victorinus 268–70
Tetricus I and II 270–3

## 'The British Empire'

Carausius 286–93
Allectus 293–6

# DIO AND TACITUS ON THE SIZE OF THE ARMY

I t is impossible to say how many legions and auxiliary units there were in the Roman army at any point in Roman history. Nor is it possible to say how many soldiers there were at any given moment. Two Roman historians, Cassius Dio and Tacitus, both provide outline descriptions, but as their accounts make clear, they were unable to supply exact information. The reference to a legion being stationed in any one province does not mean the soldiers were all together at any given moment. Legions were continually being split into vexillations, and individual soldiers were liable to be seconded to all sorts of different duties which could include a centurion being sent to another province to take over temporary command of an auxiliary unit or detachments of soldiers sent on building projects.

*Cassius Dio on the size of the army in AD 5 under Augustus and in his own time in the early third century (55.23–24):*

At this time 23, or, as others say, 25, legions of citizen soldiers were being supported. At present only 19 of them still exist, as follows: the II (Augusta), with its winter quarters in Britannia Superior; the three IIIs – the Gallica in Phoenicia, the Cyrenaica in Arabia, and the

Augusta in Numidia; the IIII (Scythica) in Syria; the V (Macedonica) in Dacia; the two VIs, of which the one (Victrix) is stationed in Britannia Inferior, the other (Ferrata) in Judaea; the VII (generally called Claudia) in Moesia Superior; the VIII (Augusta) in Germania Superior; the two Xs in Pannonia Superior (Gemina) and in Judaea; the XI (Claudia) Moesia Inferior (for two legions were thus named after Claudius because they had not fought against him in the rebellion of Camillus*); the XII (Fulminata) in Cappadocia; the XIII (Gemina) in Dacia; the XIIII (Gemina) in Pannonia Superior; the XV (Apollinaris) in Cappadocia; the XX (called both Valeria and Victrix) in Britannia Superior. These latter, I believe, were the troops which Augustus took over and retained, along with those called the XXII who are quartered in Germany – and this in spite of the fact that they were by no means called Valerians by all and do not use that name any longer. These are the legions that still remain out of those of Augustus; of the rest, some were disbanded altogether, and others were merged with various legions by Augustus himself and by other emperors, in consequence of which such legions have come to bear the name Gemina.

55.24. Now that I have once been led into giving an account of the legions, I shall speak of the other legions also which exist today and tell of their enlistment by the emperors subsequent to Augustus, my purpose being that, if any one desires to learn about them, the statement of all the facts in a single portion of my book may provide him easily with the information. Nero organized Legio I, called the Italica, which has its winter quarters in Moesia Inferior; Galba the I (Adiutrix), with quarters in Pannonia Inferior, and the VII (Gemina), in Spain; Vespasian the II (Adiutrix), in Pannonia Inferior, the Fourth (Flavia), in Moesia Superior, and the XVI

---

* Lucius Arruntius Camillus Scribonianus was a senator who led a short-lived rebellion against Claudius in 42. He had been touted as a successor to Caligula, but the Praetorian Guard's swift appointment of Claudius prevented Camillus' elevation.

(Flavia), in Syria; Domitian the I (Minervia), in Germania Inferior; Trajan the II (Aegyptia) and the XXX (Germanica), both of which he also named after himself; Marcus Antoninus the II, in Noricum, and the III, in Raetia, both of which are called Italica; and Severus the Parthicae – the I and III, quartered in Mesopotamia, and the II, quartered in Italy. This is at present the number of the legions of regularly enrolled troops, exclusive of the city cohorts and the praetorian guard; but at that time, in the days of Augustus, those I have mentioned were being maintained, whether the number is 23 or 25, and there were also allied forces of infantry, cavalry, and sailors, whatever their numbers may have been (for I cannot state the exact figures). Then there were the praetorians, 10,000 in number and organized in ten divisions, and the watchmen of the city, 6,000 in number and organized in four divisions;* and there were also picked foreign horsemen, who were given the name of Batavians, after the island of Batavia in the Rhine, inasmuch as the Batavians are excellent horsemen. I cannot, however, give their exact number any more than I can that of the evocati. These last-named Augustus began to make a practice of employing from the time when he called again into service against Antony the troops who had served with his father, and he maintained them afterwards; they constitute even now a special corps, and carry rods, like the centurions.

## TABLE 1

### The legions in the early third century

This table is based on the list provided by Cassius Dio for the number of legions in existence in his time in the early third century, but with additional information from other sources. A number of legions had been disbanded or lost in the preceding two centuries, such as the

---

* These watchmen were an earlier organization and preceded the formation in AD 6 of the Cohortes Vigilum, with which they should not be confused.

XVII, XVIII and XVIIII in AD 9, V Alaudae by the late first or early second century, and VIIII Hispana in the early second century. They are not mentioned here.*

| Number | Title | Main province in which stationed | Raised by |
|--------|-------|----------------------------------|-----------|
| I | Italica | Moesia Inferior | Nero |
| I | Flavia | Germania Inferior | Domitian |
| I | Adiutrix | Pannonia Inferior | Galba |
| I | Parthica | Mesopotamia | Septimius Severus |
| II | Adiutrix | Pannonia Inferior | Vespasian |
| II | Traiana | Egypt | Trajan |
| II | Augusta | Britannia Superior† | Augustus |
| II | Italica | Noricum | Marcus Aurelius |
| II | Parthica | Italy | Septimius Severus |
| III | Augusta | Numidia | Republic or Augustus |
| III | Cyrenaica | Arabia | Republic |
| III | Gallica | Phoenicia | Julius Caesar |
| III | Italica | Raetia | Trajan |

---

* Note that Septimius Severus divided Britain into two provinces, Superior (south) and Inferior (north), to prevent any more governors using the large garrison to mount a challenge to become emperor after Clodius Albinus tried to do so in 193.

† Anyone wishing to research the history and fate of any of the individual legions should start by consulting Pollard and Berry (2015).

| Number | Title | Main province in which stationed | Raised by |
|--------|-------|----------------------------------|-----------|
| III | Parthica | Mesopotamia | Septimius Severus |
| IIII | Flavia | Moesia Superior | Vespasian |
| IIII | Scythica | Syria | Mark Antony |
| V | Macedonia | Dacia | Julius Caesar |
| VI | Ferrata | Judaea | Julius Caesar |
| VI | Victrix | Britannia Inferior | Republic |
| VII | Claudia | Moesia Superior | Julius Caesar |
| VII | Gemina | Spain | Galba |
| VIII | Augusta | Germania Superior | Republic |
| X | Gemina | Pannonia | Julius Caesar |
| X | Fretensis | Judaea | Republic |
| XI | Claudia | Moesia Inferior | Republic |
| XII | Fulminata | Cappadocia | Julius Caesar |
| XIII | Gemina | Dacia | Republic |
| XIIII | Gemina | Pannonia Superior | Republic |
| XV | Apollinaris | Cappadocia | Augustus |
| XVI | Flavia | Syria | Vespasian |
| XX | Valeria Victrix | Britannia Superior | Augustus |
| XXII | Primigenia | Germania Superior | Caligula |
| XXX | Germanica | Germania Inferior | Trajan |

*Tacitus on the army in* AD *23 (Annals 4.5):*

Italy, on either seaboard, was protected by fleets at Misenum and Ravenna; the adjacent coast of Gaul by a squadron of fighting ships, captured by Augustus at the victory of Actium and sent with strong crews to the town of Forum Julium. Our main strength, however, lay on the Rhine – eight legions ready to cope indifferently with the German or the Gaul. The Spains, finally subdued not long before, were kept by three. Mauretania, by the national gift, had been transferred to King Juba. Two legions held down the remainder of Africa; a similar number, Egypt: then, from the Syrian marches right up to the Euphrates, four sufficed for the territories enclosed in that enormous reach of ground; while, on the borders, the Iberian, the Albanian, and other monarchs, were secured against alien power by the might of Rome. Thrace was held by Rhoemetalces and the sons of Cotys; the Danube bank by two legions in Pannonia and two in Moesia; two more being posted in Dalmatia, geographically to the rear of the other four, and within easy call, should Italy claim sudden assistance – though, in any case, the capital possessed a standing army of its own: three urban and nine praetorian cohorts, recruited in the main from Etruria and Umbria or Old Latium and the earlier Roman colonies. Again, at suitable points of the provinces, there were the federate warships, cavalry divisions and auxiliary cohorts in not much inferior strength: but to trace them was dubious, as they shifted from station to station, and, according to the exigency of the moment, increased in number or were occasionally diminished.

# ROMAN NAMES

A Roman citizen normally had three names: the *praenomen*, the *nomen* and the *cognomen* (the *tria nomina*). The best-known name today is Gaius Julius Caesar. The *nomen* was the name of the *gens*, to which the closest modern equivalent is our word 'clan' or 'house'. It identified different families united by a common descent. The actual family or branch was indicated by the *cognomen*, and the individual person by the *praenomen*, thus 'Gaius of the Julian clan, family Caesar'. The *cognomen* could be derived from any one of many sources, including the individual's or even a forebear's nickname. 'Caesar' in fact means 'hairy', though Julius Caesar himself was bald.

Sons often bore the same name as their fathers, resulting in confusingly identical names applied to several people. One Roman solution when referring to larger clans was to use two cognomens and thus assist in distinguishing one family or branch from another. Publius Cornelius Scipio acquired the additional name Africanus in honour of his defeat of Carthage at Zama. However, his adoptive descendant Publius Cornelius Scipio Aemilianus was also awarded the name Africanus after his destruction of Carthage in 146 BC.

In Roman law adoption was indistinguishable in legal validity from normal parentage. A boy or man who was adopted by a Roman citizen acquired his adoptive father's name. A freed slave retained his old slave name as the *cognomen* but acquired his former Roman citizen master's *praenomen* and *nomen*. Thanks to this process it is quite possible to find people in remote provincial frontier locations with names that hark back to great Roman families, or even to the emperor.

A man whose name began 'Marcus Ulpius' had been enfranchised by Trajan (Marcus Ulpius Trajanus), for example, or his father or another forebear had been. 'Gaius Julius' linked the man or his fore-bears to enfranchisement by Julius Caesar, Augustus or Caligula, for example Gaius Julius Apollinarius. However, the names could also be acquired by freedmen from their former masters.

Women who were the daughters of Roman citizens were generally known by the female version of the *nomen* without any other appella-tion. Thus, women of the Julian clan were normally known as 'Julia'.

Slave names included those which alluded to the slave's place of origin, mythology, personal characteristics or appropriate Roman virtues or terms, amongst others. The Legio XX centurion Marcus Favonius Facilis had a freedman called Verecundus, which means 'bashful', 'modest' or 'shy'. His other freedman was called Novicius, a word that means 'novice', and in this context a novice to slavery. Their names appear on his tombstone found at Colchester.

# GLOSSARY OF TERMS

*ala* (plural *alae*): cavalry wing, usually of auxiliary troops. An *ala quin-genaria* had 512 troopers, an *ala milliaria* had 768 troopers. The totals are nominal. Everyday strength varied

*aquilifer*: standard-bearer who bore legion's eagle standard

*aureus*: the standard Roman gold coin, weighing about 7.3 g and 17–18 mm in diameter. It was equivalent to 25 silver *denarii* (q.v.)

*auxilia*/auxiliary troops: provincial troops hired to serve in the Roman army, normally organized into infantry *cohortes* or cavalry *alae*

*beneficiarius* (plural *beneficiarii*): a soldier who benefited from the favour of the commanding officer by being allocated to special duties that exempted him from mundane soldiering

*bucinator*: horn blower/trumpeter

*canabae*: the 'hutments', the civilian town that grew up round a legionary fortress

*censor*: a Roman magistracy, held sometimes by emperors, in charge of determining social status and morals

*centuria* (plural *centuriae*): a division of 80 men within a cohort. A double-century had 160. The letters of Apion and

Apollinarius, recruits to the fleet, show that the word was also more casually used for any 'company' of men, in their case the crews of ships

*centurio* (centurion): soldier commanding a century in a legion or auxiliary unit

*cohors* (plural *cohortes*): an infantry division of a legion or a standalone auxiliary infantry unit consisting of 480 men divided into six *centuriae*. The first cohort of some legions in the imperial period consisted of ten *centuriae* and had 800 men. *Vigiles* (q.v.) cohorts seem to have been a nominal 1,120 made up of seven double-centuries each by 205, if not from their inception under Augustus,

*cohors equitata* (plural *cohortes equitatae*): an auxiliary infantry cohort with a mounted component. A *cohors equitata quingenaria* probably had 480 infantry plus 120 cavalry, a *cohors equitata milliaria* 800 infantry plus 240 cavalry. Numbers are uncertain and strength only nominal anyway. Everyday strength varied

*Cohortes Praetorianae*: the 'praetorian cohorts' – the imperial Praetorian Guard, formally founded by Augustus but which had existed for some time as the elite bodyguards of the generals of the late Republic

*colonia* (plural *coloniae*): a settlement of Roman legionary veterans, incorporated as a normal Roman town and sometimes installed in conquered territory as an example of Roman civic life and to act as a trained reserve

*comitatenses*: mobile units, mainly cavalry, of the late Roman army from Diocletian on

*concordia*: 'harmony', in a military context meaning the harmony of the emperor and army, sometimes commemorated on coins with a pair of clasped hands

*consul* (plural *consules*): the most senior magistracy in the Roman system. Two were elected annually. By imperial times they were succeeded later in the same year by a series of pairs of *consules suffecti* to supply an increased demand for qualified men who could govern provinces. Men who had served as consuls were eligible to govern the richest or most militarily demanding provinces

*consularis*: a man of consular rank, sometimes used to describe a man sent by the emperor to be a legate of a province

*contubernium*: a tent-party of eight soldiers who shared a tent or barrack room

*cornicen*: hornblower

*cornicularius*: clerk on the governor's or a senior officer's staff.

*curator*: someone placed in a position of management or charge

*decurio* (plural *decuriones*): cavalry officer commanding a *turma* (32 troopers). The literal meaning was a 'commander of ten', because originally there were three per *turma*, arranged in order of seniority.[30] It should not be confused with the civic *decurio* ('town councillor')

*delectus* (or *dilectus* – both spellings appear in our sources): a levy of troops, a 'choosing' or 'selection' of troops

*denarius*: the Roman standard silver coin introduced in the Second Punic War. It weighed about 4.5 g and was 17–18 mm in diameter, remaining in production until the third century AD when it was supplanted by a double-denarius, the *antoninianus*. During the 200s both coins were steadily debased

discharge = see *missio*

*duoviri* see *duumvir*

*duplicarius*: a soldier on double pay

*duumvir* (plural *duoviri*): senior civic magistrate, one of two elected annually in an incorporated city and modelled on the *consules* in Rome

*eques/equites*: a trooper, cavalryman. The term *equites* was also applied to irregular cavalry units, mainly in the fourth century

*equitata*: applied to the name of an auxiliary infantry unit that also had a mounted component

*evocatus*: from Augustus' time onward, a soldier who had completed his term of service, been honourably discharged (q.v.) with all the privileges a veteran enjoyed, and then subsequently voluntarily signed on again. An *evocatus* was an *immunis* (q.v.) and generally of higher rank than the other legionaries, many being made centurions

*exercitus*: the army, 'the disciplined/exercised ones'

*fabrica*: workshop where armaments, nails, tools and other essential items were manufactured

*fabriciensis*: armourer

*frumentarius*: originally a soldier responsible for purchasing or acquiring grain and fodder. By the second century the word had become incongruously applied to soldiers, usually *beneficiarii* (q.v.), who were acting as imperial spies, probably because *frumentarii* had turned out to be in a good position to pick up intelligence

*immunis* (plural *immunes*): term applied to any job which exempted the soldier concerned from normal fatigues and duties. These included soldiers who worked as engineers, made artillery and other weapons, or served in clerical and administrative capacities, among many others[31]

*ius iurandum*: the oath of loyalty sworn by soldiers, compulsory from 216 BC

Latin citizenship = a form of legal rights and privileges originally granted to certain Italian communities until the Social War, after which they were made Roman citizens, and which continued to be awarded to provincial communities and individuals. The status was inferior to being a Roman citizen but superior to being an ordinary provincial

*legatus*: the emperor's delegate, a man of senatorial status. Applied to the governor of the province, *legatus Augusti*, or to the commanding officer of a legion, *legatus legionis*

*librarius consularis*: secretary to the governor

*liburna*: a lightly-built warship, designed for speed

*limitanei*: static frontier troops from the reign of Diocletian on

*manipulus* (plural *manipuli*): a Republican term applied to a pair of centuries. By imperial times it had become a colloquial term used to refer to being in the army, e.g. 'among the maniples'

*medicus*: doctor, or if the word *veterinarius* was added, a vet

*mensor*: a measurer; either a surveyor or someone who measured commodities like grain

*miles*: soldier

*miles gregarius*: common soldier

*milliaria*: applied to a large auxiliary unit meaning '1,000', but in reality a unit made up of ten centuries of 80 men, producing a 'milliary cohort' of 800 infantry, or 768 troopers in a cavalry wing made up of 24 *turmae* (q.v.) of 32 men each

*missio*: discharge from the army, under three categories – *honesta* (honourable), *causaria* (honourable, but early due to wounds), *ignominiosa* (dishonourable)

*mithraeum*: small basilican-form, semi-subterranean, temple dedicated to the observance of Mithraic rites

*modius*: a unit of measure for dry goods, equivalent to 8.73 litres

*municipium*: a town awarded Latin status, an intermediate level between full Roman citizenship and provincial where the inhabitants lived by their own laws and rights. They were regarded as a form of Roman citizens without the obligations full Roman citizens were bound to, or the full benefits

*navarchos*: naval officer, next in seniority after a *praefectus classis* (q.v.)

*nomos* (nome): an administrative district of Egypt inherited from the era of the Greek Ptolemaic pharaohs. Each was governed by a *strategos*, all of whom reported to one of three *epistrategoi*, overseen by the prefect of Egypt, an equestrian official appointed by the emperor

*numerus*: irregular infantry unit of hired provincials. Such men had no discharge rights like other auxiliaries

*optio*: second in command of a *centuria* after a centurion

*ordinarius*: 'normal' or 'customary', but in a military context often applied to jobs or positions equivalent to the centurionate and thereby carrying the same pay

Phrygian cap: also known as a 'Liberty' cap, worn by Liberty on some US coinage

*praefectus*: applied to a litany of administrative and military commands held by men of equestrian status from the *praefectus praetorio* (prefect of the Praetorian Guard) to a *praefectus* commanding an auxiliary infantry *cohors* (see also *tribunes* q.v.)

*praefectus castrorum*: prefect of the camp. Third in command of a legion, the highest equestrian post and usually filled by a man who had risen to the top of the centurionate. A *praefectus castrorum* could be left in sole charge of a legion if there was an interregnum between legates

*praefectus classis*: commanding officer of a fleet

*praetor*: a senior Roman magistracy, one level below consul (q.v.). Typically a man who had served as a praetor would go on to command a legion (for example Agricola)

Praetorian Guard: see *Cohortes Praetorianae*

*praetorium*: a commanding officer's house (or tent on campaign) from where he directed the unit. The literal meaning was 'place of the foremost man'

*princeps praepositus*: a centurion placed in command of a unit or detachment

*princeps praetorii*: centurion in charge of the headquarters

*principia*: the headquarters building in a camp or fort

*proconsul*: a man who had served as a consul (q.v.)

*procurator*: financial administrator, for example of a province

*protectores*: bodyguard troops first introduced in the mid-third century, and which replaced the Praetorian Guard under Constantine I

*quaestor*: the entry-level magistracy in a senatorial career, with responsibility for the management of the state treasury and accounts

*quingenaria*: applied to an auxiliary cavalry or infantry unit nominally 500 in strength (in practice usually 480 for infantry or 512 for cavalry)

*quinquereme*: warship with three banks of oars, two operated by a pair of oarsmen each and one by a single oarsman. See also *trireme* (q.v.)

*Saturnalia*: the midwinter feast, during which war was banned[32]

*scholae palatinae*: late Roman mounted imperial bodyguards whose creation marginalized the Praetorian Guard from Diocletian on until the latter's abolition in 312

*sesquiplicarius*: soldier on 1½ times normal pay

*sestertius*: the largest Roman base-metal coin. It weighed about 24 g and was 33–35 mm in diameter. Four were equivalent to one silver *denarius* (q.v.)

*signifer* (plural: *signifer*): generic term applied to any standard-bearer, or standard-bearers in general

*speculator*: scout

*subpraefectus*: an under-prefect

*tesserarius*: a soldier in each *centuria* (q.v.) whose job was to organize the nightly guard rota, liaising with the unit's officers for each day's watchword

*tetradrachm*: applied to a number of large silver coins about 12 g in weight and 23 mm in diameter, struck in the Eastern Empire, especially Alexandria. The silver content was always low and by the late third century the coins were bronze and reduced in size

*tiro* (plural: *tirones*): a new recruit. The word was also used for beginners of any sort

*tractatorus*: 'masseur', but incongruously later 'inspector' or 'accountant of finances'

tribune(*tribunus*): applied to a range of military posts, from the senatorial military tribune and five equestrian tribunes in a legion (these are described in more detail in Chapter 1) to the equestrian tribune commanding a milliaria auxiliary cohort or cavalry wing

tribune of the plebs (*tribunus plebis*): a Republican elected office. A tribune of the plebs could veto senatorial legislation and propose his own legislation, but could be vetoed by another tribune. During his one-year term of office his person was inviolate. The emperors were voted a tribune's powers and privileges by the Senate annually but did not hold the office. These allowed the emperor to pose as the protector of the plebeians against the Senate and provided a pseudo-constitutional basis for their power. Not to be confused with the military tribune (q.v.)

*trierarchos*: naval officer, the third in seniority below a *praefectus classis* (q.v.), and commander of a *trireme* (q.v.), but became also applied to a commander of a quinquereme (q.v.)

*trireme*: warship with three banks of oars, each operated by a single oarsman. See also quinquereme (q.v.)

*turma*: a sub-division (squadron) of a cavalry wing, consisting of around 32 troopers.

*valetudinarium*: fort hospital; 'place of health/sickness'

*vexillarius*: standard-bearer who carried the *vexillum* standard of the legion which bore its name and emblem

*vexillatio*: a detachment of soldiers, for example two cohorts drawn from a legion

*vexillum* (plural: *vexilla*): a banner or flag which was erected in the ground to mark the presence of soldiers and the legion or cohort to which they belonged *vicus*: the unofficial civilian settlement that grew up round a fort, usually along the roads radiating from it

*vigiles*: the night watch/fire brigade of Rome, the *Cohortes Vigilum* or *Vigiles Urbani*

*voluntarius*: volunteer, for example a veteran who volunteered to fight again after retirement

# NOTES

## 1: INTRODUCTION

1   There are many books on this topic. Goldsworthy (2003), and Bishop (2016) and (2017) are excellent examples. They all contain detailed discuission of armour, equipment, forts, buildings, tactics and so on.
2   Tomlin (1998), 58.
3   Vegetius, *Military Affairs* 1.1.
4   Appian, *Preface* 7.
5   Florus, *Epitome of Roman History* 2.64, 1.1–2, 8. The date has had to be corrected. Florus makes an approximation but 29 BC is the year he means. Florus was greatly influenced by the works of Livy.
6   Formerly attributed to Ausonius. See Ausonius, *Appendix to Ausonius* 5, lines 45–57 (Loeb edition vol. II, p. 287). The date of composition is therefore uncertain.
7   Appian, *Civil War* 3.43.1.
8   Recorded on an inscription summarizing his career and found at his home town of Como. *ILS* 2927, *CIL* 5.5262.
9   Southern (2007), 18ff, provides a good summary of the main literary sources for the Roman army.
10  Plutarch, *Sulla* 28.8. Sulla's autobiography is not extant.
11  Vegetius, *preface* to Book 1.
12  Tacitus, *Annals* 1.2.1.
13  Livy 28.43.11. He was later acclaimed as Scipio Africanus for his victory over Carthage at Zama.
14  Dio 55.32.5 (Loeb vol. VI, p. 277); Scullard (1982), 213–14; Tacitus, *Annals* 1.2.1.
15  Tacitus, *Annals* 3.74.4.

16    Tacitus, *Annals* 1.9.5.

17    Vegetius 3.10.

18    Virgil, *Aeneid* 6.843: *duo fulmina belli.*

19    Livy 22.36.1.

20    Tacitus, *Annals* 4.5.3.

21    *ILS* 2288, *CIL* 6.3492, Campbell (1994), no. 144.

22    Holder (2003), 120, supplies 217,624 though this is an error because his subtotals for infantry (143,200) and cavalry (74,624) come to 217,824; Spaul (2000), 526, estimated the total at 180,800.

23    See for example Kehne (2011), 325.

24    Dio 55.23.1–7, 24.1–8 (Loeb vol. VI, pp. 453–7).

25    de la Bédoyère (2017), 28–9, 91–3.

26    This figure is based on nine cohorts at 480 men each plus a double-strength first cohort of 800 (5,120). However, there is some debate about when the double cohort was introduced. It was certainly in existence by Flavian times but may not have been used for all legions at all times thereafter (see Plan 2 Inchtuthil). Without a double first cohort, the nominal infantry total of a legion of imperial date was 4,800 plus 120 cavalry (though the only source for the latter is Josephus, *Jewish War* 3.120.

27    Vegetius 2.6–7.

28    *SHA* (Hadrian) 10.2. See also Tacitus, *Annals* 1.20.1 for a specific use of the term to refer to working parties of soldiers sent out to build roads and bridges, 'Meanwihile the maniples . . .'

29    *SHA* (Commodus) 6.1.

30    *ILS* 2662, *CIL* 14.2523. An engraving of the monument is reproduced by Cowan (2014), 44.

31    Kennedy (1983).

32    Keppie (1998), 78, 81.

33    Dio 38.47.2 (Loeb vol. III, p. 399), 'this was the way the legions of the Republic were named, according to the order of their enrolment'.

34    *ILS* 2224, *CIL* 3.12280. Keppie (1998), 78, states that this is evidence Legio XIIX was in Cilicia. There is nothing on the inscription that either says or implies this.

35    de la Bédoyère (2017), 28–30, 91–3, 166–7, and in tabular form 276–9 (amongst other references).

36    Dio 38.47.2 (Loeb vol. III, p. 399); Frontinus also refers to Caesar's favouring of the legion, *Stratagems* 1.11.3.

37    Suetonius, *Caesar* 24; Pliny the Elder, *Natural History* 11.44.121.

38    See for example a letter found in Egypt from a soldier of this legion in Adamson (2012), 83.

39    Listed on a diploma recording the discharge of one of the men involved, Sapia Anazarbus of the Syrian Cohors I Antiochenses. *ILS* 9054, Campbell (1994), no. 324.

40 Tacitus, *Annals* 4.5. See Appendices. For gladiators see Tacitus, *Histories* 1.11.

41 Though Tacitus, *Histories* 1.59, refers to 'eight cohorts of Batavians' in 69. It is possible the Revolt of Civilis of 69–70 involved some Batavian units whose numbers were abandoned after the Revolt collapsed.

42 Kennedy (1983), 216.

43 Southern (1989) explains the problems involved in understanding the *numeri*.

44 Smallwood (1967), no. 297, Campbell (1994), no. 337.

45 Dio 40.18.1–2 (Loeb vol. III, p. 431).

46 Pliny the Elder 10.16.

47 Dio 40.18.1–2 (Loeb vol. III, p. 431); for the fellow aquilifers see AE 1976.641, also cited by Campbell (1994), no. 38. Found at Byzantium (Constantinople, Istanbul).

48 Josephus, *Jewish Antiquities* 18.55, describing when Pontius Pilate brought his army into Jerusalem with the standards, causing immense offence to the Jews.

49 *ILS* 2581, *CIL* 3.3256. Found at Novi Slankamen (Acumincum). See Chapter 2 for a discussion of their ethnicity.

50 *ILS* 2349, *CIL* 13.7753. Found at Niederbieber. A number of units, legions and auxiliaries, are attested in the area. For standard-bearers in general see D'Amato et al. (2018) and (2020).

51 Dio 40.18.3 (Loeb vol. III, pp. 431–3).

52 Tacitus, *Annals* 1.65.3–6.

53 Josephus, *Jewish War* 6.402.

54 Tacitus, *Annals* 1.20.1.

55 *SHA* (The Gallieni) 8.1–7.

56 *SHA* (Probus) 23.1–3.

57 Hebblewhite (2017) provides an excellent up to date survey of how the army changed from the reign of Maximinus up to the death of Theodosius I in 395. However, at the time of writing the book is prohibitively expensive.

58 Zosimus 1.20.

59 Rivet and Smith (1979), 219.

60 Ammianus Marcellinus 31.13.19.

61 Jerome, *Letters* 60.17. The reference is to the fact that the Huns were mounted.

## 2: STRENGTH AND HONOUR

1 *RIB* 109 (Cirencester). The Frisiavones lived in what was by then the province of Gallia Belgica.

2 Tacitus, *Histories* 1.5; Seneca, *On Providence* 4.5.1 and 3, observed that it was for the common good to have the best men serving in the army, even if that meant 'perverts and wastrels' were able to remain safe in Rome.

3   Livy 10.17.4.

4   Livy 42.34.1–15, 35.1–2.

5   Varro, *Latin Language* 5.87.

6   Polybius 6.20–21, 26.

7   Aulus Gellius, *Attic Nights* 16.4.3–4, citing as his source the now lost work *On Military Science* by Cincius.

8   Polybius 6.21, 24, 25.

9   Polybius 6.22–23.

10  Polybius 6.26.

11  Polybius 6.19.

12  Plutarch, *Tiberius Gracchus* 8.4.

13  Stockton (1992), 4.

14  Plutarch, *Marius* 7.3–4, 9.1.

15  In 100 BC. Appian, *Civil Wars* 1.29.

16  Plutarch, *Marius* 35.4.

17  Augustus, *Res Gestae* 3.

18  Tacitus, *Annals* 4.4.2–3.

19  Vegetius 2.3.

20  Vegetius 1.5; equivalent to 5ft 8in–5ft 10in.

21  Vegetius 1.5–6, 8.

22  Plutarch, *Crassus* 8.2. See Chapter 7 for a description of the revolt.

23  Dio 79.1.3 (Loeb vol. IX, p. 341).

24  *SHA* (Maximus and Balbinus) 9.4, 15.6.

25  Vegetius 1.6.

26  Arrius Menander 1, Campbell (1994), no. 7.

27  P. Oxy 1022, Campbell (1994), no. 9.

28  Campbell (1994), nos. 2, 5 (the latter involving a Christian and dating to 295).

29  Vegetius 1.7.

30  Vegetius 2.3.

31  Valerius Maximus, *MDS* 6.3.4.

32  Livy 22.57.11.

33  Valerius Maximus, *MDS* 7.6.1a; *SHA* (Marcus Aurelius) 21.6 – see below in this section. Macrobius 1.11.31 refers to the same occasion and adds (1.11.32) that Caesar once accepted slaves from his friends as recruits, though this is mentioned by no other source.

34  Dio 37.35.4 (Loeb vol. III, p. 157).

35  Dio 55.31.1 (Loeb vol. VI, p. 473).

36  Dio 56.23.1–4 (Loeb vol. VII, p. 51).

37  Tacitus, *Annals* 13.35. Tacitus says the extra legion came from Germany, but the legion attested in Syria after this date is Legio VI Scythica from Moesia.

38  Tacitus, *Histories* 2.16.

39    Suetonius, *Vitellius* 15.1.

40    Tacitus, *Histories* 4.14

41    *SHA* (Hadrian) 10.8.

42    *SHA* (Marcus Aurelius) 21.6–8; *CIL* 3.1980, dated to 169–70, records the original names of the two new legions. The slaves who volunteered after Cannae are recorded by Livy at 22.57.11. See above in this section.

43    *SHA* (Caracalla) 6.2–3; Dio 78.22 (Loeb vol. IX, p. 335).

44    *CIL* 3.6627, cited by Alston (1995), 42 and Table 3.1.

45    Sauer (1999), 62, note 28.

46    Gore (1984), 572–3.

47    A number of praetorians are attested in the area. See de la Bédoyère (2017), 127.

48    Van Lommel (2013), 165.

49    Herodian 5.4.8 (referring to the praetorians in the reign of Macrinus, but as a general observation).

50    Vinnius Valens' story can be found in Pliny the Elder's *Natural History* 7.82, Maximinus' in the *Historia Augusta* (Maximinus) 3.5.

51    'Dawn of the warrior-emperor', Hebblewhite (2012), 22.

52    Aulus Gellius, *Attic Nights* 5.6 *passim*, writing around the middle of the second century AD.

53    *ILS* 2272, *CIL* 5.4365.

54    Velleius Paterculus 2.112.1–2.

55    Josephus, *Jewish War* 7.13–16.

56    Josephus, *Jewish War* 7.17–18; for the defeat of Legio XII see *Jewish War* 2.540–58, and Chapter 7.

57    *SHA* (Probus) 11.5.1–4.

58    Millar (1981), 61.

59    *RIB* 293 (Wroxeter, Salop, Britain).

60    *RIB* 294 (Wroxeter, Salop, Britain).

61    *RIB* 255, 256, 258 (Lincoln, Britain). Gaius Saufeius' city of origin, Heraclea, cannot be specifically identified though modern sources state one or other as a fact, without substantiation. Szombathely was founded as colony in 45. Vilanius Nepos was therefore probably a veteran's son.

62    *ILS* 2127.

63    *ILS* 2658, *CIL* 8.217, Campbell (1994), no. 86.

64    *ILS* 2038, *CIL* 8.21021.

65    *RIB* 814.

66    Edwell (2010), 138.

67    *RIB* 2144.

68    *ILS* 2225, *CIL* 10.3886.

69    *ILS* 2341, *CIL* 13.6901 (Mainz). The tombstone is illustrated by Pollard and Berry (2015), 191.

70    *ILS* 2339, *CIL* 5.3374, 3375, 3747.

71   Goldsworthy and Haynes (1999), 166.

72   Dio 55.24.7 (Loeb vol. VI, p. 457).

73   Sherk (1988), no. 76. See also Rankov (2004), who treats the labels 'Batavian' and 'German' as synonymous.

74   *ILS* 2581, *CIL* 3.3256. Birley (1979), 103.

75   Ivleva (2012), 162, noting that attested Britons in 'British units' are even rarer and specifically referring to the evidence for Virssuccius and Bodiccius as being exceptional. The doubled 'tt' is a common variant in the spelling of Britannica/Brittannica. For Crescentinus, see *AE* 1968.31.

76   *AE* 1950.86. The reading rests on the expansion of NB as *n(atione) B(essi)*.

77   *RIB* 109 (Cirencester, Gloucestershire, spelled Ala Trhaec(um) (*sic*). Sextus Valerius Genialis' troop commander was also called Genialis. Stylistically, the stone could be as late as the early second century, but this is difficult to square with the burial of a serving soldier unless he was passing through, since the fort was given up by the late first century. The unit is attested in Britain in 103 (*CIL* 16.48) and 124 (*RIB* 2401.6).

78   *AE* 1926.67.

79   Campbell (1994), no. 149. The original papyrus gives Ala Maurorum, 'the Ala of Moors'. This is recognized now to be a colloquial term for the Mauretanian Ala I Thracum as Campbell explains. The original text can be accessed at http://papyri.info/ddbdp/bgu;15;2492

80   *SHA* (Probus) 15.7.

81   Ivleva (2012), 176.

82   *ILS* 2562, *CIL* 13.7041. A cohort of the same name is later mainly attested in Syria, and Dacia after 109.

83   'Caesar', *African War* 20.1 (the authorship is uncertain).

84   *Select Papyri*, no. 112 (Loeb edition vol. I, pp. 305–7). Letter viewable at: http://berlpap.smb.museum/privatbrief-eines-soldaten/?lang=en

85   Campbell (1994), no. 338; see Virgil, *Aeneid* 6.89 for a favourable reference to Achilles.

86   Aristides, *Regarding Rome* 74–5.

87   *Select Papyri*, no. 111 (Loeb edition vol. I, pp. 303–5). P.Mich.inv. 4528 recto (P.Mich. VIII 491). Letter viewable at: https://www.lib.umich. edu/online-exhibits/exhibits/show/from-trace-to-text--highlights/ item/1737?exhibit=89&page=313

88   Suetonius, *Augustus* 24.1.

89   Valerius Maximus, *MDS* 6.3.3c.

90   *CJ* 51.7.3.

91   Theodosius Law Code 7.13.10.

92   Pliny the Younger, *Letters* 10.29–30.

93   Campbell (1994), no. 1 (Fayum, Egypt).

94   Ovid, *Ex Ponto* 2.5.61.

95   Frontinus, *Stratagems* 4.1.4.

96   Aulus Gellius, *Attic Nights* 6.18.1.
97   Vegetius 2.5.
98   Seneca, *On Anger* 2.9.3.
99   Appian, *Punic Wars* 17.115 (Loeb vol. I, p. 605).
100  Dio 76.14.7 (Loeb vol. IX, p. 231).
101  Dio 77.4.5, 6.3 (Loeb vol. IX, pp. 247, 251).
102  *ILS* 2487, Campbell (1994), no. 17.
103  Cicero, *Letters to Atticus* 10.16.4.
104  Appian, *Civil Wars* 2.35.
105  *ILS* 2671, *CIL* 5.923.
106  ILS 2054, *CIL* 6.2672.
107  Tacitus, *Histories* 2.68.
108  *CIL* 10.788, *ILS* 6363b. The statue plinth recording this man's career is
     illustrated at de la Bédoyère (2010) (ii), 27, fig. 2.3. *Praefectus fabrum* could
     also mean 'skilful (or ingenious) prefect'. There are many other examples
     of these titles in civilian civic contexts.

## 3: GLORIA EXERCITUS

1    Josephus, *Jewish War* 3.72–3.
2    Vegetius 1.1.9.
3    Josephus, *Jewish War* 3.102.
4    See Gibbon 1.36; also Varro, *Latin Language* 5.87.
5    Vegetius 1.9.
6    Vegetius 1.1.
7    Vegetius 2.5.
8    Appian, *Spanish Wars* 14.86, 15.95–98.
9    Seneca, *Letters* 18.1.6.
10   Josephus, *Jewish War* 3.72–5.
11   *ILS* 2416, *CIL* 2.4083.
12   Tacitus, *Annals* 1.20.1.
13   Tacitus, *Annals* 1.23.3–6.
14   Tacitus, *Annals* 1.31.3–5, 32.1, 35.1.
15   Cicero, *Letters to his brother Quintus* 3.7.4 (Loeb edition Letter 27.4).
16   Frontinus, *Stratagems* 2.7.2–3.
17   Frontinus, *Stratagems* 2.8.13; Plutarch, *Caesar* 56.2 supplies an alternative
     version of this story in which Caesar runs through the soldiers' ranks
     challenging them not to shame him.
18   Tacitus, *Agricola* 5, 7.3, 9.5.
19   Dio 68.23 (Loeb vol. VIII, p. 401).
20   *SHA* (Hadrian) 2.1–6, 3.6.
21   *SHA* (Clodius Albinus) 6.1–2.
22   *SHA* (Clodius Albinus) 10.6–8.

23 Tacitus, *Agricola* 19.2. The claim was at least partly rhetorical. Tacitus was of course trying to paint his father-in-law in the best possible light.

24 Pliny the Younger, *Letters* 10.86B, 87. This man was probably called Fabus Valens. For Pliny's request that Nymphidius Lupus join his staff, see Chapter 15.

25 Dio 74.3.1 (Loeb vol. IX, p. 127). For the murder of Pertinax, see Chapter 11.

26 Pliny the Younger, *Letters* 3.8, 10.87; *Tab. Vindol.* III.660.

27 *Tab. Vindol.* II.250.

28 *P. Mich.* 8.466. II 18. Accessed online at http://papyri.info/ddbdp/p.mich;8;466. The text of course is in Greek and gives the Greek equivalent for consularis, ὑπατικὸς (hypatikos). Normally, a legion based in Egypt had to be commanded by an equestrian but its presence in Syria at the time obviated that. For the legal protection see *Digest* 50.6.7.

29 Tarruntenus Paternus, a jurist, detailed the jobs that earned a soldier immunity. See Campbell (1994), no. 35.

30 *P. Mich.* 9.562. Accessed online at http://papyri.info/ddbdp/p.mich;9;562

31 *ILS* 9100.

32 *RIB* 19.

33 *AE* 1977.811; for speculatores as executioners, see Millar (1981), 61.

34 *CIL* 3.6108.

35 In the original text the word is in the accusative plural, *frumentarios*. *SHA* (Hadrian) 11.4.

36 *ILS* 2368, *CIL* 3.433, 'agens curam carceris'.

37 *AE* 1933.256 (Sardis, Asia).

38 *CIL* 6.3357.

39 *SHA* (Macrinus) 12.4–6.

40 *P. Mich.* Inv. 1804. Accessed online at https://quod.lib.umich.edu/a/apis/x-1531

41 Ibid.

42 *AE* 1993 no. 1575. The Latin is ambiguous: *discenti(s) lanciari(orum)*, from *disco*, 'I learn'. Here it could mean 'with learning of spears' which could either refer to his having knowledge of the skills and thius able to teach them, or that he was still training.

43 *ILS* 2344/*CIL* 8.2988.

44 *ILS* 2696/*CIL* 3.6809.

45 Valerius Maximus, *MDS* 7.8.6.

46 Fronto, *Stratagems* 4.1.16. Cato's relevant work is lost. The sentence could be commuted to being bled at the headquarters building if the soldier was not thought to have been of sound mind; see Aulus Gellius, *Attic Nights* 10.8.1–3.

47 *Digest* 49.16.3.1–22; Josephus, *Jewish War* 3.102–3.

48 Livy 22.61.14.

49   Livy 22.53.3–4.
50   Livy 23.31.1–2.
51   Appian, *Punic Wars* 17.115–16.
52   Valerius Maximus, *MDS* 5.8.4; Frontinus, *Stratagems* 4.1.13.
53   Plutarch, *Sulla* 6.9.
54   Livy 2.59.11; Frontinus, *Strategems* 4.1.34.
55   Appian, *Illyrian Wars* 26 (Loeb vol. II, p. 97).
56   Appian, *Civil War* 3.43.1.
57   Appian, *Civil War* 3.44.1.
58   Frontinus, *Stratagems* 4.1.21.
59   *SHA* (Avidius Cassius) 4.4–6.
60   *SHA* (Avidius Cassius) 6.1–4.
61   *SHA* (Macrinus) 12.1–2, 7–9, 14.1; Dio 79.40.2 (Loeb vol. IX, p. 431);
     Herodian 5.4.1–13.
62   *CJ* 6.26.13.
63   *SHA* (Severus Alexander) 52.1, 53–54.
64   Polybius 3.114, 117.
65   Livy 22.57.10.
66   Josephus, *Jewish War* 3.96–7.
67   Josephus, *Jewish War* 3.113.
68   *RIB* 2425.2.
69   *RIB* 2425.3.
70   Tomlin (1998), 58.
71   *SHA* (Hadrian) 10.5–6. Written *c.* 293–305.
72   Quintilianus 10.1.11; see Petronius, *Satyricon* 130, for the use of *ferrum* for
     sword.
73   *SHA* (Aurelian) 5.2.
74   Aulus Gellius, *Attic Nights* 10.25.1–3, supplying *machera* (*sic*). The word was
     Greek in origin, μάχαιρα.
75   *P. Mich.* 8.467. Accessed online at http://www.papyri.info/ddbdp/c.
     ep.lat;;141
76   See Bishop (2016), 13–15, for these two swords.
77   *RIB* 2427.26; and see the discussion there on the possible historical
     context of the piece relating to an attested expedition under Gallienus
     on *CIL* 13.6780, dated to 255, and another on *CIL* 3.3228.
78   *RIB* 2427.2.
79   *RIB* 2428.4. See Seneca, *Letters* 18.1 for a comment on how the feast seemed
     to go on indefinitely by his time.
80   *CIL* 13.10026. The objects are displayed in the British Museum, London.
81   Sumner and D'Amato (2014), 146.
82   Second-century date, *RIB* 2445.16. Sollius Julianus is attested on
     Hadrian's Wall (*RIB* 3454) on a stretch known to have been built by Legio
     VI Victrix.

83 *CIL* 13.7507.

84 *AE* 1926.67.

85 Lewis and Reinhold (1990), no. 152, p. 515.

86 *RIB* 2501.1.

87 Polybius 6.37.12–13.

88 Exact year uncertain. *P.* Giss. 47, also given as *P. Giss. Apoll.* 6. Acceesed online at http://papyri.info/ddbdp/p.giss.apoll;;6. The key word (σα) μσειϱα is at line 11. Text and German translation also at https://bookandsword.com/armour-in-texts/greek-and-roman-inscriptions-and-papyri/. Bishop (2016), 45, provides a brief reference to the papyrus but without any contextual detail. For σαμσειϱα, see Liddell & Scott, vol. II, 1582, citing this papyrus.

89 See Chapter 15 for a veteran called Gentilius Victor who dealt in swords.

90 *RIB* 2426.1.

91 Frontinus, *Stratagems* 4.1.5. Note that this resembles a story that only survives in a summary of one of Livy's books; see Livy, *Periochiae* 57 where the issue seems to have been the shield's excessive size.

92 Dio 67.10.1 (Loeb vol. VIII, p. 339). This recommendation also appears in Vegetius 2.18.

93 Tacitus, *Histories* 1.38.

94 Josephus, *Jewish War* 3.95; Vegetius 1.19.

95 Appian, *Syrian Wars* 21 (Loeb vol. II, p. 141).

96 Appian, *Spanish* Wars 14.85 (Loeb vol. I, p. 273); Frontinus, *Stratagems* 4.1.1.

97 Frontinus, *Stratagems* 4.1.7.

98 Plutarch, *Marius* 13.2.

99 *SHA* (Avidius Cassius) 5.3; Onasander, *The General* 10.5.

100 See Zienkiewicz (1986), pp. 117–27.

101 Henig (1999), 153.

102 Kent and Painter (1977), 26–7, nos. 16 and 17.

103 *SHA* (Hadrian) 10.5.

104 *SHA* (The Two Gallieni) 20.3–5.

105 Sutherland (1974), 46. See also the Appendix on Roman Money.

## 4: GOLD AND SILVER

1 Tacitus, *Annals* 1.17.

2 Livy 4.36.2–3.

3 Livy 4.59.11 (406 BC), 5.7.12 (402 BC). No figure is specified in either case.

4 Polybius 6.39.12 for the mid-second century BC rates. He used the word drachma, a very approximate equivalent to the denarius, with which he and his Greek readers were familiar.

5 Dio 54.25.6 and 55.23.1 (Loeb vol. VI, p. 349, 453).

6 Tacitus, *Annals* 1.17.4.

7   Suetonius, *Domitian* 12.1.

8   Speidel (2009), 350; Dio 67.3.5 (Loeb vol. VIII, 325); Suetonius, *Domitian* 7. These sources use various denominations, so here the denarius equivalent is given for simplicity (1 denarius = 4 sestertii, 1 gold aureus = 25 denarii).

9   See Chapter 7. Tacitus, *Annals* 1.16–17, 32.4, 35.1–2.

10  Tacitus, *Annals* 1.37.1–3.

11  Tacitus, *Annals* 1.45.1–2, 1.48–9.

12  Tacitus, *Annals* 1.17.6, Dio 67.3.5 (Loeb vol. VIII, p. 323–5), and also Suetonius, *Domitian* 7, who cites the equivalent in gold. The rise was three gold *aurei*, equivalent to 75 denarii; Dio 56.32.2 (Loeb vol. VII, p. 73).

13  *SHA* (Aelius) 3.3 and 6.3, citing '300 million sestertii', the equivalent of 75 million denarii. At the time legionary pay was 300 denarii per annum, 300 denarii x 250,000 men = 75 million denarii.

14  *SHA* (Antoninus Pius) 7.1, 10.2; (Marcus Aurelius) 7.9. The figure of 50 million denarii is based on the assumption that there were about ten milliaria cohorts of praetorians, as there certainly were fifty years later; see Dio 55.24.6 (Loeb vol. V, p. 457).

15  Dio 77.1.1 (Loeb vol. IX, p. 239).

16  Tacitus, *Annals* 13.51.1.

17  Dio 55.23.1 (Loeb vol. V, p. 453).

18  Augustus, *Res Gestae* 17.2.

19  Tacitus, *Annals* 12.27.1.

20  Herz (2011), 308.

21  Dio 77.15.2 (Loeb vol. IX, pp. 271–3).

22  Tacitus, *Histories* 1.46.

23  *SHA* (Hadrian) 10.7.

24  Speidel (2009), 351–3, considers all the evidence but concludes that the position for auxiliaries cannot be 'proven beyond cavil'.

25  Speidel (2009), 353ff.

26  Vegetius 2.20.

27  Tacitus, *Annals* 1.17.4 (listed as grievances during the Pannonian mutiny of 14).

28  Polybius 6.39.

29  Plutarch, *Gaius Gracchus* 5.1; Tacitus, *Annals* 1.17.4.

30  Vegetius 2.20.

31  Campbell (1994), no. 24, also quoted by Webster (1996), 258. Julius Proculus was a name with mythological connections. See earlier in this chapter.

32  Tomlin (1992), 147.

33  *CJ* 51.21.5.

34  *RIB* 526.

35  Thirty-four were found originally, and three more seem to have been

found separately. Robertson (2000), 6, no. 22. A gold *aureus* was the highest value coin normally available in the Roman world.

36 Dio 60.20.1–4 (Loeb vol. VII, p. 417–19). For Bredgar the hoard and its possible explanation, see Bland (2018), 35, and Moorhead and Stuttard (2012), 44–5, and for Narcissus' role see Dio 60.19.2 (ibid., 415). For both the Bredgar and the Caerleon hoards see Robertson (2000), 6, no. 22, and 65, no. 316.

37 Robertson (2000), no. 132.

38 Edwell (2008), 176.

39 *CJ* 51.61.3.

40 Sear (2000), nos. 1832, 1834, 1845, 1847.

41 Sear (2000), no 1799.

42 Sear (2000), no. 1957.

43 For example Nerva *RIC* 80; Sear 3042.

44 Sellars (2013), 42; Butcher and Ponting (2015), 24 ff.

45 Sear (2002), nos. 6297–6304.

46 Sear (2005), nos. 10252–75 (Gallienus); 11139–52 (Victorinus).

47 Sear (2005), nos. 12279 and 12386.

48 For example, Sear (2000), no. 2325, struck in 71.

49 For example, Sear (2000), no. 2765, struck in 88–9.

50 Sear (2002), nos. 5826, 6443 (among others).

## 5: A SOLDIER'S LIFE

1 Vegetius 1.21.

2 Josephus, *Jewish War* 3.89–92.

3 Johnson (1983), 241.

4 Polybius 6.26.10.

5 Polybius 6.31.10.

6 Caesar, *Gallic Wars* 8.8–10.

7 Caesar, *Gallic Wars* 5.40. See Lepper and Frere (1988), 262–3, for other details of camps.

8 Josephus, *Jewish War* 3.76–88.

9 *De Munitionibus Castrorum* ('On the fortifications of a military camp').

10 Johnson (1983), 27–30, provides an excellent summary of Polybius and Pseudo-Hyginus.

11 Suetonius, *Domitian* 7.3. The occasion was the rebellion of Lucius Antonius Saturninus.

12 Wilson (2002), 298; for the fortress excavation see Pitts and St Joseph (1985), and also Maxwell (1989, especially 100–105). Mint condition coins of Domitian struck in 86/7 in demolition layers date the fortress's abandonment.

13 *ILS* 9184, *CIL* 13.6592.

14 *RIB* 605. The unit's title demonstrates that the inscription was made

during the reign of Postumus. His name had been deleted from the beginning of the inscription after his regime collapsed.

15  *RIB* 1912.

16  Fields (2006) provides a convenient overview of the Saxon Shore forts.

17  Accessed (2019) at https://www.sumoservices.com/brancaster-roman-fort-case-study

18  But see Kennedy and Bewley (2004), 171 for the late fort at Qar Bshir in Jordan and an inscription recording its construction under the Tetrarchy 293–305. For Brancaster's unit, see Rivet and Smith (1979), 219 (Notitia Dignitatum 28.16).

19  *RIB* 2091.

20  *CIL* 13.8082.

21  *CIL* 13.7946.

22  *RIB* 1024. The altar is lost.

23  *ILS* 474.

24  *AE* 2014.1137. Found at Municipium Montanensium (Montana), north-west Bulgaria.

25  *CIL* 8.261. The inscription has required some restoration of the text. [Geometr]ae is only probable rather than a certainty.

26  *AE* 1979.89.

27  For Birdoswald see Biggins and Taylor (2004); for Zugmantel see Hanel (2011), 411, and Hermann (1983). Similar work at the Saxon Shore Fort of Brancaster has produced extensive traces of a vicus. See https://www.sumoservices.com/brancaster-roman-fort-case-study

28  See Edwell (2008), 119–120, fig. 4.17, for a plan of the layout of known camp buildings at Dura-Europos.

29  Campbell (1994), no. 183; British Museum Papyrus 2851; Ott (1995), 144. The document had perhaps been taken to Egypt in the archives of a veteran. Its presence in Egypt is otherwise hard to explain.

30  Pliny the Younger, *Letters* 10.27, 28.

31  Pliny the Younger, *Letters* 10.21, 22.

32  Campbell (1994), no. 182.

33  Tacitus, *Histories* 1.46.

34  *CJ* 2.51.3.

35  Papyrus Genève Lat. 1.4.B. See also Goldsworthy (2003), 91.

36  *ILS* 5863, and *AE* 1973.473. Both texts supplied by Campbell (1994), nos. 199–200.

## 6: LIVING OFF THE LAND

1  Livy 34.9.12–13.

2  Tacitus, *Annals* 1.35.1; Hanson (1978) discusses the type of timber, size of posts, supply and logistics involving the Roman army in Britain.

3  Pitts and St Joseph (1985).

4  *Britannia* 17 (1986), 450, no. 84; see also Brewer (1987), 16. For 150 ha, see Boon (1987), 18.

5  *Britannia* 27 (1996), 449, no. 14.

6  *CIL* xiii.11781. The exact reading of the text is uncertain but wood of some sort is a certainty.

7  Rook (1978); Reece (1997) makes the 5,600 ha estimate but points out that coal was available from the seams in the nearby Forest of Dean.

8  *RIB* 1008.

9  de la Bédoyère (2010), 62–3, plate 12.

10  Pollard (2000), 242; *CIL* 3.14396f. Known as Zeugma in antiquity.

11  *AE* 1984.898,

12  *CIL* 3.25.

13  Alston (1995), 80.

14  Murphy et al (2018).

15  Boon (1984), 51. This publication covers Legio II Augusta's tile production at Caerleon in considerable detail.

16  *RIB* 2491.96.

17  Josephus, *Jewish War* 3.120; and see Dixon and Southern (1992), 27 on whether this was always the case. In the fourth or early fifth century Vegetius 2.6 said each legion had 726 cavalry (see Chapter 1).

18  Holder (2003), 120, provides figures that suggest a ratio of cavalry to infantry of 1.93:1 under Hadrian, involving about 74,624 cavalry.

19  Kehne (2011), 325, who also adds that the army in Syria required *c.* 22.4 tons of wheat per day pus 6.9 tons of barley for the horses. Erdkamp (2011), 102, in Erdkamp (ed.) (2011), provides an estimate of 4,000 horses and 3,500 pack animals for an army of 40,000 though it is not clear what proportion of this army is infantry or cavalry.

20  Plutarch, *Sulla* 12.2.

21  *AE* 1968.31.

22  Kovács and Szabó (2009), 2.604.

23  *SHA* (Probus) 8.1–7.

24  King (2011), 139 –48.

25  *RIB* 1463.

26  *RIB* 1049.

27  *RIB* 1060.

28  *SHA* (Hadrian) 10.2.

29  Onasander, *The General* 6.13.

30  Livy 23.19.13–17; Pliny the Elder, *Natural History* 8.222; Valerius Maximus, *MDS* 7.6.3.

31  Livy 25.20.1–2.

32  Livy 29.25.6–7, and 25.2.

33  Livy 34.9.12–13.

34  Appian, *Macedonian Affairs* 13 (Loeb vol. II, p. 41).

35  Sallust, *Jugurthine War* 44–45.

36  Appian, *Civil Wars* 1.109.

37  Onasander, *The General* 10.7–8; Dio 39.40.1–2 (Loeb vol. III, p. 367).

38  Tacitus, *Annals* 14.24.1–2.

39  Tacitus, *Histories* 4.15.

40  Polybius 6.31; *CIL* 8.18224, *ILS* 2415.

41  *Tab. Vindol.* III.596.

42  *Tab. Vindol.* III.581.

43  *Tab. Vindol.* III.301. The word for expenses (*souxtum*) is uncertain as well as being unrecognizable in the form written. It may refer to a commodity, using a local native word in garbled Latin spelling that is unknown to us.

44  *Tab. Vindol.* II.343.

45  *Tab. Vindol.* II.190. It is also worth mentioning that a document found in Frisia was originally thought to be a unique record of the sale of a cow to a military food supplier, and witnessed by two centurions (see Lewis and Reinhold, 1990, p. 516). That reading has been completely overturned by Bowman, A.K., Tomlin, R.S.O., and Worp, K. (2009), '*Emptio bovis Frisica*: the Frisian ox sale reconsidered', *Journal of Roman Studies* 99, 156–9, demonstrating it to be a record of a debt owed to a slave. There is no mention of a bovide of any sort.

46  *RIB* 2445.6.

47  *Britannia* 42 (2011), 452, no. 17.

48  *Britannia* 42 (2011), 454, no. 20.

49  See de la Bédoyère (1991), 212–13, and 221 for reconstruction drawings and plans.

50  *RIB* 2492.7.

51  *Britannia* 46 (2015), 395, no. 14.

52  Frontinus, *Stratagems* 2.11.7.

53  *SHA* (Avidius Cassius) 4.1.

54  Herodian 7.3.6.

55  *CIL* 3.12336; *IGRR* 1.674. Accessed at: http://judaism-and-rome.cnrs.fr/gordian-iii-and-imperial-petition-skaptopara

56  Tacitus, *Germania* 37.

## 7: IGNOMINY AND DEFEAT

1  Suetonius, *Augustus* 23.2. The translation usually offered is 'Varus, give me back my legions'. There is however no word for 'my' or 'me' in the Latin text, though either could be understood. Such translations also say Augustus bashed his head against doors. The word in the Latin original is clearly that for corridors.

2  Valerius Maximus, *MDS* 7.2.2.

3   Livy 3.83.9; Livy 22.5.1–2.

4   Livy 22.4.1–7, 5.1–2; Polybius 3.84.

5   Polybius 3.84.5–6; Livy 22.6.3–4. See also Salimbeti and D'Amato (2014),
    44–5. The Insubrians were a Gaulish tribe that had been forced to
    become allies of Rome, but rebelled when the Second Punic War broke
    out.

6   Polybius 3.84.13–14.

7   Livy 22.7.1; Polybius 3.85.

8   Livy 22.7.14, 8.1.

9   Polybius 3.106.

10  Polybius 3.107.

11  Polybius 3.110.1–3.

12  Livy 22.44.4.

13  Polybius 3.111.

14  Polybius 3.113–14; Livy 22.45.5.

15  Valerius Maximus, *MDS* 7.4.ext. 2.

16  Polybius 3.115–16: Livy 22.47.1ff, 49.9–13.

17  Polybius 3.116–17; Livy 22.49.14–16, 54.7, 56.4. Other ancient sources
    supplied similarly large numbers. Polybius wrote closest to the time.

18  Polybius 3.118.

19  Valerius Maximus, *MDS* 6.4.1a.

20  Valerius Maximus, *MDS* 1.1.16.

21  Livy 25.5.10.

22  Livy 25.5.10–7.4.

23  Livy 29.24.11–13.

24  Appian, *Civil Wars* 1.116.1 tells us Spartacus was originally a soldier in the
    Roman army. Plutarch, *Crassus* 8.1 does not mention this but tells us more
    about the gladiatorial school.

25  Appian, *Civil Wars* 1.117.

26  Appian, *Civil* Wars 1.118; Plutarch, *Crassus* 10.2– 3.

27  Appian, *Civil Wars* 4.116.1–121.1; Pompey, *Crassus* 8 –12.

28  Valerius Maximus, *MDS* 1.6.11, Plutarch, *Crassus* 23.1–2.

29  Plutarch, *Crassus* 23–32.

30  Dio 40.27.3–4 (Loeb vol. III, p. 447).

31  Suetonius, *Augustus* 21.3, *Tiberius* 9.1; Ovid, *Fasti* 5.584–98; Velleius
    Paterculus 2.91.1.

32  Macrobius 3.14.15.

33  Strabo 3.4.17, 18 (see Loeb vol. 2, pp. 111, 115).

34  Dio 54.11.3–6 (Loeb vol. VI, p. 311) covers the story; for the identity of
    Legio I see Pollard and Berry (2015), 61. It then disgraced itself in the
    mutiny of AD 14, see Tacitus, *Annals* 1.42.3.

35  Dio 56.18.1–2 (Loeb vol. VII, p. 39).

36  Beeson (2018) (ii), 38.

37  Dio 56.18.4–19.5 (Loeb vol. VII, p. 41–3); Velleius Paterculus 2.118.4.

38  Velleius Paterculus 2.119.2.

39  Dio 56.20.1–2 (Loeb vol. VII, p. 43).

40  Frontinus, *Stratagems* 3.15.4. Cutting off prisoners' or hostages' hands was a long-established practice. Scipio Aemilianus ordered the hands of 400 young men of the city of Lutia in Spain to be cut off in 133 BC. Appian, *Spanish Wars* 15.94 (Loeb vol. I, p. 289).

41  Velleius Paterculus 2.119.3–5.

42  Tacitus, *Annals* 1.61–2.

43  Tacitus, *Annals* 1.60.3; 2.25.1–2; Dio 60.8.7 (Loeb vol. VII, p. 389).

44  *CIL* 13.8648. On this inscription the legion's number is given as XIIX, an alternative version of XVIIII.

45  Tacitus, *Annals* 2.52.1–2.

46  Tacitus, *Annals* 3.20.1–2.

47  Tacitus, *Annals* 3.21.1–2. The line Tacitus used about the revival of an ancient practice was also used by Plutarch when describing how Crassus decimated his army in the Spartacus War, at *Crassus* 10.2–3.

48  Tacitus. *Annals* 15.10–15.

49  Tacitus. *Annals* 15.26.1.

50  Josephus, *Jewish War* 2.487–93.

51  Josephus, *Jewish War* 2.494–98.

52  Josephus, *Jewish War* 2.499–510.

53  Josephus, *Jewish War* 2.511–16.

54  Josephus, *Jewish War* 2.516–26.

55  Josephus, *Jewish War* 2.527–45.

56  Josephus, *Jewish War* 2.546–55.

57  Josephus, *Jewish War* 2.555, 7.18; Pollard and Berry (2015), 165.

58  Josephus, *Jewish War* 3.132–4.

59  Josephus, *Jewish War* 6.359–62.

60  Josephus, *Jewish War* 6.186–9.

61  Dio 77.13.1–4 (Loeb vol. IX, pp. 265–7); Herodian 3.14.5–10.

62  The Homer quote is *Iliad* 6.57–59, cited by Dio 77.15.1 (Loeb vol. IX, p. 271). Severus changed the third line, which in Homer refers to the extermination of the whole population.

63  *ILS* 2089, CIL 6.2464.

64  Tacitus, *Annals* 1.16.2.

65  Tacitus, *Agricola* 26.1.

66  *RIB* 665.

67  N. Lewis, *The Documents from the Bar Kokhba Period in the Cave of Letters. Greek Papyri* (1991) nos. 16 (giving his *praenomen* as Titus), 23, 25–6; Birley (1981), 238.

68  *CIL* 3.87 (Petra).

## 8: I CAME, I SAW, I CONQUERED

1   Tacitus, *Annals* 14.37.
2   Polybius 15.15.
3   Josephus, *Jewish War* 3.99, 107, 115ff.
4   Hope (2003), 81.
5   *RIB* 1051.
6   Livy 30.6.8–9.
7   Polybius 14.5.
8   Livy 30.10.20.
9   Polybius 15.2–3.
10  Polybius 15.6–9.
11  Polybius 15.12.
12  Polybius 15.13.
13  Livy 30.34.12.
14  Polybius 15.14.
15  Polybius 15.16.
16  Polybius 15.18–19; Livy 30.42–3.
17  Livy 30.45.1–2, 6–7.
18  Appian, *Syrian Wars* 9–10 (Loeb vol. II, pp. 119–21).
19  Sallust, *Jugurthine War* 92–4; Frontinus, *Stratagems* 3.9.3.
20  Caesar, *Gallic Wars* 4.22–25.
21  Frontinus, *Stratagems* 2.8.4.
22  Appian, *Civil Wars* 4.110–13; Plutarch, *Brutus* 43.7. Valerius Maximus, *MDS* 9.9.2 says Cassius killed himself, contradicting another reference of his at 6.8.4 which echoes Plutarch.
23  Appian, *Civil Wars* 4.115–16.
24  Valerius Maximus, *MDS* 6.4.5.
25  Appian, *Civil Wars* 4.125–32.
26  Tacitus, *Annals* 13.35.3–4.
27  Tacitus, *Annals* 13.36.
28  Tacitus, *Annals* 13.38–41.3
29  Tacitus, *Annals* 14.23–26.
30  These are the legion's names and titles as they were at the time the Boudican Revolt started.
31  Tacitus, *Annals* 14.29.
32  Tacitus, *Annals* 14.30.
33  Tacitus, *Annals* 14.31; for the loans see Dio 62.2.1 (Loeb vol. VIII, p. 83).
34  Tacitus, *Annals* 14.32.
35  Tacitus, *Annals* 14.32.3.
36  Dio 62.2.3 (Loeb vol. VIII, p. 85).
37  Tacitus, *Annals* 14.33.
38  Tacitus, *Annals* 14.34–35.

39   Dio 62.2.2 (Loeb vol. VIII, p. 85).

40   Tacitus, *Annals* 14.36–37. There is a discrepancy here with his earlier reference to the infantry being wiped out. If so, Legio VIIII would have needed more than twice as many reinforcements.

41   Tacitus, *Annals* 14.38.

42   Tacitus, *Histories* 2.11.

43   Tacitus, *Histories* 1.79.

44   Tacitus, *Histories* 2.24. The legions were I Adiutrix and XIII Gemina.

45   Dio 64.10 (Loeb vol. VIII, p. 213).

46   Frontinus, *Stratagems* 3.9.2.

47   Frontinus, *Stratagems* 3.7.3.

48   Caesar, *Gallic Wars* 4.17–18.

49   Caesar, *Gallic Wars* 4.19.

50   *CIL* 3.13580.

51   Dio 71.3 (Loeb vol. IX, p. 7).

52   Vegetius 1.10.

53   'Caesar', *Spanish War* 32. The author is unknown but was not Caesar himself. The story is also at Valerius Maximus, *MDS* 7.6.5.

54   Vitruvius 10.10–12 (artillery), 13–15 (siege machines).

55   Dio 39.4.1–4 (Loeb vol. III, pp. 313–15).

56   Livy 42.34.10.

57   Valerius Maximus, *MDS* 8.14.5.

58   Campbell (1994), no. 90, *ILS* 2648, *CIL* 11.395, dated to AD 66.

59   *ILS* 9200.

60   Josephus, *Jewish War* 6.161–3.

61   Josephus, *Jewish War* 6.169–76.

62   Josephus, *Jewish War* 6.56–7.

63   Zonaras 7.21.

64   Rich and Shipley (1993), 50.

65   Livy 39.6–7.

66   Plutarch, *Life of Aemilius Paullus* 32–34.

67   Sear (2000), no. 156.

68   *CIL* 6.1303. For this war see Scullard (1982), 40, and particularly n. 39 discussing the problems in the sources.

69   Valerius Maximus, *MDS* 6.9.4.

70   Richardson (1975), 62.

71   Suetonius, *Caesar* 54.3.

72   Plutarch, *Caesar* 28.8–10; Dio 40.41.2–3 (Loeb vol. III, p. 469).

73   Suetonius, *Caesar* 37.1–2.

74   Suetonius, *Caesar* 49.4.

75   Josephus, *Jewish War* 7.123–31.

76   Josephus, *Jewish War* 7.132–8.

77  Josephus, *Jewish War* 7.139–47.
78  Josephus, *Jewish War* 7.148–57.
79  Martial, *Epigrams* 1.4.
80  Dio 59.25.1–5 (Loeb vol. VII, p. 339); Suetonius, *Caligula* 46. For Augustus, see Dio 53.22.5 (Loeb vol. VI, p. 253).
81  Dio 60.19.1–5 (Loeb vol. VII, pp. 415–17).
82  Ammianus Marcellinus 19.6.
83  Ammianus Marcellinus 24.6.1–16. Julian subsequently pulled back. He was wounded in a battle en route, and died.

## 9:  LIVING BY THE SWORD

1  Josephus, *Jewish War* 6.400ff.
2  Tacitus, *Histories* 1.39.
3  *SHA* (Probus) 14.2.
4  Polybius 1.32.
5  Polybius, *Histories* 10.15.
6  Valerius Maximus, *MDS* 8.7. ext. 7; Livy 25.31.9–11.
7  Livy 31.34.4–5.
8  Bishop (2017), 61 (the body is illustrated), and also (2016), 36.
9  Valerius Maximus, *MDS* 9.2.1.
10  Appan, *Civil Wars* 1.109.
11  Suetonius, *Tiberius* 37.2–3.
12  *CIL* 13.7255 (Mainz).
13  *RIB* 201 (Colchester).
14  *CIL* 13.7029.
15  Catalogued at the Hunterian Museum Objects Database GLAHM F1969.22 (Glasgow); *RIB* 3507.
16  Caesar, *Gallic Wars* 7.25–28.
17  Caesar, *Gallic Wars* 8.43–44.
18  The story of the Varian disaster can be found in Velleius Paterculus 2.119–120, and in Dio 56.20–22 (Loeb vol. VII, pp. 43–7.
19  Tacitus, *Annals* 1.49–51.
20  Tacitus, *Annals* 1.56.
21  Caesar, *Gallic Wars* 5.44. These men later appeared as the two main characters in the BBC/HBO series *Rome*, albeit largely fictionalized, but the series opens with the action described here.
22  Suetonius, *Augustus* 14.
23  Tacitus, *Histories* 2.66.
24  Petronius, *Satyricon* 82.
25  Juvenal *Satires* 16.7–25.
26  Tacitus, *Histories* 2.42.
27  Tacitus, *Histories* 2.70. This is a good example of a story where the

description is rhetorical in tone; as such it reinforced the moral lesson about the destructive futility of civil war.

28 Tacitus, *Histories* 3.15.

29 Tacitus, *Histories* 3.27.

30 Dio 64.15.1–2 (Loeb vol. VIII, p. 245).

31 Tacitus, *Histories* 3.23.

32 Tacitus, *Histories* 3.25. Legio VII was formed by Galba in Spain to support his campaign to topple Nero. Galba was the first emperor after Nero but was himself toppled and killed in 69, to be succeeded by Otho, then Vitellius and afterwards Vespasian (69–79).

33 Appian 4.42.1.

34 Appian 5.68.

35 Josephus, *Jewish War* 2.177.

36 Josephus, *Jewish War* 6.400ff.

37 Josephus, *Jewish War* 7.9.

38 Dio 62.7.2–3 (Loeb vol. VIII, p. 95).

39 Tacitus, *Histories* 1.63.

40 Tacitus, *Histories* 1.66–68.

41 Tacitus, *Histories* 2.12.

42 Livy 21.53.2.

43 Appian, *Civil Wars* 2.77–82. See Von Lommel (2013) for an interesting discussion about the impact of war on Roman soldiers.

44 Dio 76.7.2–3 (Loeb vol. IX, p. 211).

45 Dio 76.9.4–11.2 (Loeb vol. IX, pp. 219–23).

46 *RIB* 755.

47 Smallwood (1967), no. 279; Campbell (1994), no. 33.

48 Appian, *Civil Wars* 4.19–20. The sum of money was equivalent to roughly the same number of Roman denarii.

49 Suetonius, *Tiberius* 22; Tacitus, *Annals* 1.6.1–2.

50 Suetonius, *Tiberius* 53.1–2.

51 Suetonius, *Tiberius* 60.1.

52 Suetonius, *Caligula* 32.1.

53 *SHA* (Elagabalus) 16.2–4. Ulpian was later recalled and made praetorian prefect by Severus Alexander.

54 The most detailed account of the assassination of Caligula and his wife and daughter is at Josephus, *Jewish Antiquities* 19.84–201.

55 Plutarch, *Marius* 44.4; Valerius Maximus, *MDS* 8.9.2.

## 10: QUINQUEREMES AND TRIREMES

1 Plutarch, *Antony* 68.1.

2 Polybius 1.20–21.

3 Polybius 1.22. For the recovery of these rams see Natanson (2011).

4   Livy 22.57.7.

5   Livy 29.25.6–7.

6   Appian, *Syrian Wars* 22 (Loeb vol. II, p. 145).

7   Plutarch, *Pompey* 24–25.1; Appian, *Mithridatic War* 92 (Loeb vol. II, pp. 413–15).

8   Plutarch, *Pompey* 25.1–29.

9   Dio 39.40.2–43.5 (Loeb vol. III, pp. 367–77).

10  Appian *Civil Wars* 2.102.

11  Josephus, *Jewish War* 3.522–31.

12  Plutarch, *Antony* 64–68.

13  Suetonius, *Augustus* 17.2 –3.

14  Suetonius, *Augustus* 8.1, 16.1.

15  Murray (1989), 34, 56, 86; Plutarch, *Antony* 68.1; Suetonius, *Augustus* 18.2.

16  *AE* 1925.93. He was a friend of Octavian's, see Appian, *Civil Wars* 5.102.

17  Pliny the Younger, *Letters* 6.16.13–20.

18  *RIB* 66.

19  *SHA* (Caracalla) 5.8.

20  *CIL* 12.2412, *ILS* 2909.

21  Saddington (2011), 201–2, citing an inscription from Cos (*IGRR* 1.843).

22  *CIL* 3.43 (Memnon 38).

23  *AE* 2015.1344.

24  *ILS* 2816, *CIL* 6.32775.

25  Philp (1981), and https://historicengland.org.uk/listing/the-list/list-entry/1012478.

26  Hingley and Unwin (2018), 131.

27  Fields (2006), 43.

28  Found at Castellamare di Stabia, *ILS* 1986, *CIL* 16.1.

29  Wesch-Klein (2011) in Erdkamp (2011), 439. A fleet diploma of 118–19 cites 26 years (*CIL* 16.66). Diplomas postdating 206 give 28 years. See Roxan (1994), no. 189, note 6.

30  Tacitus, *Histories* 4.16.

31  Tacitus, *Histories* 4.79.

32  *RIB* 1340.

33  *RIB* 3036. The expansion of the name is uncertain.

34  *RIB* 2481.1–119.

35  Brodribb and Cleere (1988), 266.

36  Brodribb (1979), 149.

37  *CIL* 13.7719.

38  *SHA* (Commodus) 15.6. The reference implies this was a routine and established job.

39  *CIL* 6.3094. Webster (1996), 157, says that fleet soldiers were in Rome and based at the Castra Praetoria but provides no examples or evidence to substantiate the statement. But see Josephus, *Jewish Antiquities* 19.253.

40   Suetonius,*Vespasian* 8.3.
41   Campbell (1994), discussion on p. 201.
42   *AE* 1956.249.
43   Tomlin (2018), 458 (f); and Tomlin and Pearce (2018). The diploma is the first fleet veteran's to be found in Britain.
44   *CIL* 10.3593.
45   *RIB* 258.
46   *RIB* 653.
47   Tacitus, *Annals* 14.3.3.
48   Tacitus, *Annals* 14.8.2–5.
49   Tacitus, *Histories* 3.57.
50   Adams (2008), 54.
51   *CIL* 6.3123.
52   *RIB* 2432.1.
53   Sear (2011), no. 13636.
54   *CIL* 6.3145. He was a member of the Thracian Bessi tribe.
55   Saddington (2011, 214) states the raft as an attested fact but it is qute clear the inscription's reading is far from certain (*AE* 1964, 103).
56   Josephus, *Jewish War* 3.522–31. This may be the event commemorated by Vespasian's VICTORIA NAVALIS coins.
57   Dio 78.18.1 (Loeb vol. IX, pp. 325–7).
58   Frontinus, *Stratagems* 2.7.14. The story goes unmentioned in Plutarch and Livy.
59   Tacitus, *Annals* 1.70.
60   The dramatic story of the storm that hit Germanicus' fleet is at Tacitus, *Annals* 2.22–24.
61   Valerius Maximus, *MDS* 1.8 ext. 19 (nb the Loeb note is in error here, giving the date as 266), reporting the story from Livy whose main text at this point is lost, surviving only in the summary at *Periochiae* 18.

## 11: MUTINEERS AND REBELS

1   Tacitus, *Histories* 1.80.
2   Valerius Maximus, *MDS* 9.7. *mil. Rom.* 1, 2 (Loeb vol. II, p. 347); Plutarch, *Marius* 8 –9. This all took place in Italy.
3   Livy, *Periochae* 75; Plutarch, *Sulla* 6.9; Valerius Maximus, *MDS* 9.8.3.
4   Dio 41.96.1–2, 28.1, 31.1–3 (Loeb vol. IV, pp. 47, 49, 55).
5   Tacitus, *Annals* 1.16.2, and 19.4-5.
6   On collecting the cohorts from Aquileia, see Keppie (1998), 113. For the praetorian cavalry and German troops see Tacitus, *Annals* 1.24.2, presumably what Dio (55.24.6–7; Loeb vol. VI, p. 457) called the Batavian cavalry. For the mutiny, Tacitus, *Annals* 1.16.1–30.5 *passim*.
7   Tacitus, *Histories* 1.25.

8   Tacitus, *Histories* 1.42–43.

9   Tacitus, *Histories* 1.44–46.

10  Tacitus, *Histories* 1.56.

11  Tacitus, *Histories* 1.80–82, 4.13.

12  Tacitus, *Germania* 29.2.

13  Tacitus, *Histories* 4.12.

14  Tacitus, *Histories* 4.13.

15  Tacitus, *Histories* 4.14

16  Tacitus, *Agricola* 28.

17  Dio 66.20.1–3 (Loeb vol. VIII, pp. 301–3).

18  Suetonius, *Domitian* 7.3; Dio 67.11.1–3 (Loeb vol. VIII, p. 341).

19  Herodian 1.10. See also *SHA* (Pescennius Niger) 3.4 for Niger's role in pursuing Maternus in Gaul.

20  Tacitus, *Histories* 1.51.

21  *SHA* (Pertinax) 15.7, Dio 74.1.2 (Loeb vol. IX, p. 123), though the sources differ over whether he paid all of it. The former claims that only 50 per cent was ever produced. Dio 74.8.3 (Loeb vol. IX, p. 137) compares the amounts of the donative in 161 and 193.

22  Herodian 2.2.5.

23  Dio 74.1.3 (Loeb vol. IX, p. 123); *SHA* (Pertinax) 6.6 states that Pertinax confirmed all these concessionary practices.

24  Dio 74.5.4, 8.1–9 (Loeb vol. IX, pp. 133, 137).

25  Dio 74.9.1–10.3 (Loeb vol. IX, pp. 139–41).

26  Herodian 2.3.4, 2.4.4, Dio 74.1.2, 5.4, 8.4 (Loeb vol. IX, pp. 123, 133, 137), Herodian 2.5.1–9.

27  Herodian 2.6.4.

28  Dio 63.9.1 (Loeb vol. VIII, p. 209) where Dio describes the loathing felt for Otho because he had shown the Empire was for sale, and that the praetorians could make or break an emperor.

29  *SHA* (Didius Julianus) 2.5–6.

30  Dio 74.12.1-4 (Loeb vol. IX, p. 145).

31  *SHA* (Didius Julianus) 3.1. This source also states that Julianus upped his promise to 30,000 sestertii which conflicts with Herodian's more convincing claim that he failed to pay anything due to a lack of resources; for the coinage aimed at the soldiers see for example Sear (2002), no. 6072; Herodian 2.7.1–2; Dio 74.13.1 (Loeb vol. IX, p. 147). *SHA* (Didius Julianus) 7.1–2.

32  *SHA* (Didius Julianus) 3.7. Dio 74.13.4–5 (Loeb vol. IX, p. 149); for Severus' appointment as consul, see Dio 73.12.4 (Loeb vol. IX, p. 97).

33  Dio 74.14.3 (Loeb vol. IX, p. 151); 74.15.1–2 (Loeb vol. IX, p. 153).

34  Dio 74.16.2 (Loeb vol. IX, p. 155); Herodian 2.2.9. The problem was the same as in the Civil War of 68–9; see Tacitus, *Histories* 2.18.

35  Dio 74.17.1–5 (Loeb vol. IX, p. 159). Herodian 2.12.4.

36  *SHA* (Severus) 6.11. Herodian 2.13.2–4; Dio 75.1.1 (Loeb vol. IX, p. 161).
37  Herz (2011), 318.
38  Herodian 2.13.5–12; Dio 75.1.2 (Loeb vol. IX, p. 161).
39  *CIL* 6.2578 and 210; Campbell (1994), nos. 65–66.
40  *SHA* (Caracalla) 7.1.
41  Nixon and Rodgers (1994) discuss this topic in extensive detail.
42  de la Bédoyère (1998).
43  *RIB* 2291.
44  Sear (2011), nos. 13767–74.
45  The various items of literary and numismatic evidence for the Carausian story are most easily found in Casey (1994), though this was published before the discovery of the Fourth Eclogue references. For the latter, see de la Bédoyère (1998). The imperial panegyrics are in Nixon and Rodgers (1994).

## 12: PEACETIME DUTIES

1  Pliny the Younger writing to the emperor Trajan *c.* 110, *Letters* 10.77.
2  *SHA* (Probus) 9.3–4.
3  From *They Shall Not Grow Old* (2018), dir. Peter Jackson, at 48:12.
4  Dio 56.19.1 (Loeb vol. VII, p. 41).
5  Tacitus, *Histories* 1.4; *Annals* 1.77.
6  Tacitus, *Annals* 3.72.2–4.
7  Dio 59.4.1 (Loeb vol. VII, p. 287).
8  Tacitus, *Annals* 4.72.1–4.
9  Suetonius, *Caligula* 40.
10  Josephus, *Jewish Antiquities* 19.28.
11  Herodian 5.6.10.
12  Tacitus, *Annals* 13.48.
13  *CIL* 4.3081; Tacitus, *Annals* 14.17.
14  Pliny the Younger, *Letters* 10.19, 20.
15  Tacitus, *Annals* 4.15.2–3. Asia was a senatorial province and therefore was overseen by the Senate. Pliny's governorship was a special appointment by Trajan to deal with problems arising from corrupt previous governors.
16  Pliny the Younger, *Letters* 10.74.
17  Bassaeus Rufus came from modest origins and reached the praetorian prefecture via the prefecture of Egypt, but Macrinius Vindex's earlier career is unknown. Rossignol (2007), 142. See also Dio 72.5.3 (Loeb vol. IX, p. 19) for Bassaeus Rufus' modest origins.
18  Suetonius, *Augustus* 32; *ILS* 2051, *CIL* 3.6085.
19  *CIL* 9.2438, from Saepinum (Sepino), cited in Lewis and Reinhold (1990), 186.
20  *RIB* 152.

21   Egypt was divided into 26 such nomes (Strabo 17.1.3); document at *P. Mich.* Inv. 2979. Accessed online at https://quod.lib.umich.edu/a/apis/x-1724

22   *P. Mich.* Inv. 6127. Accessed online at https://quod.lib.umich.edu/a/apis/x-2641

23   *P. Mich.* Inv. 2798. Accessed online at https://quod.lib.umich.edu/a/apis/x-1634

24   Campbell (1984), 253.

25   Lewis and Reinhold (1990), p. 401; Lewis (1954).

26   Lewis and Reinhold (1990), p. 402.

27   Mark 15.39, 44, 45; Luke 23.47

28   Petronius, *Satyricon* 111.

29   Acts 21.32, 22.25, 26; see also 2.23.

30   Acts 27, 28 *passim*.

31   Pliny the Younger, *Letters* 10.77.

32   Pliny the Younger, *Letters* 10.78.

33   Campbell (1994), no. 188.

34   *CIL* 3.6628.

35   *ILS* 9142, *AE* 1911, 121.

36   Alston (1995), 81.

37   See Alston (1995), 199, for a plan.

38   *O. Florida* 2. Ostrakon text accessed at http://papyri.info/ddbdp/o.florida;;2. Herennius' rank is given as a δεκάδαρχης, 'commander of ten', for which the Latin equivalent is decurion, but this appears to be understood rather than being in the text. See also Alston (1998), 81–2.

39   *ILS* 9073, Campbell (1994), 118, no. 187.

40   Plutarch, *Marius* 15.1–3.

41   Tacitus, *Annals* 11.20.3–4.

42   Polybius 34.9.8–11 (Strabo 3.5.7 C172).

43   Livy 45.18.7. Strabo 3.2.10 suggests that the mines at Cartagena and others had ceased to be state property by his time and had passed to 'private ownership'.

44   Strabo 3.2.8.

45   Strabo 3.2.11.

46   *RIB* 2401.1.

47   *RIB* 2404.2 (assuming the reading is correct).

48   Jones (1976) discusses military involvement in the region.

49   Tallet and Zivie-Coche (2012), 449.

50   Tacitus, *Annals* 13.53. 2–3.

51   Tacitus, *Annals* 11.20. 1–2.

52   Sherk (1988) no. 85a.

53   For Caesarea, see Olami and Ringel (1975), 148–50. For Leptis Magna's aqueduct, see *AE* 1973.573.

54　Campbell (1994), 126, no. 206. The ancient and modern names of the town are the same.

55　Strabo, *Geography* 5.3.5; Suetonius, *Claudius* 18, 20; Pliny, *Natural History* 9.15.

56　Tacitus, *Annals* 12.56.1-4.

57　Tacitus, *Annals* 14.15, Dio 62.20.3 (Loeb vol. VIII, p. 79).

58　Tacitus, *Annals* 16.5.2.

59　Tacitus, *Annals* 14.1ff.

60　Josephus, *Jewish War* 3.540.

61　Suetonius, *Nero* 19.2.

62　*SHA* (Probus) 19.3–8.

63　*SHA* (Probus) 21.2–4.

64　Pliny the Elder, *Natural History* 6.181.

65　Pliny the Elder, *Natural History* 5.35–6.

66　Dio 58.9.5–6 (Loeb vol. VII, p. 211).

67　Dio 55.8.7 (Loeb vol. VI, p. 401).

68　Suetonius, *Augustus* 25.2, Strabo 5.3.7, Dio 55.26.4 (Loeb vol. VI, p. 463) tells us the Vigiles were organized into seven cohorts, but not their size, or the total number of men. See Rainbird (1976), 216ff and (1986) *passim*, but especially 150–1 for a discussion of the evidence for the size of the cohorts, derived from *CIL* 5.1057.

69　Herodian 1.14.1–6.

70　Claridge (2010), 406–7.

71　Dio 58.12.2 (Loeb vol. VI, p. 219); for Silvanus, see Sherk (1988) no. 49A.

72　*ILS* 9180, *CIL* 6.32709a.

73　Suetonius, *Claudius* 18.2, 25.2.

74　*ILS* 9494, Campbell (1994), no. 61.

75　Dio 62.17.1–18.1 (Loeb vol. VIII, p. 115).

76　Tacitus, *Annals* 13.35.1–2.

77　Tacitus, *Histories* 2.18–19.

78　*SHA* (Hadrian) 10–11.2.

79　*RIB* 1820. There are numerous other examples. See Chapter 13 for how the Wall's real name was discovered.

80　Onasander, *The General* 9.3.

81　Fronto, *Letters* 'To Lucius Verus', 19, Loeb vol. II, p. 149–51, and 'Preamble to History' to Lucius Verus 11, ibid. p. 209; Laelianus' career is listed on an inscription: *CIL* 6,1497.

## 13: LEISURE AND LEAVE

1　*RIB* 1041. The verb used was *praedare*, which is the origin of the English 'predate' or 'predation'. Here it means 'to make (or turn) [the boar] into booty', as if it was a prize of war.

2　Cicero, *Letters to his brother Quintus* 2.16.4 (Loeb edition Letter 20.4).

3 Cicero, *Letters to his brother Quintus* 3.1.13 (Loeb edition Letter 21.13). As it happens the author of this book has in his possession a volume of the poems of Catullus, taken to France on 5 June 1917 by a British officer, Captain Thomas Byrnand Trappes-Lomax of the Scots Guards (1895–1962), who no doubt found it a similar distraction. The author's father-in-law took the Oxford Greek text of Homer with him to sea while on the crew of HMS *Liverpool* in 1942 for the same reason.

4 Caesar, *Gallic Wars* 4.33.

5 Suetonius, *On Poets: Horace* lines 5–6.

6 Horace, *Odes* 2.7.9–12, 3.4.25–8.

7 Horaces, *Epistles* 2.2.47–50. He was befriended by Augustus' literary friend Maecenas.

8 Martial, *Epigrams* 9.45.

9 *ILS* 2926.

10 The Aquino inscription is *CIL* 10.5382, *ILS* 2926. Juvenal, *Satires* 2.161.

11 *SHA* (Clodius Albinus) 11.8.

12 Virgil, *Aeneid* 12.646.

13 *Aeneid* 10.830.

14 *SHA* (Carus, Carinus, Numerian) 13.3–5; Terence, *The Eunuch* 426.

15 *P. Mich.* 8.466 II 18, accessed at http://papyri.info/ddbdp/p.mich;8;466.

16 *Tab. Vindol.* II.233.

17 *RIB* 1041.

18 *RIB* 3219. The inscriptions usually start DEL . . . but refer to the Dalmatae.

19 Arrian, *Tactica* 34.

20 See Breeze and Bishop (2013) for the Crosby Garrett helmet; Kaminski and Sim (2019) provide a useful description and analysis of the Ribchester helmet.

21 *RIB* 323.

22 *RIB* 3149; *CIL* 13.6728. The legion's number on the Mainz inscription is incomplete: X[ . . . ].

23 Pollard (2000), 243; AE 1937,239. The Dura arena was 105 × 85 ft (32 × 26 m). Chester's by comparison was 335 × 299 ft (102 × 91 m). See also Baird (2018).

24 Tacitus, *Histories* 1.46.

25 *CJ* 2.13.9.

26 *CJ* 4.6.5.

27 *CJ* 3.37.2.

28 *CJ* 2.28.1.

29 *CJ* 3.42.1.

30 Quoted in Lewis (1983), 138.

31 *P. Wisc.* Inv. 2.70. Accessed online from the Advanced Papyrological Information System, University of Michigan.

32 *Tab. Vindol.* II.174, 175. For the Raetian soldiers' request, see Bowman *et alia* (2019), 242–43.

33   Suetonius, *Galba* 6.3.
34   Tacitus, *Annals* 15.9.2.
35   Bowman *et alia* (2019), 237–39.
36   Alston (2002), 97.
37   See Boon, G.C., 'A Strigil Inlaid with the Labours of Herakles', in Zienkiewicz (1986), 156–66.
38   Beeson (2018) (i), 14.
39   Sauer (1999), 53, in Goldsworthy and Haynes (1999).
40   For Titius Alexander, see *ILS* 2092, *CIL* 6.20, Campbell (1994), 104, no. 167. For Harmodius see *AE* 1952, 143.
41   *AE* 1937.180.
42   *AE* 1987.952.
43   *ILS* 9174, *CIL* 3.14537.
44   *ILS* 9169.
45   *ILS* 2437, *CIL* 8.2563.
46   *ILS* 9071.
47   *Tab. Vindol* II.181.
48   *CIL* 3.11215.
49   Celsus, *De Medicina* 7.5 *passim*.
50   Tacitus, *Histories* 2.93.
51   Campbell (1994), no. 182.
52   *RIB* 1209.
53   Dio 77.13.2 (Loeb vol. IX, p. 267); Galen, *Method of Medicine* 19.17–18.
54   *SHA* (Marcus Aurelius) 17.2.
55   *SHA* (Lucius Verus) 8.2. In the mid-1800s London's cholera epidemic was routinely blamed on the miasma emanating from the polluted Thames. It took the construction of sewers for the discovery that it was waterborne.
56   *RIB* 143–4.
57   *RIB* 156.
58   *RIB* 2034.
59   See Breeze (2012) for an extended description of the pans. See also Tomlin (2004), 344, note 43. The present author was responsible for the original identification of *rigore vali* as two words rather than a place name, which meant that it was possible to associate *Aeli* with *vali*.

## 14: WIVES AND LOVERS

1   *ILS* 2590 (Asia). Campbell (1994), no. 51.
2   Plautus, *Miles Gloriosus* 1.1.
3   No women soldiers are attested. Women, however, are attested occasionally as gladiators. See for example Suetonius, *Domitian* 4.1.
4   Grant (1974), 78.

5    Silver tridrachms of Trajan depicting legionary standards, including an aquila, struck between 112 and 117, commemorate the legion's first sojourn at Bostra.

6    Tacitus, *Annals* 1.40–44 (mutiny), 69 (bridge). Dio 57.5.6–7 (Loeb vol. VII, pp. 125–7).

7    *CIL* 8.26498 (Thugga), and later appeared on coins. See for example Sear (2002), no. 6595.

8    *SHA* (Marcus Aurelius) 26.8; Dio 72.10.5 (Loeb vol. IX, p. 33).

9    *CIL* VI.2149. See Friggeri et al (2012), 311. IVLIAE.AVG(VSTAE). DOMINAE.MATRI.CAS.IT.IM(M)VNIS. Here the name/title Domna is given in its expanded form of Domina ('Mistress').

10   Caesar, *Gallic Wars* 5.14.

11   For example on the dedication inscription of the Dura amphitheatre in 216. *AE* 1937.239.

12   Scheidel (2011), 417–8.

13   Campbell (1994), no. 257 (iii).

14   *Select Papyri* vol. II (Loeb edition), p. 89, no. 213.

15   Campbell (1994), no. 257 (iv).

16   Tacitus, *Histories* 2.80.

17   *Fragmenta Vaticana* no. 195. Accessed online at http://www.ancientrome.ru/ ius/library/vatican/FragVat.htm#195

18   *CJ* 2.52.2.

19   Dio 60.24.3 (Loeb vol. VII, p. 429).

20   Campbell (1994), no. 257; see also Phang (2010), 23. The ban referred to is that thought to have been introduced by Augustus.

21   Herodian 3.8.5.

22   Scheidel (2011), 417–18.

23   *ILS* 2254; McDermott (1970), 189; Bruun and Edmondson (2014), 574. In this context the word *spurius* (illegitimate) should not be confused with the common Roman praenomen Spurius. Despite the identical spelling, the two were not etmylogically connected.

24   Pliny, *Letters* 10.106, 107. The unit was presumably an infantry cohort with a cavalry contingent.

25   *Tab. Vindol.* II.291.

26   Beeson (2018) (i), 13. The miniature sword is just over 7 in (180 mm) long.

27   *CIL* 4.4360, a dedication to Diana by Brocchus as prefect of the unit c. 103–115. No other names or information were included.

28   Virgil, *Aeneid* 9.473, *Tab. Vindol.* II.118; Virgil, *Aeneid* 1.1–2, *Tab. Vindol.* II.453.

29   *P. Mich* 8.464. Accessed online at http://papyri.info/ddbdp/p.mich;8;464

30   See, for example, Rowlandson and Takahashi (2009).

31   *ILS* 2658, *CIL* 8.217, Campbell (1994), no. 86.

32   *CIL* 8.2623.

33   *ILS* 3092, *CIL* 3.4363.

34  *RIB* 1600.
35  CIL 16.161.
36  *ILS* 2460, *CIL* 9.4684. The text also names three slaves freed by Gaius Julius Longinus, and refers to the deified Vespasian thus dating it to late 79 or afterwards.
37  *AE* 1922, 135; Adams (2008), 58–63. Note that Tomlin and Pearce (2018), who refer to this tablet in passing, have Memmius Montanus selling the slave girl. However, Adams' full transcription and extensive discussion of the document, which was written in Latin but using Greek characters, makes it quite clear that Memmius was the purchaser. Tomlin (personal communication) confirms he and Pearce were in error.
38  Martial, *Epigrams* 4.13, 6.58.10, 11.53. Pudens' unit is unknown.
39  It is assumed here in that Sarapias' mother Cleopatra was the same Cleopatra as the one named as Longinus Castor's heir. For the text see Campbell (1994), no. 375.
40  See for example *Select Papyri* no. 32 (Loeb edition vol. I, pp. 97–9) concerning the sale of a slave woman in Egypt in 120.
41  *RIB* 1065. For the Barates at Corbridge, see *RIB* 1171.
42  *CJ* 5.16.2.
43  *ILS* 2653, *CIL* 8.2877, Campbell (1994), no. 87.
44  Scheidel (2011), 423.
45  *Gnomon of the Ideos Logos* 53. Text accessed at https://droitromain.univ-grenoble-alpes.fr/Negotia/Idiologi_riccobono.gr.htm
46  Seneca, *On Firmness* 18.4 recounts how the nickname was acquired but also how the adult Caligula disliked it.
47  Suetonius, *Caligula* 8–9.
48  Illustrated in Ägyten. Dauer und Wandel. Symposium anlässlich des 75jährigen bestehens des Deutschen Archäologischen Instituts Kairo am 10 und 11 Oktober 1982. (Deutsches Archäologisches Institut Kairo Sonderschrift 18), p. 101. Text of inscription accessed at http://db.edcs.eu/epigr/epi_ergebnis.php catalogued under EDCS-ID: EDCS-58200215.
49  *CIL* 3.12056.
50  *RIB* 690.
51  *CIL* 14.2289.
52  *RIB* 1920.
53  *RIB* 860.
54  *CIL* 8.3022.
55  *AE* 1906.113. The stone is badly damaged but once carried a carved portrait of the deceased.
56  *CIL* 3.8719.
57  *CIL* 3.5814.
58  *ILS* 2661, *CIL* 2.4461.
59  *Tab. Vindol.* II.310.

60    *ILS* 2632, *CIL* 13.7054.

61    *CIL* 13.6951a.

62    *RIB* 369.

63    *RIB* 363, 373.

64    *CIL* 3.1124. The expansion of the abbreviation for *signif(erorum)*, 'of
      the standard-bearers' is not certain. He may have been an optio and
      progressed to being a signifer.

65    Adamson (2012) provides the full text, translation and discussion. He
      refers to Poleion as 'Polion' but Poleion is a better phonetic transliteration
      of the Greek. There is no suggestion here of an Egyptian brother-sister
      marriage as in the case of Julius Terentianus, but Poleion's evident
      worries suggest his concern went beyond the purely personal. The letter
      is damaged and incomplete in the second half, obscuring that part's exact
      meaning.

66    Grubbs (2002), 217; Baird (2018), 82; text accessed at http://papyri.info/
      ddbdp/p.dura;;32

67    *CJ* 3.2.2.3.

68    Ammianus Marcellinus 24.4.27.

69    Polybius 10.19.3.

70    Valerius Maximus, *MDS* 6.7.1.

71    Pliny the Younger, *Letters* 6.31.4–6.

72    *SHA* (Macrinus) 12.4–6, 10.

73    *SHA* (Aurelian) 7.4. The source is not wholly reliable and the story
      possibly entirely apocryphal.

74    Strong (2016), 10, 42, citing *CIL* 9.2029, 116 and 238.

75    Seneca, *Natural Questions* 7.31.2.

76    *CIL* 4.8356 (found in the House of the Menander, Pompeii). For the
      praetorian base near Nocera see de la Bédoyère (2017), 42.

77    Strong (2016), 39, citing *IGR* 1.1183/*OGIS* 2.674.

78    *CIL* 4.1751.

79    Suetonius, *Caligula* 40.

80    Phang (2010), 251.

81    *Digest* 29.1.41.1.

82    *Digest* 34.9.14 (Papinian).

83    Valerius Maximus, *MDS* 7.3.7.

84    Livy *Periochiae* 57; also at Appian, *Spanish Wars*, 14.85, Frontinus, *Stratagems*
      4.1.1. See too McGinn (1998), 40.

85    Valerius Maximus, *MDS* 2.7.1.

86    Phang (2010), 248–51.

87    Phang (2010), 263ff.

88    Plautus, *Pseudolus* 4.7 (lines 1180–1).

89    Dio 68.7.4 (Loeb vol. VIII, p. 373).

90    Phang (2010), 280.

91 Polybius 6.37.9. The word for the punishment given by Polybius is *bastinado*. It was also known as *fustuarium*.

92 Valerius Maximus, *MDS* 6.1.11.

93 Plutarch, *Marius* 14; Valerius Maximus, *MDS* 6.1.12.

94 Valerius Maximus, *MDS* 6.1.10.

95 *RIB* 1064.

96 The tunic was also associated with Carthaginians so Scipio's words may have been inspired in part by disdain for his old enemies. Aulus Gellius, *Attic Nights* 6.12.4–7.

97 Sallust, *Catiline War* 51.9; Phang (2010), 267.

98 Tacitus, *Histories* 3.33.

99 Tacitus, *Histories* 4.14

100 Dio 67.11.4–5 (Loeb vol. VIII, p. 343).

101 *SHA* (Elagabalus) 5.1.

102 *SHA* (Elagabalus) 12.1–2. Elagabalus' grandmother, Julia Maesa, was the sister of Caracalla's mother Julia Domna, the empress of Septimius Severus. Severus Alexander's mother Julia Mamaea was the sister of Elagabalus' mother Julia Soaemias. The cousin relationships were thus all through the female line.

## 15: VETERANS

1 *ILS* 2338, *CIL* 5.5832. Publius Tutilius' tombstone is unusual in supplying consular dates for his birth and death. These have been replaced here with our equivalent dates. Since he lived to the age of seventy-two, he had obviously served far longer than he need have though it is not entirely clear whether he died still in post.

2 Wesch-Klein (2011), 440; the classes of military discharge are itemized in Justinian's *Digest* 3.2.2.2.

3 Scheidl (2011), 426–7.

4 *ILS* 9059; Campbell (1994), no. 341.

5 Campbell (1994), no. 330.

6 For Diocletian and Maximianus see *CJ* 7.64.9. The ruling appears to apply to legionaries and auxiliaries.

7 Wesch-Klein (2011), 443, provides a useful survey of all the privileges a veteran enjoyed.

8 *Digest* 49.18.2, 4.

9 Livy 22.5.4.

10 Dio 37.30.5 (Loeb vol. III, p. 147).

11 The story is told by Suetonius, *On Grammarians* 9. Macrobius 2.6.4 repeats the court story but says it was Galba, father of the later emperor of the same name, who asked Orbilius Pupillus what he did for a living.

12 Cicero, *Letters to Brutus* 1.8.2 (Loeb edition Letter 15.2). The use of Crete

as a pirate base made it part of Pompey's naval war against pirates. The Roman army in the Cretan War was led by Quintus Caecilius Metellus, later honoured as Creticus, and an opponent of Pompey. Nothing else is known about Nasennius or his connections.

13 Caesar, *Gallic Wars* 8.8. This book was written by Aulus Hirtius, one of Caesar's legates, and it is he who made this observation about Legio XI being outclassed by veterans, in spite of its ability and experience.

14 *ILS* 2249, *CIL* 8.14697; McDermott (1970), 186–7; Caesar, *African War* 1.5.

15 Augustus, *Res Gestae* 3, 15.

16 *ILS* 2243, *CIL* 5.2501; *ILS* 2336, *CIL* 5.2503.

17 Augustus, *Res Gestae* 16.1, 3.3, 15.3.

18 Tacitus, *Annals* 1.35.2, 36.3.

19 Suetonius, *Tiberius* 48.2.

20 Suetonius, *Nero* 9, 32.1.

21 Campbell (1994), no. 331.

22 See Campbell (1994), nos. 327–8, and his discussion on p. 201.

23 *RIB* 2401.1.

24 Frere et al (1990), 1–2.

25 *RIB* 2401.11.

26 Campbell (1994), no. 329; also Lewis and Reinhold (1990), 525–6.

27 Dio 53.25.1–5 (Loeb vol. VI, p. 259).

28 Tacitus, *Annals* 12.32.

29 *AE* 2013.1116.

30 *ILS* 2476, *CIL* 3.1008.

31 *CIL* 3.7773.

32 *ILS* 2474, *CIL* 3.6166. Called here *magistri* rather than *duoviri*.

33 *SHA* (Hadrian) 20.10, noted by McDermott (1970), 195. The building was probably a temple. For Hadrian's tribuneship with Legio V, see *ILS* 308, *CIL* 3.550 found in Athens.

34 *CIL* 10.3903. One suggestion is that he was retired as early as 36 BC and that the dedication came after Augustus' death in AD 14 (see Campbell (1994), no. 366). This makes for an implausible retirement of at least fifty years, which is patently unlikely.

35 *CIL* 10.3803.

36 *RIB* 685.

37 Tacitus, *Annals* 1.17.

38 Tacitus, *Histories* 4.65, and see also *Annals* 12.27. Tacitus made a similar observation about serving soldiers becoming intertwined with provincials at *Histories* 2.80.

39 *RIB* 252.

40 *ILS* 2462, *CIL* 3.4057.

41 *ILS* 7531, *CIL* 13.1906.

42 *ILS* 2472, *CIL* 13.6677.

43  Edwell (2008), 78; Baird (2018), 82. Greek text at http://papyri.info/ddbdp/p.dura;;26

44  *CIL* 3.3114.

45  *CIL* 5.5269.

46  *CIL* 9.4682.

47  *ILS* 5795, *CIL* 8.2728.

48  *Britannia* 45 (2014), 434, no. 4. The name is unique.

49  'Gemellus' means 'twin', so it is likely he was born as one of a pair of twins. P. Fay. 118. Alston (1999), 185–6; Elliott (2004), 264. Accessed at http://papyri.info/ddbdp/p.fay;;118

50  P. Fay. 117 accessed at http://papyri.info/ddbdp/p.fay;;117

51  Waebens (2018), 2, citing the archive of Apollinarius at *BGU* II.180. See also Campbell (1994), no. 339. Strategos here is assumed.

52  Suetonius, *Galba* 6.3.

53  *ILS* 2034, *CIL* 6.2725, Campbell (1994), no. 63.

54  Tacitus, *Histories* 2.94.

55  *CIL* 6.2725. The tombstone is displayed in the Vatican Museums in Rome.

56  *ILS* 2283; *ILS* 7076, *CIL* 13.6797 from Mainz attests Vibius Rufinus as governor in 43–4.

57  *ILS* 2259; Campbell (1994), no. 32.

58  Tacitus, *Histories* 1.46.

59  Pliny the Younger, *Letters* 10.87. For the recommendation of the son see Chapter 3.

60  Dio 68.14.2 (Loeb vol. VIII, p. 387).

61  *AE* 1984.721.

62  Cicero, *On Oratory* 1.175; Valerius Maximus, *MDS* 7.7.1. Neither names any of the parties.

63  Dio 67.13.1 (Loeb vol. VIII, p. 347).

64  *RIB* 359.

65  Tacitus, *Annals* 14.27.1–3, and compare with *Histories* 2.80.

## 16: JUPITER'S MEN

1  *RIB* 1600.

2  Polybius 6.56.

3  *AE* 1964.261; *SHA* (Caracalla) 6.6. See Cowan (2002), 142–3, for a discussion of the epigraphic evidence for the legion's involvement in the campaign.

4  *CIL* 3.3344.

5  Livy 8.9.1–8.10.10.

6  Livy 10.41–42, 46.8; the story also appears in Valerius Maximus, *MDS* at 7.2.5.

7  Suetonius, *Tiberius* 2.2.

8   Livy 22.3.10–14.

9   Polybius 10.11.5–9.

10  Frontinus, *Stratagems* 1.12.1.

11  Aulus Gellius, *Attic Nights* 15.22.1–2.

12  Dio 41.14.1–3 (Loeb vol. IV, p. 27). The eclipse is dated to 21 August 49 BC
    on astronomical dating, which includes a year 0; this equates to 50 BC by
    the normal calendar which does not have a year 0. For the eclipse, see
    https://eclipse.gsfc.nasa.gov/SEsearch/SEsearchmap.php?Ecl=-00490821

13  https://eclipse.gsfc.nasa.gov/SEsearch/SEsearchmap.php?Ecl=-00500307.
    This eclipse is dated to 7 March 50 BC on astronomical dating. See
    previous note.

14  There are other instances where Dio exaggerated stories about eclipses,
    for example that of 30 April 59 which he implies was total in Rome but
    which was in fact about 75 per cent (see de la Bédoyère 2018, 257–8).

15  Dio 72.8–9 (Loeb vol. IX, pp. 27–31).

16  *RIB* 1426.

17  *CIL* 6.2256.

18  Pliny the Younger, *Letters* 10.100–103.

19  The text of the Feriale Duranum is in Campbell (1994), no. 207.

20  For this possible clock/calendar from Vindolanda and other examples,
    see Meyer (2019).

21  Herodian 6.8.4.

22  Herodian 6.8.2–3.

23  Herodian 6.8.8.

24  Herodian 6.9.7.

25  Herodian 8.5.9.

26  *RIB* 816.

27  Edwell (2010), 138.

28  Cato, *On Agriculture* 141.1–4; Keppie and Arnold (1984), 27, no. 68.
    Note that Keppie and Arnold refer to the third animal as a 'bull'. In
    fact the ruminant shown is the same size as the others, and is more
    commensurate with Cato's description of the ritual.

29  *RIB* 916.

30  *ILS* 4312, *CIL* 13.8201. A similar anonymous one is at Mainz, *CIL* 13.6804.

31  Jones and McFadden (2015), especially 105–34.

32  Tacitus, *Histories* 3.24.3.

33  For Saturninus, see *RIB* 1530 –1; for Cosconianus, see *RIB* 1534.

34  Stoll (2011), 466.

35  For Clodius Marcellinus, see *ILS* 4780, *CIL* 13.8021. For Calpurnius Proclus
    see *AE* 1930.24, and Domitia Regina *AE* 1930.27.

36  *AE* 1931.13.

37  *AE* 1931.16.

38  Seyrig (1933), 68–71.

39 Published and illustrated in 'Glimpses', a catalogue published by ArtAncient (2014), no. 48. Said to be held in a private collection in the Far East.

40 Jerome, *Letters* 107.

41 Tertullian, *De Corona* 15.3-4

42 *AE* 1979.425.

43 https://artgallery.yale.edu/collections/objects/6746

44 Rostovtzeff (1939), 83; see also Edwell (2008), 125ff.

45 *AE* 1940.220. Illustrated in Edwell (2008), 140, fig. 4.26.

46 *RIB* 1544–46. The altars displayed in the mithraeum today are replicas. The originals are in the Great North Museum: Hancock, Newcastle.

47 *RIB* 1545.

48 Henig (1999), 161.

49 *CIL* 3.1122.

50 *ILS* 4311, *CIL* 3.11135.

51 *CIL* 8.2630.

52 *P. Mich.* 8.466 II18, accessed at http://papyri.info/ddbdp/p.mich;8;466. For Apion, see Chapter 1. For the second letter see *P. Mich.* 8.501 II17, accessed online at http://papyri.info/ddbdp/p.mich;8;501. The text is incomplete, meaning that Stoll's description of the text as representing thanks to Sarapis for the trip relies on a speculative restoration.

53 *RIB* 658.

54 *CIL* 3.882, *ILS* 4361; Antigonus *CIL* 3.881. See also Takacs (2015), 201.

55 *CIL* 3.1342. Takacs (2015), 198; *SIRIS* 679a, ibid., 164.

56 *Acts of the Apostles* 10.

57 *Acta Maximiliani* 1–3 (St Maximilianus of Tebessa). Accessed at https://www.ucc.ie/archive/milmart/Maximilian.html

58 *Acta Marcelli* 1 –17. Accessed at https://www.ucc.ie/archive/milmart/Marcellus.html

59 Eusebius, *Life of Constantine* 28.

## EPILOGUE

1 *ILS* 2238, *CIL* 3.6825; McDermott (1970), 188. Antioch in Pisidia was made a colony early on in imperial times.

2 *CIL* 3.2014. Legio VII Claudia received the title Pia Fidelis in 42 and was moved to Moesia in 57. This provides the time period in which Caesius Bassus must have died.

3 See Nick Squires, 'Chinese villagers descended from Roman soldiers', *Daily Telegraph*, 23 November 2010, citing one authority dismissing it as a 'fairy tale'.

4 Sear (2005) no. 9404.

5 Sear (2005) no. 12258, noting how incompatible this was with Numerian's nature.

6   Caesar, *Gallic Wars* 1.22; Virgil (example), *Aeneid* 2.63.

7   Livy 28.15.7 (Loeb vol. VIII, pp. 64–5). Note that Nixon and Rodgers (1994), 18, observed the similarity of Livy 28.44.8 to Panegryic XII.15.6.

8   *SHA* (Probus) 2.3.4, and 7.

9   Nixon and Rodgers (1994), 24.

10  *Notitia Dignitatum, Western Sector* 28 (II Augusta), 34 (X Gemina).

11  For example Vegetius 1.4, 8, 9, 13, 15, 2.3, 3.10.

12  *CJ* 3.44.5.

13  Hope (2003), 85.

14  Sallust, *Jugurthine War* 4.5.

15  For the Parentalia/Feralia, see Ovid, *Fasti* 2.533ff. See Hope (2003), for more on this fascinating topic.

16  Appian, *Civil Wars* 2.82.

17  Appian, *Civil War* 1.43. One might consider how much long-term impact the visibility of First and Second World War cemeteries has, by contrast, had in more recent times on public support for war and the preference of powers such as the United States and the United Kingdom to assert their power through the use of unmanned equipment like drones. For cremation, see Pliny, *Natural History* 7.54.187.

18  Tacitus, Annals 1.62.1; Hope (2003), 91.

19  Tacitus, *Histories* 2.45, 70. Photographs taken, for example, at Antietam of corpses after the battle there in 1862 during the American Civil War provide an idea of what this scene may have resembled.

20  *ILS* 2089, CIL 6.2464.

21  *RIB* 200.

22  *CJ* 6.21.7.

23  Jones and McFadden (2015), 22 (fig. 1.6), 23. Now in Luxor Museum.

24  *RIB* 3121.

25  *RIB* 685. For more detail on this tombstone see Chapter 15. The coffin was inscribed for a woman but since it contained the body of a man it had also clearly been reused.

26  *RIB* 1172.

27  *CIL* 3.435.

28  *ILS* 2558, *CIL* 3.3676; Campbell (1994), no. 47. Note that 'Batavian' here does not mean ethnically, but is referring to the name of a unit labelled 'Batavian'. See Chapter 2 for this aspect of auxiliary unit names.

29  *RIB* 292. Tartarus was a place for sinners in the afterlife.

30  Polybius 6.25; Varro, *Latin Language* 91; see also Strobel (2011), 274. There remains some confusion about the exact arrangements, which undoubtedly changed over time.

31  See *Digest* 50.6.7.

32  Macrobius 1.10.1.

# ABBREVIATIONS AND BIBLIOGRAPHY

*(NB online references are subject to change)*

*AE* = *L'Annee épigraphique* (Paris, 1888– ). Some of this material is now available online from the same source as *ILS* below

*BGU* = *Berliner griechische Urkunden*, Berlin State Museums (Berlin 1895– )

*CJ* = *Codex of Justinian*. Accessed at: https://droitromain.univ-grenoble-alpes.fr/Anglica/codjust_Scott.htm

*CIL* = *Corpus Inscriptionum Latinarum* (Berlin, 1863– ) in 16 volumes. Some of this material is now available online from the same source as *ILS* below

*Digest* = *Digest of Justinian*. Accessed at: https://droitromain.univ-grenoble-alpes.fr/Anglica/digest_Scott.htm

*IGRR* = *Inscriptiones Graecae ad Res Romanas Pertinentes et impensis Adademiae inscriptionvm et litterarvm hvmaniorvm collectae et editae* (Paris, 1901 and later)

*ILS* = Dessau, H. (1892–1916), *Inscriptionum Latinae Selectae*, Berlin, 3 vols (now available online with full search facilities at http://db.edcs.eu/epigr/epi_en.php

*JRS* = *Journal of Roman Studies*, published by the Society for the Promotion of Roman Studies

*MDS* = Valerius Maximus, *Memorable Doings and Sayings*, available in the Loeb Classical Library Series, Harvard University Press, Vols I and II

*O. Florida* = Bagnall (1976)

*P. Giss.* = *Griechische Papyri im Museum des Oberhessischen Geschichtsvereins zu Giessen.* Accessed online from the papyri.info website

*P. Mich.* = papyri at the University of Michigan's Papyrology Collection, accessed online through the Advanced Papyrological Information System (APIS) at: https://quod.lib.umich.edu/cgi/i/image/image-idx?c=apis&page=search

*P. Oxy* = *Papyri Oxyrhynchus.* See Grenfell and Hunt (1898)

*RIB* = *Roman Inscriptions of Britain*: see Collingwood and Wright (1965) (revised edition 1995 with Addenda and Corrigenda by R.S.O. Tomlin) For inscriptions reported 1995–2006 see Tomlin, R.S.O., Wright, R.P., and Hassall, M.W.C. (2009; for non-stone inscriptions see Frere, S.S., et al., eds (1990 and later; see Bibliography). The superb and indispensable online resource at https://romaninscriptionsofbritain.org/ is highly recommended and will eventually feature all of Britain's Roman inscriptions

*SHA* = *Scriptores Historiae Augustae*, available in the Loeb Classical Library Series, Harvard University Press, Vols I and II (also in Penguin translation, *The Lives of the Later Caesars*, trans. A. Birley, Penguin)

*SIRIS* = *Sylloge inscriptionum religionis Isiacae et Sarapiacae*, ed. L. Vidman (Berlin, 1969)

*Tab. Vindol.* = Bowman, A.K., and Thomas, J.D. (1994 and 2003). The tablets are also online at http://vindolanda.csad.ox.ac.uk/

## 1. Ancient Sources

Ancient written sources are by far and away the most important source of information used for this book. There are frequently serious issues with meaning, details, ambiguities and reliability and these are often noted in the text; however, those interested in pursuing the complexities of interpreting ancient sources would do well to pursue some of the more specialist titles listed below under Modern Sources, as these are largely beyond the scope of this book. Ancient sources (as opposed to inscriptions) used in this book, and listed below, are principally to be found in the Loeb Classical Library Series, now published by Harvard University Press. These feature parallel texts of Greek or Latin on the left and English on the right, and include the vast majority of surviving writings of Greek and Roman history and literature, as well as some documents, including Egyptian papyri. However, they are prohibitively expensive for most readers and are normally only found in specialist libraries or in dedicated private collections. Some of the public domain texts in the Loeb volumes can be found online at:

http://penelope.uchicago.edu/Thayer/E/Roman/Texts/
Subscription access to the whole collection is available at:
https://www.loebclassics.com/

Note that in this book as far as Cassius Dio is concerned I have always referred to the volume and page number in the Loeb series. This is due to an extraordinarily confusingand absurd numbering system adopted by the editor which has resulted in different numbers being given for the same book at the beginning each and in the running headers. The result has been that the authorities referring to his text have used different numbers, causing tedious confusion.

Some of the key ancient texts are also to be found in English translation in the Penguin Classics series, and in Oxford University Press's World's Classics series. The sources of other texts, such as papyri, are indicated in the associated notes.

The principal (not all) sources used in this book were:

Ammianus Marcellinus (325–95)
Appian (Appianus Alexandrinus) (*c.* 95–165)
Arrian (Arrianos) (*c.* 86–160?)
Aulus Gellius (*c.* 123–70)
Caesar (100–44 BC)
Cato the Elder (234–149 BC)
Cicero (Marcus Tullius Cicero) (106–43 BC)
Codex of Justinian (compiled on Justinian's orders) (530–3)
Digest of Roman Law (compiled on Justinian's orders) (530–3)
Dio (Cassius Dio) (*c.* 150–235)
Eusebius of Caesarea (263–339)
Frontinus (Sextus Julius Frontinus) (35–103)
Fronto (Marcus Cornelius Fronto) (*c.* 100–67)
Horace (Quintus Horatius Flaccus) (65–8 BC)
Josephus (Titus Flavius Josephus) (*c.* 37–100)
Juvenal (Decimus Junius Juvenalis) (*fl.* early second century AD)
Livy (Titus Livius) (59 BC–AD 17)

Macrobius (Macrobius Ambrosius Theodosius) (*fl.* early fifth century AD)
Onasander (*fl.* mid-first century AD)
Ovid (Publius Ovidius Naso) (43 BC–AD 17/18)
Pliny the Elder (Gaius Plinius Secundus) (23–79)
Pliny the Younger (Gaius Plinius Secundus) (*c.* 61–113)
Polybius (*c.* 200–118 BC)
Pseudo-Hyginus (third century AD)
Sallust (Gaius Sallustius Crispus) (86–34 BC)
*Scriptores Historiae Augustae* (various authors 292–323)
Suetonius (Gaius Suetonius Tranquillus) (*c.* 69–123)
Tacitus (Cornelius Tacitus) (*c.* 55–120)
Valerius Maximus (*fl.* 14–37)
Varro (Marcus Terentius Varro) (116–17 BC)
Vegetius (Publius Flavius Vegetius Renatus) (*fl.* early fifth century AD)
Velleius Paterculus (*c.* 19 BC–AD 31)
Virgil (Public Vergilius Maro) (70–19 BC)
Zonaras (Joannes Zonaras) (1074–1130)

## 2. Modern sources

The following books and articles were useful either as sources of specific information or for general context. They are referenced appropriately in the endnotes. The literature on the Roman army is astronomical and it is only possible to provide here a representative and realistic selection. The bibliographies in many of the following titles will take the reader even further into the details of the Roman army's history, organization and equipment.

Please note that as far as coins are concerned I have chosen not to make reference to the *Roman Imperial Coinage* (*RIC*) classification series used by scholars. These volumes are effectively unobtainable outside a highly specialized library and even then are arcane and difficult to use. I have therefore used David Sear's outstanding collector handbooks which are readily found and include the *RIC* references should the reader need them.

Adams, J.N. (2008), *Bilingualism and the Latin Language*, Cambridge University Press, Cambridge

Adamson, G. (2012), 'Letter from a Soldier in Pannonia', *Bulletin of the American Society of Papyrologists*, vol. 49, 79–94

Alston, R. (1995), *Soldier and Civilian in Roman Egypt: A Social History*, Routledge, London

Alston, R. (1999), 'Ties that bind: soldiers and societies', in Goldsworthy and Haynes (1999)

Bagnall, R.S. (1976), *The Florida Ostraka (O. Florida): documents from the Roman army in Upper Egypt*, Greek, Roman and Byzantine Monographs no. 7, Duke University, Durham, North Carolina

Baird, J.A. (2018), *Dura-Europos*, Bloomsbury Academic, London

Beeson, A. (2018) (i), 'A Hoard of Roman Cavalry Weapons from Vindolanda', *ARA News* 39, Spring 2018, Association for Roman Archaeology, Swindon

Beeson, A. (2018) (ii), 'Superb Gilded Bronze Horse's Head goes on display at Saalburg', *ARA News* 40, Autumn 2018, Association for Roman Archaeology, Swindon

Biggins, J.A., and Taylor, D.J.A. (2004), 'Geophysical survey of the *vicus* at Birdoswald Roman fort, Cumbria', *Britannia* 35, 159–78

Birley, A. (1979), *The People of Roman Britain*, Batsford, London

Birley, E. (1951), 'The Prefects and their Altars', in Richmond and Gillam (1951)

Bishop, M.C. (2016), *The Gladius: The Roman Short Sword*, Osprey, Oxford

Bishop, M.C. (2017), *The Pilum: The Roman Heavy Javelin*, Osprey, Oxford

Bland, R. (2018), *Coin Hoards and Hoarding in Roman Britain AD 43–c. 498*, British Numismatic Society Special Publication No. 13, Spink, London

Boon, G.C. (1984), *Laterarium Iscanum: The Antfixes, Brick & Tile Stamps of the Second Augustan Legion*, National Museum of Wales, Cardiff

Boon, G.C. (1987), *The Legionary Fortress of Caerleon – Isca*, Roman Legionary Museum, Caerleon

Bowman, A.K. (2004), *Life and Letters on the Roman Frontier: Vindolanda and its People*, British Museum Press, London

Bowman, A.K., and Thomas, J.D. (1994), *The Vindolanda Writing-tablets (Tabulae Vindolandenses II)*, British Museum Press, London (the most important letters also cited in Bowman, 2004)

Bowman, A.K., and Thomas, J.D. (2003), *The Vindolanda Writing-tablets (Tabulae Vindolandenses III)*, British Museum Press, London (the most important letters also cited in Bowman, 2004)

Bowman, A.K., Thomas, J.D., and Tomlin, R.S.O. (2019), 'The Vindolanda Writing Tablets (*Tabulae Vindolandenses* IV, Part 3): New Letters of Iulius Verecundus, *Britannia* 50, 225–51

Breeze, D., ed. (2012), *The First Souvenirs: Enamelled Vessels from Hadrian's Wall*, Cumberland and Westmorland Archaeological Society

Breeze, D.J., and Bishop, M.C. (2013), *The Crosby Garrett Helmet*, Armatura Press, Pewsey

Brewer, R. (1987), *Caerleon – Isca: The Roman Legionary Museum*, Roman Legionary Museum, Caerleon

Brodribb, G. (1979), 'Tile from the Roman Bath House at Beauport Park', *Britannia* 10, 139–56

Brodribb, G., and Cleere, H. (1988), 'The *Classis Britannica* Bath-house at Beauport Park, East Sussex', *Britannia* 19, 217–74

Bruun, C., and Edmondson, J. (2014), *The Oxford Handbook of Roman Epigraphy*, Oxford University Press, Oxford

Butcher, K., and Ponting, M. (2015), 'The Reforms of Trajan and the End of the Pre-Neronian Denarius', *Annali dell'Istituto Italiano di Numismaticai (Rome)*, 61, 21 –42

Campbell, J.B. (1984), *The Emperor and the Roman Army 31 BC–AD 235*, Clarendon Press, Oxford

Campbell, J.B. (1994), *The Roman Army 31 BC–AD 337: A Sourcebook*, Routledge, London

Casey, P.J. (1994), *Carausius and Allectus: The British Usurpers*, Batsford, London

Claridge, A. (2010), *Rome. An Oxford Archaeological Guide*, Oxford University Press, Oxford

Collingwood, R.G. and Wright, R.P. (1965), *The Roman Inscriptions of Britain. Volume I, Inscriptiions on Stone*, Clarendon Press, Oxford (second edition 1995, with Addenda and Corrigenda by R.S.O. Tomlin, Alan Sutton, Stroud)

Cowan, R. (2002), *Aspects of the Severan Field Army: The Praetorian Guard, Legio II Parthica, and Legionary Vexillations, AD 193–238*, PhD thesis, University of Glasgow

Cowan, R. (2014), *Roman Guardsmen 62 BC–AD 324*, Osprey, Oxford

D'Amato, R., and Dennis, P. (2018), *Roman Standards and Standard-Bearers (1), 192 BC–AD 192*, Osprey, Oxford

D'Amato, R., and Negin, A.E. (2020), *Roman Standards and Standard-Bearers (2), AD 192–500*, Osprey, Oxford de la Bédoyère, G. (1991), *The Buildings of Roman Britain*, Batsford, London

de la Bédoyère, G. (1998), 'Carausius and the Marks RSR and I.N.P.C.D.A.' in *Numismatic Chronicle*, 79–88

de la Bédoyère, G. (2010) (i), *Hadrian's Wall: A History and Guide*, Amberley, Stroud

de la Bédoyère, G. (2010) (ii), *Cities of Roman Italy*, Bristol Classical Press, London

de la Bédoyère, G. (2015), *The Real Lives of Roman Britain*, Yale University Press, London

de la Bédoyère, G. (2017), *Praetorian: The Rise and Fall of Rome's Imperial Bodyguard*, Yale University Press, London

de la Bédoyère G. (2018), *Domina: The Women who Ruled Imperial Rome*, Yale University Press, London

Dixon, K.R., and Southern P. (1992), *The Roman Cavalry; From the First to the Third Century AD*, Batsford, London

Edmondson, J. (2014), 'Roman Family History', in Bruun and Edmondson, eds (2014), 559–81

Edwell, P. (2008), *Between Rome and Persia:The Middle Euphrates, Mesopotamia, and Palmyra under Roman control*, Routledge, Abingdon

Elliott, J.K., ed. (2004), *The Collected Biblical Writings of T.C. Skeat*, Novum Testamentum Supplements, Brill, Leiden

Erdkamp, P. (2011), 'War and State Formation', in Erdkamp, P., ed. (2011), 96–113

Erdkamp, P., ed. (2011), *A Companion to the Roman Army*, Blackwell, Oxford

Fields, N. (2006), *Rome's Saxon Shore: Coastal Defences of Roman Britain AD 250–500*, Osprey, Oxford

Frere, S.S., Roxan, M., and Tomlin, R.S.O. (1990), *The Roman Inscriptions of Britain. Volume II, Fascicule 1: The Military Diplomata; Metal Ingots; Tesserae; Dies; Labels; and Lead Sealings*, Alan Sutton Publishing, Gloucester

Frere, S.S., and Tomlin, R.S.O., eds (1991), *The Roman Inscriptions of Britain. Volume II, Fascicule 2: Weights, Metal Vessels etc (RIB 2412–2420)*, Alan Sutton Publishing, Stroud

Frere, S.S., and Tomlin, R.S.O., eds (1991), *The Roman Inscriptions of Britain. Volume II, Fascicule 3: Jewellery, Armour etc (RIB 2421–2441)*, Alan Sutton Publishing, Stroud

Frere, S.S., and Tomlin, R.S.O., eds (1992), *The Roman Inscriptions of Britain.*

    *Volume II, Fascicule 4. Wooden barrels, tile stamps etc (RIB 2442–2480)*, Alan Sutton Publishing, Stroud

Frere, S.S., and Tomlin, R.S.O., eds (1993), *The Roman Inscriptions of Britain. Volume II, Fascicule 5: Tile Stamps of the Classis Britannica; Imperial, Procuratorial and Civic Tile-stamps; Stamps of Private Tilers; Inscriptions on Relief-patterned Tiles and graffiti on Tiles (RIB 2481–2491)*, Alan Sutton Publishing, Stroud

Frere, S.S., and Tomlin, R.S.O., eds (1995), *The Roman Inscriptions of Britain. Volume II, Fascicule 7: Graffiti on Samian Ware (Terra Sigillata) (RIB 2501)*, Alan Sutton Publishing, Stroud

Frere, S.S., and Tomlin, R.S.O., eds (1995), *The Roman Inscriptions of Britain. Volume II, Fascicule 8: Graffiti on Coarse Pottery; Stamps on Coarse Pottery; Addenda and Corrigenda to Fascicules 1–8 (RIB 2502–2505)*, Alan Sutton Publishing, Stroud

Friggeri, R., Cecere, M.G.G., and Gregori, G.L. (2012), *Terme di Diocleziani. La Collezione Epigrafica*, Electa, Milan

Gibbon, E. (1776), *Decline and Fall of the Roman Empire* (numerous editions)

Goldsworthy, A.K. (1996), *The Roman Army at War 100 BC–AD 200*, Clarendon Press, Oxford

Goldsworthy, A.K. (2003), *The Complete Roman Army*, Thames and Hudson, London

Goldsworthy, A.K., and Haynes, I., eds (1999), *The Roman Army as a Community*, Journal of Roman Archaeology Supplementary Series 34, Portsmouth, Rhode Island

Gore, R. (1984), 'The Dead Do Tell Tales', *National Geographic* vol. 165, no. 5 (May), Washington DC

Grant, M. (1974), *The Army of the Caesars*, Scribner's Sons, New York

Grenfell, B.P., Hunt, A.S. *et alia* (1898), *The Oxyrhynchus Papyri*, Egypt Exploration Fund, London

Grubbs, J.E. (2002), *Women and the Law in the Roman Empire: A Sourcebook on Marriage, Divorce and Widowhood*, Routledge Sourcebooks for the Ancient World, Routledge, London

Hanel, N. (2011), 'Military Camps, *Canabae*, and *Vici*, The Archaeological Evidence', in Erdkamp, ed. (2011), 395–416

Hanson, W.S. (1978), 'The organisation of Roman military timber-supply', *Britannia* 9, 293–305

Haynes, I. (1999), 'Military service and cultural identity in the *auxilia*', in Goldsworthy and Haynes (1999), 164–74

Hebblewhite, M. (2017), *The Emperor and the Army in the Later Roman Empire AD 235–395*, Routledge, London

Henig, M. (1999), 'Artistic patronage and the military community in Britain', in Goldsworthy and Haynes (1999), 150–64

Hermann, F.R. (1983), *Kastell Zugmantel und der Limes bei Orlen. Archäologische Denkmäler in Hessen 33*. Abteilung für Vor- und Frühgeschichte im Landesamt für Denkmalpflege Hessen, Wiesbaden

Herz, P. (2011), 'Finances and Costs of the Roman Army', in Erdkamp, ed. (2011)

Hingley, R., and Unwin, C. (2018), *Londinium. A Biography: Roman London from its Origins to the Fifth Century*, Bloomsbury Academic, London

Holder, P.A. (2003), 'Auxiliary Deployment in the Reign of Hadrian', *Bulletin of the Institute of Classical Studies, Supplement no. 81, Documenting the Roman Army: Essays in honour of Margaret Roxan*, 101–145, Oxford University Press, Oxford

Holder, P.A. (2006), 'Roman Military Diplomas V', *Bulletin of Classical Studies* 88

Hope, V.M. (2003), 'Trophies and Tombstones: Commemorating the Roman Soldier', *World Archaeology* 35.1, 79–97, Taylor and Francis, Abingdon

Ivleva, T.A. (2012), *Britons Abroad: The Mobility of Britons and the Circulation of British-made Objects in the Roman Empire*, University of Leiden dissertation (accessed at http://hdl.handle.net/1887/20136)

Johnson, A. (1983), *Roman Forts*, A&C Black, London

Jones, R.F.J. (1976), 'The Roman Military Occupation of North-West Spain', *Journal of Roman Studies* 66, 45–66

Jones, M., and McFadden, S., eds (2015), *Art of Empire: The Roman Frescoes and Imperial Cult Chamber in Luxor Temple*, American Research Center in Egypt, Yale University Press, New Haven

Kaminski, J., and Sim, D.N. (2019), 'Interpreting the Ribchester Helmet', *Arms & Armour* vol. 16, issue 1, 1–26

Keenan, J.G. (1989), *Roman Criminal Law in a Berlin Papyrus Codex*, Faculty Publications, Loyala University, Chicago

Kehne, P. (2011), 'War- and Peacetime Logistics', in Erdkamp, ed. (2011), 323–38

Kennedy, D.L. (1983), 'Cohors XX: An Alternative Explanation of the Numeral', *Zeitschrift für Papyrologie und Epigraphik*, Bd. 53 (1983), pp. 214–216

Kennedy, D., and Bewley, R. (2004), *Ancient Jordan from the Air*, Council for British Research in the Levant, London

Kent, J.P.C., and Painter, K.S. (1977), *Wealth of the Roman World A D 300–700*, British Museum, London

Keppie, L.J.F. (1998), *The Making of the Roman Army: From Republic to Empire*, Routledge, London

Keppie, L.J.F., and Arnold, B.J. (1984), *Corpus Signorum Imperii Romani. Great Britain*, Volume I, Fascicule 4, Scotland, British Academy, Oxford

King, A. (1999), 'Animals and the Roman army: the evidence of animal bones', in Goldsworthy and Haynes (1999) 139–48

Kovács, P., Szabó, Á. (2009) *Tituli Aquincensesi* I y II, Pytheas, Budapest

Lepper, F. and Frere, S. (1988), *Trajan's Column*, Alan Sutton, Gloucester

Lewis, N. (1954), 'On Official Corruption in Roman Egypt: the Edict of Vergilius Capito', *Proceedings of the American Philosophical Society* 98.2, 153–8

Lewis, N. (1983), *Life in Egypt under Roman Rule*, Clarendon Press, Oxford

Lewis, N. (1991), *The Documents from the Bar Kokhba Period in the Cave of Letters. Greek Papyri*

Lewis, N., and Reinhold, M. (1990), *Roman Civilization. Sourcebook II: The Empire*, Harper Torchbooks, New York

Liddell, H.G., and Scott, R. (1940), *A Greek-English Lexicon*, Clarendon Press, Oxford

Maxfield, V.A., and Dobson, B. (1995), *Inscriptions of Roman Britain*, Lactor no. 4, London Association of Classical Teachers

Maxwell, G.S. (1989), *The Romans in Scotland*, James Thin, Edinburgh

McDermott, W. (1970), 'Milites Gregarii', *Greece and Rome* vol. 17, no. 2 (October 1970), 184–96

McGinn, T.A.J. (1998), *Prostitution, Sexuality, and the Law in Ancient Rome*, Oxford University Press, Oxford

Meyer, A. (2019), 'The Vindolanda Calendrical Clepsydra', *Britannia* 50, 185–202

Millar, F. (1981), *The Roman Empire and its Neighbours*, Duckworth, London

Moorhead, S. and Stuttard, D. (2012), *The Romans Who Shaped Britain*, Thames and Hudson, London

Murphy, E., Goldfus, H., and Arubas, B. (2018), 'The Jerusalem Legio X Fretensis Kilnworks: contextualizing Ceramic Manufacture and Legionary Wares', *Oxford Journal of Archaeology*, vol. 37, issue 4, 443–66

Murray, W.M. (1989), 'Octavian's Campsite Memorial for the Actian War',

*Transactions of the American Philosophical Society*, New Series, vol. 79, no. 4, 1–172

Natanson, A. (2011), 'Roman Naval Power: Raising the Ram', *History Today* vol. 61, issue 8

Nixon, C.E.V., and Rodgers, B.S. (1994), *In Praise of Later Emperors: the Panegyricii Latini*, University of California Press, Berkeley

Olami, J., and Ringel, J. (1975), 'New Inscriptions of the Tenth Legion Fretensis from the High Level Aqueduct of Caesarea', *Israel Exploration Journal*, vol. 25, no. 2/3, 148–50

Ott, J. (1995), *Die Beneficiarier. Untersuchungen zu ihrer Stellung innerhalb der Rangordnung des Römischen Heeres und zu ihrer Funktion*, Franz Steimer, Stuttgart

Phang, S.E. (2010), *The Marriage of Roman Soldiers (13 BC–AD 235): Law and Family in the Imperial Army*, Columbia Studies in the Classical Tradition 24, Brill, Leiden, Boston and Cologne

Philp, B.J. (1981), *The Excavation of the Roman Forts of the Classis Britannica at Dover 1970–1977*, Kent Archaeological Rescue Unit, Dover

Pitts, L.F., and St Joseph, J.K.S. (1985), *Inchtuthil: The Roman Legionary Fortress*, Britannia Monograph 6, Society for the Promotion of Roman Studies, London

Pollard, N. (2000), *Soldiers, Cities and Civilians in Roman Syria*, University of Michigan Press, Ann Arbor

Pollard, N., and Berry, J. (2015), *The Complete Roman Legions*, Thames and Hudson, London

Rainbird, J.S. (1976), *The Vigiles of Rome*, unpublished PhD dissertation, University of Durham, accessed at etheses.dur.ac.uk/7455/

Rainbird, J.S. (1986), 'The Fire Stations of Imperial Rome', *Papers of the British School at Rome* 54, 147–69

Rankov, B. (2004), *The Praetorian Guard*, Osprey, Oxford

Reece, R. (1997), *The Future of Roman Military Archaeology*, The Tenth Annual Caerleon Lecture, National Museums and Galleries of Wales

Renberg, I., Persson, M.W., and Emteryd, O. (1994), 'Pre-industrial Atmospheric Lead Contamination Detected in Swedish Lake Sediments', *Nature* 368, 323–6

Rich, J., and Shipley, G. (1993), *War and Society in the Roman World*, Leicester-Nottingham Studies in Ancient History, Routledge, London

Richardson, J.S. (1975), 'The Triumph, the Praetors and the Senate in the Early Second Century BC', *Journal of Roman Studies* 65, 50–63

Richmond, I., and Gillam, J.P. (1951), *The Temple of Mithras at Carrawburgh*, Society of Antiquaries of Newcastle upon Tyne

Riggs, C., ed. (2012), *The Oxford Handbook to Roman Egypt*, Oxford University Press, Oxford

Rivet, A.L.F., and Smith, C. (1979), *The Place-Names of Roman Britain*, Batsford, London

Robertson, A.S. (2000), *An Inventory of Romano-British Coin Hoards*, Royal Numismatic Society Special Publication 20, London

Rook, A. (1978), 'The development and operation of Roman hypocausted baths', *Journal of Archaeological Science* 5, 281

Rossignol, B. (2007), 'Les Préfets du Prétoire de Marce Aurèle', *Cahiers du Centre Gustav Glotz* 18, 141–77

Rostovtzeff, M.I. (1939), *The Excavations at Dura-Europos: Preliminary Report of the Seventh and Eighth Seasons of Work*, Yale University Press, New Haven

Rowlandson, J., and Takahashi, R. (2009), 'Brother-Sister Marriage and Inheritance Strategies in Greco-Roman Egypt', *Journal of Roman Studies* 99, 104–39

Roxan, M. (1994), *Roman Military Diplomas 1985–1993*, University College Institute of Archaeology Occasional Publication 14, London

Saddington, D.B. (2011), '*Classes*. The Evolution of the Roman Imperial Fleets', in Erdkamp (2011), 201–17

Salimbeti, A., and D'Amato, R. (2014), *The Carthaginians 6th–2nd Century BC*, Elite Book 201, Osprey, Oxford

Sauer, E. (1999), 'The Augustan army spa at Bourbonne-les-Bains', in Goldsworthy and Haynes (1999), 52–79

Scheidel, W. (2011), 'Marriages, families, and survival: demographic aspects', in Erdkamp, ed. (2011)

Scullard, H.H. (1982), *From the Gracchi to Nero: A History of Rome from 133 BC to AD 68*, Routledge, London

Sear, D. (2000), *Roman Coins and their Values. Volume I. The Republic and Twelve Caesars 280 BC–AD 96*, Spink, London

Sear, D. (2002), *Roman Coins and their Values. Volume II: The Accession of Nerva to the Overthrow of the Severan Dynasty AD 96–AD 235*, Spink, London

Sear, D. (2005), *Roman Coins and their Values. Volume III: The 3rd Century Crisis and Recovery AD 235–AD 285*, Spink, London

Sear, D. (2011), *Roman Coins and their Values. Volume IV: The Tetrarchies and the Rise of the House of Constantine AD 284–AD 337*, Spink, London

Sellars, I.J. (2013), *The Monetary System of the Romans*, e-book, Lulu

Seyrig, H. (1933), 'Altar dedicated to Zeus Betylos', in P.V.C. Baur, M.I. Rostovtzeff, and A.R. Bellinger, eds, *Excavations at Dura Europos, Preliminary Report of the Fourth Season of Work, October 1930 – March 1931*, Yale University Press, New Haven, 68–71

Sherk, R., ed. (1988), *The Roman Empire: Augustus to Hadrian*, Cambridge University Press, Cambridge

Smallwood, E.M. (1967), *Documents illustrating the Reigns of Gaius, Claudius and Nero*, Cambridge University Press, Cambridge

Southern, P. (1989), 'The Numeri of the Roman Imperial Army', *Britannia* 20, 81–140

Southern, P. (2007), *The Roman Army: A Social and Institutional History*, Oxford University Press, Oxford

Spaul, J. (2000), *Cohors 2. The evidence for and a short history of the auxiliary infantry units of the Roman Imperial Army*, British Archaeological Reports (International Series), no. 841, Oxford

Speidel, M.A. (2009), 'Roman Pay Scales', in Speidel, M.A. (2009), *Heer und Herrschaft in Römischen Reich der Hohen Kaiserzeit*, Stuttgart

Starr, C.G. (1941), *The Roman Imperial Navy: 31 BC–AD 324*, Cornell Studies in Classical Philology 26, Cornell University Press

Stockton, D. (1992), *The Gracchi*, Oxford University Press, Oxford

Strobel, K. (2011), 'Strategy and Army Structure between Septimius Severus and Constantine the Great' in Erdkamp, P. (2011) (ed.), 267–85

Strong, A.K. (2016), *Prostitutes and Matrons in the Roman World*, Cambridge University Press, Cambridge

Sumner, G. (2002), *Roman Military Clothing 100 BC–AD 200*, Osprey, Oxford

Sumner, G., and D'Amato, R. (2014), *Arms and Armour of the Imperial Roman Soldier: From Marius to Commodus*, Frontline Books, London

Sutherland, C.H.V. (1974), *Roman Coins*, Barrie & Jenkins, London

Takacs, S.A. (2015), *Isis and Sarapis in the Greco-Roman World*, Brill, Leiden

Tallet, G., and Zivie-Coche, C. (2012), 'Imported Cults' in Riggs, ed. (2012)

Tomlin, R.S.O, (1992), 'The Twentieth Legion at Wroxeter and Carlisle in the First Century: the Epigraphic Evidence', *Britannia* 23, 141–58.

Tomlin, R.S.O. (1998), 'Roman manuscripts from Carlisle: the ink-written tablets', *Britannia* 29, 31–84

Tomlin, R.S.O., and Hassall, M.W.C. (2004), 'Roman Britain in 2003. Instrumentum Domesticum', *Britannia* 35, 344–5

Tomlin, R.S.O., Wright, R.P., and Hassall, M.W.C. (2009), *The Roman Inscriptions of Britain. Volume III, Inscriptions on Stone found or notified between 1 January 1955 and 31 December 2006*, Oxbow Books, Oxford

Tomlin, R.S.O., and Pearce, J. (2018), 'A Roman military diploma for the German Fleet (19 November 150) found in northern Britain', *Zeitschrift für Papyrologie und Epigraphik* 206, 207–17

Van Lommel, K. (2013), 'The Recognition of Roman Soldiers' Mental Impairment', *Acta Classica* 56, 155–84

Waebens, S. (2018), *The Archive of Gaius Julius Apollinarius, an auxiliary soldier and gentleman*, Online publication: https://www.trismegistos.org/arch/detail.php?arch_id=566

Webster, G. (1996), *The Roman Imperial Army*, Constable, London

Wesch-Klein, G. (2011), 'Recruits and Veterans' in Erdkamp, ed. (2011)

Wilson, R.J.A. (2002), *A Guide to the Roman Remains in Britain*, Constable, London

Zienkiewicz, J.D. (1986), *The Legionary Fortress Baths at Caerleon*, Cadw: Welsh Historic Monuments, Cardiff

# LIST OF ILLUSTRATIONS

*All photographs are by the author unless otherwise indicated*

## SECTION ONE

Roman silver denarii. The silver denarius was the staple Roman bullion coin and used to pay soldiers. Left to right: 1. First denarius issue, Roma with (rev) the Dioscuri, Second Punic War 211–206 BC; 2. Vespasian (deified) by Titus, Victory with shield and Jewish captive, 80; 3. Trajan, military trophy, 103–12; 4. Marcus Aurelius, Victory inscribing shield for the Parthian War, 166. All struck at Rome. Diameter *c.* 18–19 mm.

Marcus Favonius Facilis, centurion of Legio XX. Facilis is shown bare-headed as was normal on tombstones. Colchester, Britain. About 43–7.

Legio XV Apollinaris re-enactors. Roman legionary re-enactors of Legio XV Apollinaris on display at Pram in Austria. A centurion leads. In reality, equipment and armour probably varied much more among soldiers in a single unit than this image suggests. Photo: Matthias Kabel.

Auxiliary infantryman. A re-enactor portrays an auxiliary infantryman of Cohors V Gallorum.

Auxiliary cavalryman. Auxiliary trooper re-enactor photographed near the outpost auxiliary fort of Drumlanrig, Scotland. He wears mail armour, carries a shield and wears a spatha sword. Note the horse's decorative harness.

Auxiliary cavalry horse trapping. Auxiliary cavalrymen were particularly fond of elaborate display. This silver plaque is from the equipage of a cavalry horse in a unit commanded by Pliny the Elder. Found at Xanten. Mid-first century.

Auxiliary trooper's parade helmet. Auxiliary cavalrymen sometimes took part in battle demonstrations, re-enacting episodes from myth or Roman history. This example was found at Ribchester and was owned by a soldier with the Spanish name Caravius.

Legionary fortress of Legio XX, Chester. This model shows the fortress in its completed state together with its canabae civilian settlement and extramural amphitheatre. The canabae was probably much more densely packed than shown here.

Porta Praetoria (main gate) at the reconstructed auxiliary infantry fort at Saalburg, Hesse, Germany. The fort was built in the late first century but rebuilt later in stone. The modern reconstruction dates from the early twentieth century. Photo: Ekem.

Part of Legio III Augusta's principia (headquarters) at the legionary fortress, Lambaesis, Algeria. The structure is a good example of the army's ability to produce sophisticated masonry structures, even in remote places. Photo: Zinou2Go.

Luxor (Thebes), Egypt. During the fourth century AD a Roman fort was built in and around the temple courts, by then mostly over fifteen centuries old. A chapel dedicated to the imperial cult of the Tetrarchy emperors (Diocletian, Maximianus, Galerius, and Constantius I) was adapted out of a chapel originally used for the worship of the Egyptian goddess Mut. A fresco survives showing Roman officers from the garrison honouring the emperors in a way all garrisons had been doing since the reign of Augustus.

The reconstructed praetorium (commandant's house), South Shields. Although the original building was of late date (c. 300), this reconstruction gives a good idea of the sort of houses commanding officers occupied in forts from earlier times. The open courtyard and ambulatory recalled Mediterranean style houses, even on Britain's northern frontier.

Amphitheatre of Legio II Augusta, Caerleon, Wales. Large enough to seat the whole legion, the amphitheatre was constructed around 90 to provide a venue for parades, displays of military skills, and entertainments such as gladiatorial bouts. It is one of few such military buildings visible today.

Arch of Titus, Rome. Marble relief depicting Titus at the triumph in Rome in 71, described by Josephus. Titus is in a quadriga with Victory, and accompanied by soldiers bearing spears. Constructed in 80.

Legio I Italica. A stamped tile, probably from a *pila* in a heated room, manufactured by the legion and marked as its property. From Olpia Oescus, a base built by Trajan on the Danube, prior to his Dacian campaign *c.* 106. Diameter 180 mm.

Classis Britannica. Stamped *tegula* tile from the roof of the fleet baths at Beauport Park in the iron-working region in the Weald of Kent and Sussex, Britain. Second century. Height of stamp 58 mm.

Modern replica of the *gladius* (short sword).

## SECTION TWO

Lucius Cornelius Scipio Africanus (236–183 BC), the celebrated hero of the Second Punic War after his defeat of Hannibal at Zama (202 BC). He was widely admired for centuries after his death, his military exploits and anecdotes about him being constantly cited by later authorities. A bronze bust found at the Villa of the Papyri, Herculaneum. Photo: Miguel Hermoso Cuesta.

Gaius Julius Caesar (100–44 BC), the celebrated statesman, general, and dictator of Rome. Caesar's military leadership was primarily commemorated in his Commentaries on the Gallic Wars. These accounts became a highly influential textbook on military leadership in the Roman period and long afterwards. The loyalty of his men became legendary. From a portrait on a silver denarius struck at Rome, February–March 44 BC, shortly before his assassination.

Arch of Septimius Severus, Rome. One of the best-preserved of all Roman triumphal arches, Severus' arch still dominates the Forum in Rome and commemorated both his rise to power and his Parthian war. Constructed in 203.

Marcus Aurelius (161–80) and the Praetorian Guard. Marcus Aurelius is greeted by the Praetorian Guard at Rome. The armour of the praetorians is idealized. From his triumphal arch (demolished).

Marcus Caelius, primus pilus centurion of Legio XVIII (here XIIX), killed in the disaster of 9 when three legions were wiped out by German tribes. This cenotaph was set up at Xanten by his brother Publius in the hope that one day his body would be identified and given a proper burial. Photo: Agnete.

Flavius Bassus, auxiliary trooper with Ala Noricorum. He signed up at the age of twenty and died after 26 years' service at Cologne. The carving shows his scale armour and spatha cavalry sword to good effect. Probably late first century. Photo: Marcus Cyron.

The face of a Roman soldier. Aurelius Martinus, an *eques singularis Augusti* (the cavalry attached to the Praetorian Guard) portrayed on his tombstone. A Pannonian by birth, he served for 24 years before dying at Rome. Probably early third century.

Oclatius, signifer of Ala Afrorum at Neuss depicted wearing a tunic and holding his standard. His unit was originally raised in Africa but he was from the Tunger region (Belgium). Photo: Hartmann Linge.

Gaius Saufeius of Legio VIIII Hispana. This legionary signed up at eighteen and served 22 years before dying at the age of forty. He almost certainly took part in the invasion of Britain and ended up at the legion's new fortress at Lincoln, built not long after the mid-first century. Legio VIIII Hispana was later to disappear without trace in the early second century. Found at Lincoln.

Trajan's Column. The column bears a detailed 200-metre-long frieze depicting Trajan's campaigns in Dacia and is an enormously important visual resource for the Roman army. In this panel soldiers can be seen engaged on fort construction tasks while others in the foreground drag a captive before the emperor for interrogation.

Aqueduct, Caesarea Maritima, Judaea. The aqueduct built by Legio VI Ferrata under Hadrian, and representative of countless other such infrastructure projects undertaken by the Roman army. Photo: Mark87.

Tombstone of Regina, the slavewoman of the British Catuvellauni tribe who was freed and married by Barates, the Palmyrene. He served at South

Shields. By the time Barates married Regina, soldiers were allowed to marry. From Augustus' reign on to the earlier part of Septimius Severus' reign, soldiers were banned from having wives. However, this was routinely flouted. Regina died aged thirty. The text adds a briefer version in Palmyrene at the bottom. Early third century.

Tombstone of Victor, a Moorish tribesman and freedman of Numerianus, a trooper with Ala I Asturum at South Shields. The depiction of Victor and the text makes it likely the two were engaged in a same-sex relationship. Early third century.

Coventina. Titus Cosconianus, prefect of Cohors I Batavorum at Carrawburgh on Hadrian's Wall, commissioned this dedication to the water nymph whose sacred spring lay close to the south-west corner of the fort. Propitiating native deities was a common practice by Roman troops who wanted to be sure such gods and goddesses were on their side. Found in the spring. Mid to late second century?

Mithraeum, Dura-Europos. The mithraeum was originally built in a private house in 168–9 for Cohors XX Palmyrenorum and went through two later phases of rebuilding. This reconstruction, using the original building components, frescoes and carving, belongs to *c.* 240 in the reign of Gordian III. The building was destroyed in the mid-250s when the fort's defences were redesigned. Two tauroctony (Mithras killing the bull and releasing the sacred life force) reliefs were displayed instead of the usual one. Other mithraea are known at Roman military sites across the Empire, especially in Britain along Hadrian's Wall. The cult, open only to men, was immensely popular with soldiers. Photo: Yale University Art Gallery.

Tombstone of Lucius Pompius Marcellinus, tribune of Cohors I Ligurum. Marcellinus died at Ephesus aged twenty-three, apparently while travelling to take up his command. The tombstone was set up by his mother Flavia Marcellina and sister Pompeia Catullina. Despite the early end to his military career, Marcellinus was depicted in a heroic pose. The tombstone is typical of the random survival of evidence for individual officers and soldiers.

Brass sestertius of Nero (54–68), struck at Rome *c.* 64, celebrating Rome's military success. Roma (personification of the city) is shown in typical martial pose, seated on a cuirass and holding a Victory. Diameter 33 mm.

# ACKNOWLEDGEMENTS

I am grateful to Richard Beswick at Little, Brown for his immediate interest in and commitment to the project to write *Gladius* and for the rapid route to the book being commissioned, as well as for his involvement in its development from start to finish. Zoe Gullen, also at Little, Brown, oversaw the book's editing and production. I would like also to thank the following: Roger Tomlin for fielding some epigraphic queries; Scott Vanderbilt for his stunning achievement in putting the Roman Inscriptions of Britain online (the best digital epigraphic source yet created) and for giving me advance access to the electronic version of *The Roman Inscriptions of Britain* Volume 3; Joann Fletcher for advice about Roman Egypt; the authors and archaeologists Roy and Lesley Adkins for their encouragement and help with suggestions on how to treat the material; Adrian Goldsworthy for assistance with the knotty issue of styling the names of Roman army units; Bill Thayer for his extraordinary website at the University of Chicago, which features so many ancient sources in accessible online form (and which was used by this author in association with the printed texts in his personal library); Kym Ramadge of KRD Graphic Design and Multimedia Communications, Melbourne, Australia, for her detailed and perceptive assistance with proofreading and style advice; and my wife Rosemary for tolerating her Roman Empire-obsessed husband for the last four decades.

## Acknowledgements

Although there are far too many to name here, I should also like to acknowledge the scholars and others who have worked tirelessly on Roman army studies and are responsible for the epic range of published resources available to anyone wishing to pursue the subject, without which this book would have been impossible to write. Many of them are listed in the Further Reading section of this book. As every author must always acknowledge, any mistakes in the book are mine and mine alone.

I would like to commemorate the late Dr Graham Webster (1913–2001), one of the great experts on the Roman army, with whom I worked when writing my first books for Batsford over thirty years ago. His pithy advice, particularly for writers of non-fiction, was much appreciated and is still remembered with gratitude by this one. So too is the late Dr Gerald Brodribb (1915–99) whose friendship was a great privilege and who gave me access on a number of occasions to the astonishing fleet bath-house of the Classis Britannica at Beauport Park (East Sussex), which he had both discovered and excavated. Finally, I should also pay tribute to my late mother, Irene de la Bédoyère (1934–2016), whose enthusiastic and proactive role in taking me to Roman sites when I was a child, including a first trip to Hadrian's Wall in 1968 aged ten, played a much more important part in my future career than she ever appreciated.

# INDEX

## Index of Names

Names are normally indexed here in their most familiar form. For example Julius Caesar is indexed under Caesar and the general Scipio Africanus is indexed under Scipio Africanus. Individual soldiers are not normally indexed. Literary sources (e.g. Tacitus) are only indexed where they are specifically discussed in the text.

# Index

# Index of Places

Places are indexed by their modern name unless the location has no or no familiar modern equivalent, for example Dura-Europos or Vindolanda. Provinces like Britain and Egypt which are mentioned in numerous instances are only indexed where major events are involved.

# General Index

# Index

rodents, used as food 131
rounding, of losses etc 6, 128, 145

*sacellum*, 376 *and see* shrine
sacrifice 34, 49, 146, 149, 185, 200, 202, 221, 290, 367, 370, 371, 380
Salassi tribe 358
samian ware 87, 137, 362
Samnite Wars 32, 343, 372
*samseira*, Persian name for a sword 87
Sarmatians 12, 49, 80, 190
Sassanids 119
Saturnalia 86, 100, 134, 204, 205, 426
savings 99, 256
scabbard 85, 215, 343
*scholae palatinae* 29, 426
scouts 15, 74, 187, 426
second jobs 27, 296, 304
Serapis 57, 381, 389, 390, 391
*sesquiplicarius* 75, 426
sestertius 97, 102, 104, 106, 254, 258, 260, 328, 330, 340, 353, 394, 426
shame 69, 79, 141, 145, 159, 163, 175, 178, 180, 254, 275, 319, 323, 343
shield 2, 35, 52, 85, 87, 88, 105, 108, 130, 131, 136, 152, 153, 156, 166, 190, 194, 197, 201, 215, 217, 298, 341, 381, 395
ships, names of 244–5
shipwreck 57, 235, 236
shrine 24, 25, 113, 118, 214, 271, 274, 290, 291, 303, 304, 308, 309, 370, 376, 380, 381, 382, 386, 387, 396
Sicarii Jewish sect, at Masada 220
sieges 48, 65, 89, 93, 128, 130, 131, 150, 153, 156, 159, 163, 176, 194, 196, 201, 202, 205, 206, 208, 209, 212, 213, 220, 245, 293
silver 86, 105, 195, 198, 201, 281–2, 308, 372, *see also* denarius
size, of the army 13–15, 39, 94, 411–12
slaves: captured 68, 132, 220; married by soldiers 327, 329; owned by soldiers 304, 318, 327, 329, 336; rebellions of 36, 148–51; recruited 42, 43, 45, 59; supervised by soldiers 280
sling-bolt 117
smallpox 45, 313
smelting 124, 240, 269, 282
Social War 58, 79, 249, 350, 398, 423
soothsayers 371, *see also* pullarii
sources, reliability and availability of xvii, 2–7
souvenirs 314, 315, 355
spas 274, 309, 313

spatha, cavalry sword xvi, 84
*speculatores*, *see* scouts
spies (*frumentarii*) 74, 75, 267, 339, 422
sport 129
stables xiii, 128
standards and standard bearers, general: 24–7, 35, 85, 108, 113, 123, 151, 153, 154, 157, 158, 171, 179, 208, 253, 292, 301, 337, 366, 374, 376; *aquila* (eagle) 24, 49, 151, 157, 158, 178, 212, 252; *aquilifer* 24, 25, 52, 53, 76, 157, 158, 178, 212, 347, 419; *imaginifer* 25; *signifer* 25, 35, 50, 53, 55, 87, 103, 300, 373, 379, 389, 401, 426; *vexillarius* 25, 367, 427
standing army, establishment of 12, 32, 38, 133, 146, 349, 416
*statio* (staging post) 280
*stationarius* (armed police) 271, 272, 273, 274
stealing 77, 138, 186, 274, 343
stone: as weapons 111, 162, 197, 219, 247; for stoning someone to death 79, 249
storms 173, 235, 246, 289, 375
strength report 14, 19, 119, 120, 121, 312
*strigil* 308–9
strongmen soldiers 47
Sturi tribe 114
Suebi tribe 192, 193, 382
*suovetaurilia*, sacrifice 380
superstition 204, 251, 371, 374, 390
supply chains 131, 133, 152, 192
surveyor 7, 73, 108, 111, 116, 117, 284, 307, 423
swamps 166, 309, 312, 361
swimming 193, 403
swords, *see* gladius

taxation 8, 94, 97, 138, 219, 269, 271, 340, 348, *see also* poll tax
teaching 325, 350, *see also* education
temple 3, 24, 32, 82, 118, 119, 164, 187, 196, 197, 199, 200, 201, 201, 219, 237, 269, 271, 276, 281, 301, 309, 326, 327, 355, 356, 360, 376, 377, 379, 380, 381, 384, 390, 424, *see also* mithraea
term of service 28, 45, 347, 348, 355, 367, 422
*tesserarius* 62, 426
*testudo* (tortoise) formation 162, 217
tetradrachm 105, 426
Tetrarchy 115, 266, 381, 395
theatres 32, 152, 200, 226, 227, 270
Thracians 12, 31, 47, 54, 55, 124, 148, 211, 245
thunderbolts 12, 26, 374, 375

505